Texts in Applied Mathematics 50

Texts in Applied Mathematics

(continued after index)

Ferdinand Verhulst

Methods and Applications of Singular Perturbations

Boundary Layers and Multiple Timescale Dynamics

With 26 Illustrations

 Springer

Ferdinand Verhulst
University of Utrecht
Utrecht 3584 CD
The Netherlands

Series Editors

J.E. Marsden
Control and Dynamical Systems, 107–81
California Institute of Technology
Pasadena, CA 91125
USA
marsden@cds.caltech.edu

L. Sirovich
Division of Applied Mathematics
Brown University
Providence, RI 02912
USA
chico@camelot.mssm.edu

S.S. Antman
Department of Mathematics
and
Institute for Physical Science
  and Technology
University of Maryland
College Park, MD 20742-4015
USA
ssa@math.umd.edu

Mathematics Subject Classification (2000): 12147

Library of Congress Cataloging-in-Publication Data
Verhulst, F. (Ferdinand), 1939–
    Methods and applications of singular perturbations : boundary layers and multiple
timescale dynamics / Ferdinand Verhulst.
      p. cm.
    Includes bibliographical references and index.

    1. Boundary value problems—Numerical solutions.   2. Singular perturbations
(Mathematics)   I. Title.
QA379.V47   2005
515'.35—dc22

                                                                    2005042479

ISBN-13: 978-1-4419-1992-2   Printed on acid-free paper.
e-ISBN-13: 978-0-387-28313-5

Printed in the United States of America.   (EB)

9 8 7 6 5 4 3 2 1

springeronline.com

Series Preface

Mathematics is playing an ever more important role in the physical and biological sciences, provoking a blurring of boundaries between scientific disciplines and a resurgence of interest in the modern as well as the classical techniques of applied mathematics. This renewal of interest, both in research and teaching, has led to the establishment of the series Texts in Applied Mathematics (TAM).

The development of new courses is a natural consequence of a high level of excitement on the research frontier as newer techniques, such as numerical and symbolic computer systems, dynamical systems, and chaos, mix with and reinforce the traditional methods of applied mathematics. Thus, the purpose of this textbook series is to meet the current and future needs of these advances and to encourage the teaching of new courses.

TAM will publish textbooks suitable for use in advanced undergraduate and beginning graduate courses, and will complement the Applied Mathematical Sciences (AMS) series, which will focus on advanced textbooks and research-level monographs.

Pasadena, California   J.E. Marsden
Providence, Rhode Island   L. Sirovich
Houston, Texas   M. Golubitsky
College Park, Maryland   S.S. Antman

Preface

Mathematics is more an activity than a theory
(Mathematik is mehr ein Tun als eine Lehre)
Hermann Weyl, after L.E.J. Brouwer

Perturbation theory is a fundamental topic in mathematics and its applications to the natural and engineering sciences. The obvious reason is that hardly any problem can be solved exactly and that the best we can hope for is the solution of a "neighbouring" problem. The original problem is then a perturbation of the solvable problem, and what we would like is to establish the relation between the solvable and the perturbation problems.

What is a singular perturbation? The traditional idea is a differential equation (plus other conditions) having a small parameter that is multiplying the highest derivatives. This covers a lot of cases but certainly not everything. It refers to boundary layer problems only.

The modern view is to consider a problem with a small parameter ε and solution $x(t, \varepsilon)$. Also defined is an "unperturbed" (neighbouring) problem with solution $x(t, 0)$. If, in an appropriate norm, the difference $\|x(t, \varepsilon) - x(t, 0)\|$ does not tend to zero when ε tends to zero, this is called a singular perturbation problem. The problems in Chapters 1–9 are covered by both the old(fashioned) definition and the new one. Slow-time problems (multiple time dynamics), as will be discussed in later chapters, fall under the new definition. Actually, most perturbation problems in this book are singular by this definition; only in Chapter 10 shall we consider problems where "simple" continuation makes sense.

This book starts each chapter with studying explicit examples and introducing methods without proof. After many years of teaching the subject of singular perturbations, I have found that this is the best way to introduce this particular subject. It tends to be so technical, both in calculations and in theory, that knowledge of basic examples is a must for the student. This view is not only confirmed by my lectures in Utrecht and elsewhere but also by lecturers who used parts of my text in various places. In this respect, Hermann Weyl's quotation which is concerned with the fundamentals of mathematics, gives us the right perspective.

I have stressed that the proposed workbook format is very suitable for singular perturbation problems, but I hope that the added flavour of precise estimates and excursions into the theoretical background makes the book of interest both for people working in the applied sciences and for more theoretically oriented mathematicians.

Let me mention one more important subject of the forthcoming chapters. There will be an extensive discussion of timescales and a priori knowledge of the presence of certain timescales. This is one of the most widely used concepts in slow-time dynamics, and there is a lot of confusion in the literature. I hope to have settled some of the questions arising in choosing timescales.

What about theory and proofs one may ask. To limit the size of the book, those mathematical proofs that are easy to obtain from the literature are listed at the end of each chapter in a section "Guide to the Literature". If they are readily accessible, it usually makes no sense to reproduce them. Exceptions are sometimes cases where the proof contains actual constructions or where a line of reasoning is so prominent that it has to be included. In all cases discussed in this book - except in Chapter 14 - proofs of asymptotic validity are available. Under "Guide to the Literature" one also finds other relevant and recent references.

In a final chapter I collected pieces of theory that are difficult to find in the literature or a summary such as the one on perturbations of matrices or a typical and important type of proof such as the application of maximum principles for elliptic equations. Also, in the epilogue I return to the discussion about "proving and doing".

To give a general introduction to singular perturbations, I have tried to cover as many topics as possible, but of course there are subjects omitted. The first seven chapters contain standard topics from ordinary differential equations and partial differential equations, boundary value problems and problems with initial values within a mathematical framework that is more rigorous in formulation than is usual in perturbation theory. This improves the connection with theory-proof approaches. Also, we use important, but nearly forgotten theorems such as the du Bois-Reymond theorem.

Some topics are missing (such as the homogenisation method) or get a sketchy treatment (such as the WKBJ method). I did not include relaxation oscillations, as an elementary treatment can be found in my book *Nonlinear Differential Equations and Dynamical Systems*. Also there are books available on this topic, such as *Asymptotic Methods for Relaxation Oscillations and Applications* by Johan Grasman.

Perturbation theory is a fascinating topic, not only because of its applications but also because of its many unexpected results. A long time ago, Wiktor Eckhaus taught me the basics of singular perturbation theory, and at about the same time Bob O'Malley introduced me to Tikhonov's theorem and multiple scales.

Many colleagues and students made remarks and gave suggestions. I mention Abadi, Taoufik Bakri, Arjen Doelman, Hans Duistermaat, Wiktor Eckhaus, Johan Grasman, Richard Haberman, Michiel Hochstenbach, James Murdock, Bob O'Malley, Richard Rand, Bob Rink, Thijs Ruijgrok, Theo Tuwankotta, Adriaan van der Burgh. I got most of section 15.5 from Van Harten's (1975) thesis, section 15.9 is based on Buitelaar's (1993) thesis.

The figures in the first nine chapters were produced by Theo Tuwankotta; other figures were obtained from Abadi, Taoufik Bakri and Hartono. Copyeditor Hal Henglein of Springer proposed the addition of thousands of comma's and many layout improvements.

I am grateful to all of them.

Corrections and additions will be posted on
http://www.math.uu.nl/people/verhulst

Ferdinand Verhulst, University of Utrecht

Contents

1

Introduction

Perturbation theory, defined as the theory of approximating solutions of mathematical problems, goes back to ancient times. An important example is the practice of measurement, where quantities such as distance and volume (such as the contents of wine barrels) have been estimated by professional people through the ages.

The theory of perturbations expanded very rapidly when mathematical analysis was founded in the eighteenth century, and many classical results in this field can be traced to Newton, Euler, Lagrange, Laplace, and others. One of the most stimulating fields of application of that time was celestial mechanics, where the controversies and excitement about Newton's gravitational theory triggered many detailed calculational studies.

The establishment of more rigorous foundations of perturbation theory had to wait until Poincaré (1886) and Stieltjes (1886) separately published papers on asymptotic series, which are in general divergent; see also the discussion on the literature at the end of Chapter 2. In the twentieth century, an additional stimulus came from other fields of application. In 1905, Prandtl published a paper on the motion of a fluid or gas with small viscosity along a body. In the case of an airfoil moving through air, the problem is described by the Navier-Stokes equations with large Reynolds number; see also Prandtl and Tietjens (1934) and, for modern developments, Van Ingen (1998). Ting (2000) and other authors discuss the boundary layer theory of fluids in a special issue of the Zeitschrift für Angewandte Mathematik und Mechanik dedicated to Ludwig Prandtl.

In this problem, there are two regions of interest: a boundary layer around the solid body, where the velocity gradient becomes large, and the region outside this layer, where we can neglect the velocity gradient and the viscosity. The mathematical analysis of the problem uses this insight to develop an appropriate perturbation theory in the case of the presence of boundary layers. Notes on the historical development of boundary layer theory are given by O'Malley (1991).

As mentioned above, the roots of classical perturbation theory, which are mainly in celestial mechanics, are quite old. A modern stimulus came from the theory of nonlinear oscillations in electronics and mechanics. The name of the Dutch physicist Balthasar van der Pol is connected with this field, for instance in the theory of relaxation oscillations. One can find historical remarks in the books by Bogoliubov and Mitropolsky (1961), Sanders and Verhulst (1985), and Grasman (1987).

We conclude this introduction by giving some examples. *Note that here and henceforth ε will always be a small positive parameter:*

$$0 < \varepsilon \ll 1.$$

Quantities and functions will be real unless explicitly stated otherwise.

Example 1.1

The first example is a series studied by Euler (1754) with partial sum

$$S_m(\varepsilon) = \sum_{n=0}^{m} (-1)^n n! \varepsilon^n.$$

It is clear that the series diverges as, denoting the terms of the series by a_n, we have

$$\left| \frac{a_n}{a_{n-1}} \right| = n\varepsilon.$$

However, the size of the terms for small values of ε does not increase much in the beginning (i.e. if $n\varepsilon \ll 1$), but growth seriously affects the partial sum for larger values of m. The question is, can we use a number of the first terms of such a divergent series to approximate a function in some sense? This looks like a wild idea, but consider the function $f(\varepsilon)$ defined by the convergent integral

$$f(\varepsilon) = \int_0^\infty e^{-t} \frac{dt}{1 + \varepsilon t}.$$

Partial integration leads to the expression

$$f(\varepsilon) = S_m(\varepsilon) + (-1)^{m+1}(m+1)! \varepsilon^{m+1} \int_0^\infty e^{-t} \frac{dt}{(1+\varepsilon t)^{m+2}}.$$

The integral on the right-hand side converges, and we estimate

$$|f(\varepsilon) - S_m(\varepsilon)| \leq (m+1)! \varepsilon^{m+1}.$$

In some sense, to be made precise later on, for ε small enough, S_m constitutes an approximation of f. To be more explicit, we give some numerical details.

ε	$f(\varepsilon)$	$S_2(\varepsilon) = 1 - \varepsilon + 2\varepsilon^2$
.05	.9543	.9550
.10	.9156	.9200
.20	.8521	.8800

To see when the divergence becomes effective, we list $S_m(.10)$ for $m = 1, \cdots, 21$. The best approximation in this case is found for $m = 9$.

In Fig. 1.1 we show the behaviour of the error $|S_m - f(.1)|$ as a function of m. It is typical for an asymptotic approximation that there is an optimal choice of the number of terms that generates the best approximation. In approximations by a convergent series, there is not such a finite optimal choice and usually we take as many terms as possible.

m	$S_m(f(.1) = .9156)$	m	$S_m(f(.1) = .9156)$
0	1	11	.9154
1	.9000	12	.9159
2	.9200	13	.9153
3	.9140	14	.9161
4	.9164	15	.9148
5	.9152	16	.9169
6	.9159	17	.9134
7	.9154	18	.9198
8	.9158	19	.9076
9	.9155	20	.9319
10	.9158	21	.8809

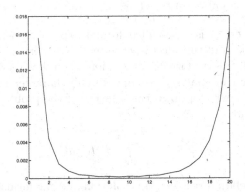

Fig. 1.1. The error $|S_m - f(.1)|$ as a function of m; this behaviour is typical for an asymptotic approximation that generally does not converge to the solution.

We shall now discuss some perturbation problems arising from differential equations.

Example 1.2
The function $\phi_\varepsilon(x)$ is defined for $x \in [0,1]$ as the solution of the differential equation

$$\frac{d\phi}{dx} + \varepsilon\phi = \cos x$$

with initial value $\phi_\varepsilon(0) = 0$.

Solving the "unperturbed" problem means putting $\varepsilon = 0$ in the equation; we find, using the initial condition,

$$\phi_0(x) = \int_0^x \cos t\, dt = \sin x.$$

To solve the problem for $\varepsilon > 0$, we might try an expansion of the form

$$\phi_\varepsilon(x) = \sum_{n=0}^\infty \varepsilon^n \phi_n(x),$$

which, after substitution in the differential equation, leads to the recurrent system

$$\frac{d\phi_n}{dx} = -\phi_{n-1}, \phi_n(0) = 0, \quad n = 1, 2, \cdots.$$

In this problem, it is natural to put all initial values for the higher-order equations equal to zero. We find for the first correction to $\phi_0(x), \phi_1(x) = \cos x - 1$, so we have

$$\phi_\varepsilon(x) = \sin x + \varepsilon(\cos x - 1) + \varepsilon^2 \cdots.$$

The expansion for $\phi_\varepsilon(x)$ is a so-called formal expansion, which leads to a consistent construction of the successive terms. (Note that it is strange that mathematicians call this a "formal" expansion when it is really "informal".) In this example, we can analyse the approximate character i.e., the validity of this expansion, by writing down the solution of the problem, obtained by variation of constants,

$$\phi_\varepsilon(x) = e^{-\varepsilon x} \int_0^x \cos t\, e^{\varepsilon t} dt.$$

We can study the relation between this solution and the formal expansion by partial integration of the integral, see Fig. 1.2. We find

$$\phi_\varepsilon(x) = \sin x + \varepsilon(\cos x - e^{-\varepsilon x}) - \varepsilon^2 \phi_\varepsilon(x)$$

so that we have

$$\phi_\varepsilon(x) = \frac{1}{1+\varepsilon^2}(\sin x + \varepsilon(\cos x - e^{-\varepsilon x})).$$

Expansion with respect to ε produces the validity of the formal approximation on $[0, 1]$.

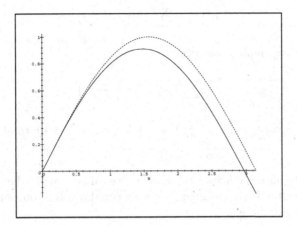

Fig. 1.2. Solution (full line) and formal expansion (dashed line) with $\varepsilon = 0.1$ in Example 1.2. Calculating higher-order approximations will improve the approximation.

Example 1.3

Suppose now that we change the interval in which we are interested in the behaviour of the solution of the equation in Example 1.2 to $[0, \infty]$. Note that the formal approximation to $O(\varepsilon)$,

$$\phi_\varepsilon(x) = \sin x + \varepsilon \cdots ,$$

still holds on the whole interval but the approximation to $O(\varepsilon^2)$ is not a formal approximation. For x in a neighbourhood of $x = 0$, we recover the formal approximation from the exact solution to $O(\varepsilon^2)$, but for x very large, we find from the exact solution the approximation

$$\sin x + \varepsilon \cos x.$$

This motivates us to be more precise in our notions of approximation; we shall return to this in Chapter 2.

Example 1.4

We consider for $x \in [0, 1]$ the function $\phi_\varepsilon(x)$ defined by the initial value problem

$$\varepsilon \frac{d\phi}{dx} + \phi = \cos x, \quad \phi_\varepsilon(0) = 0.$$

The equation is nearly the same as in Example 1.2 but, as will be apparent shortly, the different location of ε changes the problem drastically. Substituting again a formal expansion of the form

$$\phi_\varepsilon(x) = \sum_{n=0}^{\infty} \varepsilon^n \phi_n(x)$$

produces, after regrouping the terms,

$$\phi_0 - \cos x + \sum_{n=1}^{\infty} \varepsilon^n \left(\phi_n + \frac{d\phi_{n-1}}{dx} \right) = 0.$$

So we have $\phi_0 = \cos x, \phi_n = -\frac{d\phi_{n-1}}{dx}, n = 1, 2, \cdots$, and as a formal expansion

$$\phi_\varepsilon(x) = \cos x + \varepsilon \sin x - \varepsilon^2 \cos x + \cdots.$$

However, the expansion makes little sense, as we cannot satisfy the initial condition! To understand what is going on, we write down the solution obtained by variation of parameters,

$$\phi_\varepsilon(x) = \frac{1}{\varepsilon} e^{-x/\varepsilon} \int_0^x e^{t/\varepsilon} \cos t \, dt.$$

We expand the integral by partial integration to find

$$\phi_\varepsilon(x) = \cos x - e^{-x/\varepsilon} + e^{-x/\varepsilon} \int_0^x e^{t/\varepsilon} \sin t \, dt.$$

The function $\exp(-x/\varepsilon)$ is quickly varying in a neighbourhood of $x = 0$, see Fig. 1.3. For say, $x \geq \sqrt{\varepsilon}$, this term is very small; we call it "exponentially small".

To order m, we find

$$\phi_\varepsilon(x) = \sum_{n=0}^{m} (-1)^n \varepsilon^n [\cos^{(n)}(x) - e^{-x/\varepsilon} \cos^{(n)}(0)]$$
$$+ (-1)^{m+1} \varepsilon^m e^{-x/\varepsilon} \int_0^x e^{t/\varepsilon} \cos^{(m+1)}(t) dt.$$

Introducing the expansion

$$S_m(x) = \sum_{n=0}^{m} (-1)^n \varepsilon^{(n)} [\cos^{(n)}(x) - e^{-\frac{x}{\varepsilon}} \cos^{(n)}(0)],$$

we have in $[0, 1]$

$$|\phi_\varepsilon(x) - S_m(x)| \leq C \, \varepsilon^m e^{-\frac{x}{\varepsilon}} \int_0^x e^{\frac{t}{\varepsilon}} dt$$
$$\leq C \, \varepsilon^{m+1} \left(1 - e^{-\frac{x}{\varepsilon}} \right).$$

Note that $S_m(x)$ satisfies the initial condition and represents an approximation of the solution. The structure of the expansion, however, is essentially different from the formal expansion. On the other hand, the formal expansion represents the solution well outside a neighbourhood of $x = 0$.

In the next chapter, we shall make our terminology more precise. This is essential to avoid confusion and to obtain a fair appraisal of the results to be obtained by expansion techniques.

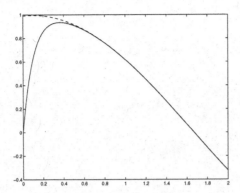

Fig. 1.3. Solution (full line) and formal expansion (dashed line) with $\varepsilon = 0.1$ in Example 1.4. The solution goes through a fast transition near $x = 0$.

2

Basic Material

When calculating series expansions for solutions of equations, it is important
to have a clear view of the meaning of these expansions. For this it is necessary
to introduce a number of definitions and concepts that will be our tools in
what follows.

2.1 Estimates and Order Symbols.

In this section, we present a number of basic concepts that are necessary to
discuss approximation theory. We would like to estimate vector fields that
depend on a small parameter ε, but we start in a simple way by considering
pairs of real continuous functions $f(\varepsilon)$ and $g(\varepsilon)$, where $0 < \varepsilon \leq \varepsilon_0 \ll 1$.

To compare these functions, we can use Landau's order symbols 0 and o.

Definition
$f(\varepsilon) = 0(g(\varepsilon))$ for $\varepsilon \to 0$ if there exist positive constants C, ε_0 such that in
$(0, \varepsilon_0]$
$$|f(\varepsilon)| \leq C|\,g(\varepsilon)| \ \text{ for } \varepsilon \to 0.$$

Example 2.1
Compare the functions
$$f(\varepsilon) = \varepsilon, \ g(\varepsilon) = 2\varepsilon,$$
$$f(\varepsilon) = \varepsilon, \ g(\varepsilon) = \sin \varepsilon - 3\varepsilon,$$
$$f(\varepsilon) = \varepsilon^2, g(\varepsilon) = \varepsilon^{\frac{1}{2}}.$$

Definition
$f(\varepsilon) = o(g(\varepsilon))$ for $\varepsilon \to 0$ if we have
$$\lim_{\varepsilon \to 0} \frac{f(\varepsilon)}{g(\varepsilon)} = 0.$$

Example 2.2
Compare the functions

$$f(\varepsilon) = \varepsilon^2, \, g(\varepsilon) = \varepsilon,$$
$$f(\varepsilon) = \sin\varepsilon - \varepsilon, \, g(\varepsilon) = \varepsilon^{3/2} + \varepsilon.$$

It is interesting to note that Landau's order symbols enable us also to compare functions that do not exist for $\varepsilon = 0$. For instance, the estimate

$$\frac{1}{\varepsilon} = o\left(\frac{1}{\varepsilon^2}\right)$$

informs us about the rate of divergence of the two functions $1/\varepsilon$ and $1/\varepsilon^2$ for $\varepsilon \to 0$.

A second remark is that, against the rules of the use of Landau's symbols, we shall usually leave out the expression "for $\varepsilon \to 0$" in our estimates as this is the case that we are *always* considering in this book.

A third remark is that an estimate $f(\varepsilon) = 0(g(\varepsilon))$ can sometimes be improved to an o-estimate and that sometimes this is not the case. For instance, $\varepsilon = 0(2\varepsilon)$ and $\varepsilon \neq o(2\varepsilon)$ but $\varepsilon^2 = 0(\varepsilon)$ and $\varepsilon^2 = o(\varepsilon)$. It is natural to compare order functions, and sometimes we have to be more precise in our estimates; two definitions are in use in the literature.

Definition of equivalence
The functions $f(\varepsilon)$ and $g(\varepsilon)$ are equivalent, $f(\varepsilon) \approx g(\varepsilon)$ as $\varepsilon \to 0$, if $f(\varepsilon) = 0(g(\varepsilon))$ and $g(\varepsilon) = 0(f(\varepsilon))$ for $\varepsilon \to 0$.

Although this looks like a natural definition, equivalence is a strong property and we shall hardly use it in practice.

Example 2.3
Simple examples of equivalence are

$$\varepsilon \approx 24\varepsilon,$$
$$\varepsilon \approx 6\sin\varepsilon.$$

There are many subtle difficulties with these order estimates. One might for instance think that if two functions $f(\varepsilon)$ and $g(\varepsilon)$ are not equivalent, $f(\varepsilon) \napprox g(\varepsilon)$, and while $f(\varepsilon) = 0(g(\varepsilon))$ for $\varepsilon \to 0$, we would have $f(\varepsilon) = o(g(\varepsilon))$. This is not the case, as for instance $\varepsilon\sin(1/\varepsilon) = 0(\varepsilon)$ for $\varepsilon \to 0$, and the reverse is not true because of the infinite number of zeros of $\sin(1/\varepsilon)$ as $\varepsilon \to 0$. However, $\varepsilon\sin(1/\varepsilon) \neq o(\varepsilon)$, as the $\lim_{\varepsilon \to 0} \sin(1/\varepsilon)$ does not exist.

Another definition enables us to give a sharp estimate of functions that is more useful.

Definition

$f(\varepsilon) = 0_s(g(\varepsilon))$ if $f = 0(g)$ and $f \neq o(g)$ for $\varepsilon \to 0$.

It is clear that equivalence of functions implies the sharp estimate, $f \approx g \Rightarrow f = 0_s(g)$. On the other hand, one can show that there exist pairs of functions such that $f = 0_s(g)$ while $f \not\approx g$.

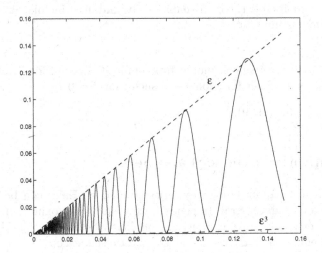

Fig. 2.1. Illustration of partial ordering of order functions by considering the functions $\varepsilon, \varepsilon^2$, and $\varepsilon \sin^2(1/\varepsilon) + \varepsilon^3$.

In what follows we shall use the $0, o$, and 0_s symbols. We should keep in mind, however, that with these tools we cannot compare all real continuous functions defined in $(0, \varepsilon_0]$. Consider for instance the functions $f(\varepsilon) = \varepsilon^2$ and $g(\varepsilon) = \varepsilon \sin^2(1/\varepsilon) + \varepsilon^3$, see Fig. 2.1. Consider two infinite sequences of numbers converging towards zero for $n = 1, 2, \cdots$

$$a_n = (2n\pi)^{-1}, \qquad \sin(1/a_n) = 0,$$
$$b_n = (2n\pi + \tfrac{1}{2}\pi)^{-1}, \quad \sin^2(1/b_n) = 1.$$

Comparison of the magnitudes of f and g on these sequences produces essentially different answers in each small neighbourhood of $\varepsilon = 0$. This example shows that the ordering of functions with respect to their magnitude introduced by the $0, o$ symbols is a partial ordering. On the other hand, this should not worry us too much, as we shall find a set of comparison functions, order functions, that will be sufficient to estimate any continuous function $f(\varepsilon)$.

Definition

$\mathcal{E}$ is the set of order functions $\delta(\varepsilon)$ defined in $(0, \varepsilon_0]$ that are real, positive, continuous, and monotonic and for which $\lim_{\varepsilon \to 0} \delta(\varepsilon)$ exists. The extended set $\bar{\mathcal{E}}$ consists of the functions $\delta(\varepsilon)$ with $\delta \in \mathcal{E}$ or $\delta^{-1} \in \mathcal{E}$.

Example 2.4
The following order functions are often used:
$\delta_n(\varepsilon) = \varepsilon^n$ or $e^{-n/\varepsilon}, n = 1, 2, \cdots$.

The following remarkable theorem enables us to find, in principle, an estimate for any continuous function.

Theorem 2.1
Suppose that $f(\varepsilon)$ is a real continuous function in $(0, \varepsilon_0)$, not identically zero, then there exists an order function $\delta \in \bar{\mathcal{E}}$ such that $f = 0_s(\delta)$.

Proof: See Eckhaus (1979).

2.2 Asymptotic Sequences and Series

We have seen in Chapter 1 how the series studied by Euler can be used to approximate a certain integral. In general, we wish to approximate functions sometimes given as integrals, or functions implicitly defined as the solution of a differential equation. First we define asymptotic sequence.

Definition
A sequence of order functions $\delta_n(\varepsilon), n = 0, 1, 2, \cdots$ is called asymptotic if for $n = 0, 1, 2, \cdots \delta_{n+1} = o(\delta_n), \varepsilon \to 0$.

Example 2.5
$\delta_n = \varepsilon^{n+1}, n = 0, 1, 2, \cdots$.

Of course, an asymptotic sequence can have a finite or infinite number of elements. It is useful to generalise this as follows.

Definition
A sequence of real continuous functions $f_n(\varepsilon), n = 0, 1, 2, \cdots$ in $(0, \varepsilon_0]$ is called asymptotic if one has $f_n(\varepsilon) = 0_s(\delta_n(\varepsilon))$ for $\varepsilon \to 0$ with the $\delta_n(\varepsilon)$ order functions and $\delta_{n+1} = o(\delta_n), n = 0, 1, 2, \cdots$ for $\varepsilon \to 0$.

The idea here is to compare the functions $f_n(\varepsilon)$, which are expected to be asymptotically ordered by using order functions $\delta_n(\varepsilon)$. This is possible according to Theorem 2.1 given in the preceding section.

Example 2.6
$\sin \varepsilon + \varepsilon^2 \sin(1/\varepsilon), \varepsilon \sin \varepsilon + \varepsilon^3 \sin(1/\varepsilon), e^{-1/\varepsilon} \cos \varepsilon$ can be compared with the order functions $\varepsilon, \varepsilon^2, e^{-1/\varepsilon}$, which form an asymptotic sequence.

We are now prepared for the following definition.

Definition

The sum $\sum_{n=0}^{m} a_n f_n(\varepsilon)$ with constant coefficients a_n is called an asymptotic series if $f_n(\varepsilon), n = 0, 1, 2, \cdots, m$ is an asymptotic sequence.

Note that we may put $m = \infty$; the existence or nonexistence of the sum has nothing to do with the asymptotic character of the series. Sometimes we call the sum an asymptotic expansion. An example is the series obtained in the introduction,

$$\sum_{n=0}^{\infty} (-1)^n n! \, \varepsilon^n.$$

It improves our insight into asymptotic sequences and series to point out one more basic difference with convergent sequences. As we know, the set of real numbers is ordered such that if we take a sequence of positive numbers converging towards zero, the terms of the sequence are ordered such that the terms become arbitrarily small *and* there exists no positive number smaller than all elements of this sequence. An analogous result does *not* hold for asymptotic sequences.

Theorem 2.2

(du Bois-Reymond)
For any asymptotic sequence $\delta_n(\varepsilon), n = 0, 1, 2, \cdots$, there exist order functions $\delta^0(\varepsilon)$ such that for all n

$$\delta^0 = o(\delta_n).$$

Remark

The theorem was introduced by du Bois-Reymond in 1887 (see Borel, 1902) to study the growth of functions near a singularity. A corollary has been given by Hardy (1910); the present formulation of the theorem is due to Eckhaus (1979). For a proof, see Section 15.1.

It is easy to find examples: take $\delta_n = \varepsilon^n, n = 0, 1, 2, \cdots$ and $\delta^0 = e^{-1/\varepsilon}$. The amazing thing about the theorem is that such a δ^0 exists for any sequence, for instance

$$\delta_1 = e^{-1/\varepsilon}, \delta_2 = e^{-1/\delta_1}, \cdots, \delta_n = e^{-1/\delta_{n-1}}, \cdots.$$

We take such a δ^0 and construct $\delta_1 = e^{-1/\delta^0}, \delta_2 = e^{-1/\delta_1}, \cdots$ and we may repeat the process ad infinitum.

2.3 Asymptotic Expansions with more Variables

Consider a function parametrised by the small parameter $\varepsilon : \phi = \phi_\varepsilon(x); x \in D \subset \mathbb{R}^n$. Take for instance the functions εx and $\varepsilon \sin x$ for $x \geq 0$. Estimating these functions with respect to ε, one notices that εx is not uniformly bounded on the domain, while $\varepsilon \sin x$ is bounded. Intuitively we would be inclined to put $\varepsilon \sin x = 0(\varepsilon)$ for $x \geq 0$; if the domain is finite, for instance $0 \leq x \leq 1$, we would also put $\varepsilon x = 0(\varepsilon)$; see Example 1.4 for such a case.

To estimate the magnitude of a function $\phi = \phi_\varepsilon(x), x \in D \subset \mathbb{R}^n$, we use a given norm $\|\cdot\|$. The choice of the norm is usually a natural one in the context of a given perturbation problem, as we shall see. Often we use the supremum (sup) or uniform norm

$$\|\phi_\varepsilon(x)\|_D = \sup_{x \in D} |\phi_\varepsilon(x)|,$$

where $|.|$ indicates the Euclidean norm of the vector function. The resulting magnitude is a scalar function of ε, which we can try to estimate in terms of order functions. So we have in the sup norm

$$\|\varepsilon \sin x\|_{\mathbb{R}+} = \sup_{x \in \mathbb{R}^+} |\varepsilon \sin x| = \varepsilon,$$

which is a simple estimate.

Definition
Consider $\phi_\varepsilon(x), x \in D \subset \mathbb{R}^n$, with ϕ_ε an element of a linear set of functions provided with some norm $\|\cdot\|_D; \delta(\varepsilon)$ is an order function. Then, for $\varepsilon \to 0$,

$$\phi_\varepsilon(x) = 0(\delta(\varepsilon)) \ \text{ in } D \text{ if } \|\phi_\varepsilon\|_D = 0(\delta(\varepsilon)),$$
$$\phi_\varepsilon(x) = o(\delta(\varepsilon)) \ \text{ in } D \text{ if } \|\phi_\varepsilon\|_D = o(\delta(\varepsilon)),$$
$$\phi_\varepsilon(x) = 0_s(\delta(\varepsilon)) \ \text{ in } D \text{ if } \|\phi_\varepsilon\|_D = 0_s(\delta(\varepsilon)).$$

In the previous examples, we have, using the sup norm,

$$\varepsilon \sin x = 0_s(\varepsilon) \in \mathbb{R}^+,$$
$$\varepsilon x = 0_s(\varepsilon) \in [0, 1].$$

Remark
It is instructive to note the influence of the choice of the norm. Consider for instance

$$\phi_\varepsilon = e^{-x/\varepsilon}, x \in [0, 1].$$

Using the sup norm we have $e^{-x/\varepsilon} = 0_s(1)$. Another norm, the L_2-norm

$$\|\phi_\varepsilon\| = \left(\int_0^1 \phi_\varepsilon^2(x)dx \right)^{\frac{1}{2}}$$

produces $e^{-x/\varepsilon} = 0_s(\sqrt{\varepsilon})$. So we may have an asymptotic estimate of a function in a particular norm that is not valid in another norm. In real-life applications it is important to have estimates in the natural norms of the problem at hand.

Remark

The definitions given here enable us to give a meaning to the expansion obtained in Example 1.4 of Chapter 1. The problem was to study the solution of

$$\varepsilon\frac{d\phi}{dx} + \phi = \cos x, \phi_\varepsilon(0) = 0, x \in [0,1].$$

We have found the expansion

$$\phi_\varepsilon(x) = \sum_{n=0}^{m} \varepsilon^n \psi_{n(x,\varepsilon)} + R_m$$

with $\psi_n = (-1)^n[\cos^{(n)}(x) - e^{-x/\varepsilon}\cos^{(n)}(0)]$ and

$$|R_m| \le C\varepsilon^{m+1}(1 - e^{-x/\varepsilon}).$$

In the sup norm, $\psi_n = 0_s(1)$ and $(1 - e^{-x/\varepsilon}) = 0_s(1)$. It is natural to interpret this expansion as an asymptotic series as follows.

Definition

$\sum_{n=0}^{m} \psi_n(x,\varepsilon)$ is an asymptotic series in D with respect to a given norm if $\psi_n(x,\varepsilon) = 0_s(\delta_n)$ and $\delta_n(\varepsilon), n = 0, 1, \cdots$ is an asymptotic sequence.

It is clear from Theorem 2.1 that any nontrivial continuous function $\phi_\varepsilon(x)$ can be estimated by an order function $\delta(\varepsilon)$ such that $\phi_\varepsilon(x) = 0_s(\delta(\varepsilon))$. This has the practical consequence that we can always rescale

$$\phi_\varepsilon(x) = \delta(\varepsilon)\bar{\phi}_\varepsilon(x)$$

with $\delta(\varepsilon)$ an order function and $\bar{\phi}_\varepsilon(x) = 0_s(1)$. Rescaling is useful in defining *asymptotic approximations* in the sense that it is natural to rescale before estimating. If we omit this, we would have a definition such as $\tilde{\phi}$ is an asymptotic approximation of ϕ if $\phi - \tilde{\phi} = o(1)$. This implies that ε would be an asymptotic approximation of ε^2; this makes little sense.

Definition

$\tilde{\phi}_\varepsilon(x)$ is an asymptotic approximation of $\phi_\varepsilon(x)$ in D with respect to a given norm if $\phi_\varepsilon(x) = 0_s(1)$ and $\phi_\varepsilon(x) - \tilde{\phi}_\varepsilon(x) = o(1)$.

The definitions of an asymptotic series and an asymptotic approximation can be combined to study asymptotic expansions, which are approximating series of the form

$$\sum_{n=0}^{m} \delta_n(\varepsilon)\psi_n(x,\varepsilon).$$

In this expression, $\delta_n, n = 0, 1, \cdots$ is an asymptotic sequence and $\psi_n(x,\varepsilon) = 0_s(1)$.

2.4 Discussion

In this section, we discuss three fundamental aspects of asymptotic expansions: nonuniqueness, convergence, and the practical significance of the du Bois-Reymond theorem. Consider a function $\phi_\varepsilon(x)$ in D. We wish to study an asymptotic expansion of the function with respect to a given norm. Note that there is no uniqueness of expansion.

Example 2.7
(non-uniqueness; Eckhaus, 1979)

$$\phi_\varepsilon(x) = \left(1 - \frac{\varepsilon}{1+\varepsilon}x\right)^{-1} \quad \text{for } x \in [0,1].$$

Using the binomial expansion rule, we have

$$\phi_\varepsilon(x) = \sum_{n=0}^{m} \left(\frac{\varepsilon}{1+\varepsilon}\right)^n x^n + O(\varepsilon^{m+1}).$$

This is an asymptotic expansion with respect to the set of order functions $\delta_n = (\frac{\varepsilon}{1+\varepsilon})^n, n = 0, 1, 2, \cdots$. However, we may also write

$$\phi_\varepsilon(x) = \frac{1+\varepsilon}{1+\varepsilon(1-x)} = (1+\varepsilon)\sum_{n=0}^{m}(-1)^n\varepsilon^n(1-x)^n + 0(\varepsilon^{m+1})$$

$$= 1 + \sum_{n=1}^{m}\varepsilon^n x(x-1)^{n-1} \quad\quad + 0(\varepsilon^{m+1}).$$

This is an asymptotic expansion of the same function with respect to a different set of order functions.

Note that the expansion is unique with respect to a given asymptotic sequence $\delta_n(\varepsilon), n = 0, 1, 2, \cdots$ with $\phi_\varepsilon(x) = 0_s(\delta_0)$. We have

$$\phi_\varepsilon(x) = \sum_{n=0}^{m}\delta_n(\varepsilon)\psi_n(x)$$

with

$$\psi_0 = \lim_{\varepsilon\to 0}\phi_\varepsilon/\delta_0,$$
$$\psi_1 = \lim_{\varepsilon\to 0}(\phi_\varepsilon - \delta_0\psi_0)/\delta_1,$$
$$\psi_2 = \lim_{\varepsilon\to 0}(\phi_\varepsilon - \delta_0\psi_0 - \delta_1\psi_1)/\delta_2, \quad\quad \text{etc.}$$

2.4.1 The Question of Convergence.

We will sometimes put

$$\delta_\varepsilon(x) = \sum_{n=0}^{\infty} \delta_n \psi_n,$$

but this is considered as an asymptotic series generally with *no* convergence; compare this for instance with the expansion of the integral in Chapter 1.

Moreover, even if the asymptotic series converges, the limit need not equal the function that we have expanded. Consider the function

$$\phi_\varepsilon(x) = \psi_\varepsilon(x) + e^{-x/\varepsilon}, x \geq 1.$$

Suppose we can expand $\psi_\varepsilon(x) = \sum_{n=0}^{\infty} \varepsilon^n \psi_n(x)$ and this series is supposed to be convergent. It is clear that

$$\tilde{\phi}_\varepsilon(x) = \sum_{n=0}^{m} \varepsilon^n \psi_n(x)$$

is an asymptotic approximation of $\phi_\varepsilon(x)$ that even converges for $m \to \infty$. The limit, however, does not equal $\phi_\varepsilon(x)$.

2.4.2 Practical Aspects of the du Bois-Reymond Theorem.

One might wonder whether the fact that there always exists an order function smaller than any term of an asymptotic sequence, as discussed in Section 2.2, has any practical importance. If one has constructed a nontrivial asymptotic approximation in some application, does this theorem matter? Remarkably enough it does, if two phenomena that are qualitatively very different are present in a problem. We shall meet examples of this in Chapter 4 and in what follows, where we have asymptotic expansions of a very different nature within one problem. We discuss another example here.

Consider a two degrees of freedom Hamiltonian system near a stable equilibrium point

$$\dot{p} = \frac{\partial H}{\partial q}, \quad \dot{q} = -\frac{\partial H}{\partial p},$$

with $H = H_0(p, q) + \varepsilon H_1(p, q)$; p and q are two-dimensional, and suppose that for $\varepsilon = 0$ the Hamiltonian system is integrable. This means that for $\varepsilon = 0$, a second independent integral exists apart from the Hamiltonian. Near stable equilibrium the energy manifold is foliated in invariant tori around the stable periodic solutions.

The KAM-theorem tells us that, under certain conditions, for ε nonzero but small enough, an infinite number of invariant tori survive. Constructing asymptotic approximations of the solutions by expansion in powers of ε, one finds these tori and periodic solutions. To any order (power) of ε, the approximating system is integrable, i.e., in the asymptotic approximation, the tori fill up phase-space near stable equilibrium.

We know however, that Hamiltonian systems with two degrees of freedom are generally nonintegrable and that chaotic phaseflow is present. There has been a lot of confusion in the literature about this, as we do not find these chaotic orbits in the perturbation expansion described above.

The explanation is that in the case of systems with *two* degrees of freedom near stable equilibrium, the measure of the chaotic orbits is $o(\varepsilon^n)$ for any n; the measure is exponentially small. An explicit example to demonstrate this has been studied by Holmes et al. (1988). A general discussion and more references can be found in Verhulst (1998, 2000).

This shows that the du Bois-Reymond theorem is of practical importance if a problem contains qualitatively different types of solutions that are characterised by different order functions of the perturbation parameter.

2.5 The Boundary of a Laser-Sustained Plasma

To illustrate the use and efficiency of asymptotic expansions, we consider a laser beam focused in a monatomic gas. If the laser intensity is high enough, a region of plasma can be created near the focal point. We consider a very simple model describing the local temperature T of the gas by an energy balance equation. In the case of equilibrium, the equation is

$$\varepsilon^2 \triangle T - T + I_0 F(r, T) H(T - 1) = 0$$

with, assuming spherical symmetry, the Laplace operator

$$\triangle = \frac{1}{r^2} \frac{\partial}{\partial r} \left(r^2 \frac{\partial}{\partial r} \right) ;$$

ε is the heat conductivity coefficient, supposed to be small; $-T$ stands for the radiation losses; I_0 is a constant indicating the laser intensity, $F(r, T)$ describes the absorption process that takes place if the temperature is high enough (i.e., $T \geq 1$). The Heaviside function $H(T - 1)$ switches off the absorption process if $T < 1$ as

$$H(T - 1) = 1, T \geq 1$$
$$= 0, T < 1.$$

The absorption coefficient takes the form

$$F(r, T) = \frac{1}{r^2 + a^2} \exp \left(- \int_r^\infty H(T(\xi) - 1) d\xi \right)$$

where a^2 represents the imperfection of the focusing device and the exponential accounts for the loss of intensity due to absorption.

Note that the trivial solution $T = 0$ satisfies the equation. We shall look for nontrivial solutions, ignited by the laser beam in a sphere around the focal point, with the properties

$$T(r) > 1, 0 \leq r < R,$$
$$T(R) = 1,$$
$$T(r) < 1, r > R,$$
$$\lim_{r \to \infty} T(r) = 0,$$

where R is the boundary of the plasma.

We start with the solution for $T(r)$ outside the plasma. The equation is

$$\varepsilon^2 \frac{1}{r^2} \frac{d}{dr} \left(r^2 \frac{dT}{dr} \right) - T = 0$$

or, with $v = Tr$,

$$\varepsilon^2 \frac{d^2 v}{dr^2} - v = 0.$$

Solving the equation produces

$$T(r) = \frac{1}{r}(Ae^{-r/\varepsilon} + Be^{r/\varepsilon}).$$

The requirement at infinity means that $B = 0$. The boundary of the plasma corresponds with $T(R) = 1$, which determines A. We find for the temperature outside the plasma

$$T(r) = \frac{R}{r} e^{-(r-R)/\varepsilon},$$

where R is still unknown. Inside the plasma, the equation becomes

$$\varepsilon^2 \Delta T - T + \frac{I_0}{r^2 + a^2} e^{-(R-r)} = 0.$$

For this equation, we have three conditions:

1. $T(R) = 1$,
2. $\lim_{r \to o} T(r)$ exists (regularity of the solution). Moreover,
3. $\frac{dT}{dr}\big|_{r \uparrow R} = \frac{dT}{dr}\big|_{r \downarrow R}$ (continuity of the heat flux across the boundary).

The third condition becomes $\frac{dT}{dr}\big|_{r \uparrow R} = -\frac{1}{\varepsilon} - \frac{1}{R}$.

Do we have too many conditions? Usually there are two conditions for a second order equation, but in this case we have the additional free parameter R. The solution describing the temperature T inside the plasma can be obtained by variation of constants; we have, using conditions (1) and (3),

$$T(r) = \frac{1}{r} \left[-\frac{I_0}{2\varepsilon} \left\{ e^{r/\varepsilon} \int_R^r e^{-t/\varepsilon} e^{-(R-t)} \frac{t\,dt}{t^2 + a^2} \right.\right.$$

$$\left.\left. -e^{-r/\varepsilon} \int_R^r e^{t/\varepsilon} e^{-(R-t)} \frac{t\,dt}{t^2 + a^2} \right\} + Re^{(R-r)/\varepsilon} \right].$$

This can be verified by differentiation and substitution. Note that condition 2 (regularity at the origin) means that the expression [   ] tends to zero as $r \to 0$. We determine R with this condition, which produces the relation

$$R = \frac{I_0}{2\varepsilon} e^{-R/\varepsilon} \left[\int_0^R e^{t/\varepsilon} e^{-(R-t)} \frac{t dt}{t^2 + a^2} - \int_0^R e^{-t/\varepsilon} e^{-(R-t)} \frac{t dt}{t^2 + a^2} \right].$$

This relation determines R implicitly as the solution of a transcendental equation, but it is not an easy one.

To get an asymptotic estimate of R, we integrate partially; we find

$$R = \tfrac{I_0}{2} \frac{R}{R^2 + a^2} + 0(\varepsilon) \quad \text{or}$$

$$R^2 = \tfrac{1}{2} I_0 - a^2 + 0(\varepsilon).$$

We conclude that a plasma solution exists whenever the intensity I_0 is such that $I_0 > 2a^2$, a result that is both simple and unexpected. See Eckhaus et al. (1985) for more results on these plasma problems.

In Chapter 3, we give an elementary introduction to the techniques estimating integrals.

2.6 Guide to the Literature

A number of the examples and basic results in this chapter can be found in Eckhaus (1979). For the fundamentals of asymptotic theory, see also Erdélyi (1956) and De Bruijn (1958).

In the nineteenth century, before the papers of Poincaré (1886) and Stieltjes (1886), the use of divergent series and asymptotics was not generally accepted. This attitude was clearly expressed by d'Alembert, who wrote around 1800: "For me, I have to state that all reasonings which are based on non-convergent series seem to be very suspect, even if the results agree with correct results obtained in another way." However, around 1900, after the papers of Poincaré and Stieltjes, divergent series became an accepted research topic.

A still very readable text also with historical remarks - among them d'Alembert's observation - is Borel (1901); see also Borel (1902, Chapter 3). The du Bois-Reymond theorem (1887) is discussed by Borel in the context of orders of infinity; see also Hardy (1910).

New developments in the basics of asymptotic series were the introduction of neutrices (Van der Corput, 1966) and nonstandard analysis. Neutrix calculus has not yet fullfilled its promise to start new theories and applications. Nonstandard analysis is flourishing; it is used to develop new concepts to characterise asymptotic behaviour together with a number of interesting applications; see for instance, Van den Berg (1987) and Jones (1997).

2.7 Exercises

Exercise 2.1 Consider again the function $\phi_\varepsilon(x) = e^{-x/\varepsilon}, x \in [0, 1]$. We have estimated this function using the sup norm and the L_2-norm; we introduce two additional norms.

$$\|\phi\| = \sup_{x \in D} |\phi| + \sup_{x \in D} \left| \frac{d\phi}{dx} \right| \text{ and } \|\phi\| = \left(\int_0^1 \left(\phi^2 + \left(\frac{d\phi}{dx} \right)^2 \right) dx \right)^{\frac{1}{2}}.$$

Estimate $e^{-x/\varepsilon}$ again using these four norms for $x \in [1, 2]$. Make the statement in Example 1.4 that this function is nearly zero for $x \geq \sqrt{\varepsilon}$ more precise.

Exercise 2.2 Compare the following order functions:

a. Show that: $\varepsilon^n = o(\varepsilon^m)$ if $n > m$.
b. Compute an order function $\delta(\varepsilon)$ such that $\varepsilon \sin \frac{1}{\varepsilon}$ is $0(\delta(\varepsilon))$; give examples in which $\varepsilon \sin(\frac{1}{\varepsilon}) = 0_s(\delta(\varepsilon))$.
c. Consider $\delta(\varepsilon) = -\varepsilon^2 \ln \varepsilon$; do we have $\delta(\varepsilon) = 0(\varepsilon^2 \ln^2(\varepsilon))$, or o, or 0_s?
d. Show that $\exp(-\frac{1}{\varepsilon}) = o(\varepsilon^n)$ with $n \in \mathbb{N}$.

Exercise 2.3 Consider the functions $f(x)$ and $g(x)$, with $x \in \mathbb{R}$:

$$f(x) = x,$$
$$g(x) = x \exp(x) + x^2.$$

a. Compare $f(x)$ and $g(x)$ for $x \to 0$. (This means compute whether $f(x) = 0(g(x))$ or $g(x) = 0(f(x))$, or o, or 0_s)
b. Answer the same question as in part (a) for $x \to +\infty$.

Exercise 2.4 Compare the functions $f(x)$ and $g(x), x \in \mathbb{R}$, for $x \downarrow 0$:

$$f(x) = x^2,$$
$$g(x) = x \sin^2(\frac{1}{x}) + x^3.$$

Exercise 2.5 Find an order function $\delta(\varepsilon)$ such that

$$\sin(\varepsilon) = \delta(\varepsilon) + 0(\varepsilon^4).$$

Does the estimate hold with 0_s instead of 0?

Exercise 2.6 Consider the following sequences:

a. $\{\delta_n(\varepsilon)\}_{n=1}^\infty$, with $\delta_n(\varepsilon) = -\varepsilon^n \ln \varepsilon$.
   Prove that this sequence is an asymptotic sequence.
b. Answer the same question as in part (a) with $\delta_1(\varepsilon) = \sin \varepsilon$ and $\delta_n(\varepsilon) = \varepsilon^n \exp(\frac{-1}{\varepsilon})$ with $n = 2, 3, 4, \cdots$.

c. Answer the same question as in part (a) with the sequences $\{\delta_n(\varepsilon)\}_{n=0}^{\infty}$, where $\delta_n(\varepsilon) = \varepsilon^n$.

Exercise 2.7 Consider the function

$$f(\varepsilon) = \sin \varepsilon + (\cos \varepsilon) \exp\left(-\frac{1}{\varepsilon}\right).$$

a. Which asymptotic sequence(s) from Exercises 2.5 and 2.6, can one use to construct approximations for $f(\varepsilon)$?
b. Are the approximations convergent or only asymptotic? Do they converge to $f(\varepsilon)$?

Exercise 2.8 Prove that if $\delta_1(\varepsilon) = 0(\delta_2(\varepsilon))$, then

a. $O(\delta_1) + O(\delta_2) = O(\delta_2)$.
b. $o(\delta_1) + o(\delta_2) = o(\delta_2)$.
c. $O(\delta_1) + o(\delta_2) = O(\delta_2)$.

Exercise 2.9 We consider expansions for $x \in \mathbb{R}^+$.

a. Consider the expansion

$$f(x;\varepsilon) = \sqrt{x + \varepsilon} = \sqrt{x}\left(1 + \frac{\varepsilon}{2x} - \frac{\varepsilon^2}{8x^2} + \cdots\right)$$

For which values of x is this an asymptotic sequence.
b. Answer the same question as in part (a) for the expansion

$$g(x;\varepsilon) = \frac{1}{1 + \varepsilon x} = 1 - \varepsilon x + \varepsilon^2 x^2 - \varepsilon^3 x^3 + \cdots.$$

Exercise 2.10 Consider the quadratic equation ($x \in \mathbb{R}$):

$$x^2 + \varepsilon x - 1 = 0,$$

with $0 < \varepsilon \ll 1$.

The solution of this equation is called x_ε; x_0 is the solution of $x^2 - 1 = 0$.

a. Is it true that $\lim_{\varepsilon \downarrow 0} x_\varepsilon = x_0$?
b. Use the explicit solutions to construct an asymptotic expansion for x_ε.
c. Substitute $x = \sum_{n=0}^{\infty} c_n \varepsilon^n$ in the equation and compute the coefficients c_n. We call this expansion a formal expansion. Is this formal expansion asymptotic? Is it convergent?
d. How do we know a priori (without using part b) that the formal expansion is asymptotic and even convergent?

Exercise 2.11 Consider the quadratic equation for $x \in \mathbb{R}$:

$$x^2 - 2x + (1 - \varepsilon) = 0.$$

a. Is it true that for the solutions of this equation we have $\lim_{\varepsilon \downarrow 0} x_\varepsilon = x_0$?

b. What is the result of the formal procedure of Exercise 2.10? Is this predictable? This phenomenon, which appears for $\varepsilon \downarrow 0$, is called a bifurcation.

Exercise 2.12 Consider the following quadratic equation for $x \in \mathbb{R}$:

$$\varepsilon x^2 + x + 1 = 0.$$

a. Which expression do we find for the roots of the quadratic equation after substituting the formal expansion $x = x_0 + \varepsilon x_1 + \varepsilon^2 x_2 + \cdots$?

b. Putting $\varepsilon = 0$, we obtain one root. Rescale $x = y/\varepsilon^p$; choose p suitable and compute the coefficients of the formal expansion $y = y_0 + \varepsilon y_1 + \varepsilon^2 y_2 + \cdots$. Determine y_0 and y_1.

c. Compare the expansion obtained in parts (a) and (b) with the exact (explicit) solutions of the equation.

Exercise 2.13 Consider a system of two bodies with Newtonian attraction and loss of mass. The variation of the eccentricity e and the angular variable E (eccentric anomaly) are given by

$$\frac{de}{dt} = \varepsilon \frac{(1 - e^2) \cos E}{(1 - e \cos E)},$$

$$\frac{dE}{dt} = \beta \frac{(1 - e^2)^{3/2}}{(1 - e \cos E)} - \varepsilon \frac{\sin E}{e(1 - e \cos E)}.$$

$0 < \beta$ is a dynamical constant independent of ε, $0 < \varepsilon \ll 1$, and $0 \le e < 1$.

a. Determine the stationary solutions $e = e_0$ (constant) and $E = E_0$ (constant).

b. Approximate the stationary solutions for e. (Construct a formal expansion in powers of (ε/β).)

For more details and references, see Verhulst (1975).

Exercise 2.14 Consider the equation $x = e^{\varepsilon x} - 1$ for $x \ge 0$. This equation has two roots, one is $x = 0$. Find an asymptotic approximation of the other root.

Exercise 2.15 Give an estimate for the integral

$$\int_0^1 e^{-\frac{x^2}{\varepsilon}} \, dx.$$

3

Approximation of Integrals

The techniques of expanding integrals have proved to be very useful in mathematical analysis, physics, fluid dynamics, and the technical sciences. There exist a number of excellent introductory books on the subject, a reason to be relatively brief here. A useful book has been written by Olver (1974); see also Erdelyi (1956), De Bruijn (1958), Jeffreys (1962), and Lauwerier (1974).

In practice, one often meets integrals for which no clear-cut theorem or result in the literature seems to be available. Therefore, instead of listing theorems, it is more important to obtain an idea of how to use the current methods such as partial integration, the method of Laplace, Fourier integrals, the method of stationary phase, the steepest descent, and the saddle point method. We discuss briefly the first three of them.

3.1 Partial Integration and the Laplace Integral

Consider the function $\phi(x)$ defined for $x > 0$ by

$$\phi(x) = \int_0^\infty e^{-xt} \frac{dt}{t+1}.$$

Note that the integral exists for positive values of x. We are interested in the behaviour of $\phi(x)$ for large values of x (i.e., for $x \to \infty$). In the context of this book, it would be natural to introduce $\varepsilon = 1/x$ and consider the behaviour of the corresponding integral for $\varepsilon \to 0$; however, we conform ourselves here to the standard formulation in the literature.

Intuitively, it is clear that the largest contribution to the value of the integral will come from a neighbourhood of $t = 0$. It is easy to obtain an expansion. From the geometric series, we have

$$\frac{1}{t+1} = \sum_{n=1}^m (-t)^{n-1} + \frac{(-t)^m}{t+1}.$$

Substitution produces

$$\phi(x) = \int_0^\infty e^{-tx} \left(\sum_{n=1}^m (-t)^{n-1} + \frac{(-t)^m}{t+1} \right) dt$$

$$= \tfrac{1}{x} - \tfrac{1}{x^2} + \cdots (-1)^{m-1} \tfrac{(m-1)!}{x^m} + R_m(x)$$

with

$$|R_m(x)| = \int_0^\infty e^{-tx} \tfrac{t^m}{t+1} dt$$

$$< \int_0^\infty e^{-tx} t^m dt = \tfrac{m!}{x^{m+1}},$$

and so we have obtained an asymptotic expansion for $\phi(x)$ with respect to the order function $(1/x)^n, n = 1, 2, \cdots$ for $x \to \infty$. Note that the series diverges; the ratio of subsequent terms yields

$$\left| \frac{a_{m+1}}{a_m} \right| = \frac{m}{x}.$$

We now generalise this result somewhat.

Theorem 3.1
Consider the Laplace integral

$$\phi(x) = \int_0^\infty e^{-xt} f(t) dt$$

with $f(t)$ uniformly bounded for $t \geq 0$; the derivatives $f^{(1)}(t) \cdots f^{(m+1)}(t)$ exist and are bounded on $[0, a], a > 0$. Then

$$\phi(x) = \sum_{n=0}^{m-1} \frac{1}{x^{n+1}} f^{(n)}(0) + 0 \left(\frac{1}{x^{m+1}} \right) + 0 \left(\frac{1}{x} e^{-ax} \right)$$

for $x \to \infty$.

Proof
The proof is simple and can be found in section 15.2.

The idea that the contribution from the integrand near $t = 0$ dominates the result can be generalised; for instance, to study integrals such as

$$\phi(x) = \int_a^b e^{xh(t)} f(t) dt,$$

one can consider the case where

$$f(t) = t^\lambda g(t)$$

with Re $\lambda > -1$ and where $g(t)$ is supposed to have an asymptotic expansion for $t \to 0$ and does not grow faster than exponentially for $t \to \infty$. Again one can expand the integral, and the corresponding result is called Watson's lemma. The lemma can be formulated for functions that are real and complex; see the references mentioned in the introduction.

3.2 Expansion of the Fourier Integral.

Consider the oscillating or Fourier integral

$$\phi(x) = \int_\alpha^\beta e^{ixt} f(t)dt$$

with α, β real; $f(t)$ is integrable and we wish to study the behaviour of the integral for $x \to \infty$. The intuitive idea is that for x large enough there are fast oscillations that tend to average out all contributions of the integrand except near the endpoints. One of the simplest results runs as follows.

Theorem 3.2
Consider the Fourier integral $\phi(x) = \int_\alpha^\beta e^{ixt} f(t)dt$ with $f(t)$ that is $(m+1)$ times continuously differentiable in $[\alpha, \beta]$; then
$$\phi(x) = -\sum_{n=0}^{m-1} i^{n-1} \frac{f^{(n)}(\alpha)}{x^{n+1}} e^{ix\alpha} + \sum_{n=0}^{m-1} i^{n-1} \frac{f^{(n)}(\beta)}{x^{n+1}} e^{ix\beta} + 0(x^{-m-1}).$$

Remark
The result remains valid if $\alpha = -\infty$ or $\beta = \infty$, but we have to require that $f^{(n)}(t) \to 0$ for $t \to -\infty$ or $+\infty$.

Proof
See section 15.2.

In a number of problems, we encounter integrals in which $f(t)$ has a singularity at $t = \alpha$ or β. Similar results can be obtained in these cases with weaker estimates for the remainder.

3.3 The Method of Stationary Phase

We consider more general oscillating integrals

$$\phi(x) = \int_\alpha^\beta e^{ixf(t)} g(t)dt$$

for $x \to \infty$; f and g are real functions.

Kelvin reasoned that the most important contribution to the integral arises from a neighbourhood of t_0 where the phase $f(t)$ is stationary (i.e., $f'(t_0) = 0$). It turns out that this intuitive reasoning leads in a number of cases to correct first terms of an expansion. Note that in Theorem 3.2, where the contributions of the endpoints dominate, there are no stationary phase points.

Example 3.1
Consider $\phi(x) = \int_{-\infty}^{-\infty} \frac{e^{ixt^2}}{1+t^2} dt$. The phase t^2 is stationary if $t = 0$, which suggests expanding the integrand around $t = 0$. A probable first asymptotic approximation of the integral would then be

$$\phi_0(x) = \int_{-\infty}^{+\infty} e^{ixt^2} dt.$$

Rotating over $\pi/4$ by $z = e^{-i\pi/4}t$ yields

$$\phi_0(x) = e^{i\pi/4} \int_{-\infty}^{\infty} e^{-xz^2} dz = e^{i\pi/4} \sqrt{\frac{\pi}{x}}.$$

$\phi_0(x)$ turns out to be an asymptotic approximation of $\phi(x)$.

A simple example of a theorem using the method of stationary phase runs as follows. We start by splitting the interval of integration into subintervals with stationary points at one of the endpoints. Consider an integral corresponding with one of the subintervals.

Theorem 3.3
Consider $\phi(x) = \int_a^b e^{ixf(t)} g(t) dt$ with $f \in C^2[a,b]$, $f'(t) \neq 0$ in $(a,b]$, $f'(a) = 0$, $f''(a) \neq 0$, and $g(t)$ is bounded in a neighbourhood of $t = a$ and integrable. If $f(a)$ is a minimum,

$$\phi(x) = \frac{1}{\sqrt{x}} \left(\frac{\pi}{2f''(a)} \right)^{\frac{1}{2}} g(a) e^{\frac{i\pi}{4} + ixf(a)} + 0(x^{-1}).$$

If $f(a)$ is a maximum, the coefficient of $f''(a)$ changes sign.

Proof
See Lauwerier (1974) or Olver (1974).

3.4 Exercises

Exercise 3.1 Consider the incomplete gammafunction

$$\gamma(a, x) = \int_0^x e^{-t} t^{a-1} dt, a > 0, x > 0.$$

a. Show that a suitable series for calculation is given by

$$\gamma(a, x) = \sum_{n=0}^{\infty} \frac{(-1)^n x^{n+a}}{(a+n)n!}, \quad \text{if } x > 0 \text{ is small.}$$

b. Is the expansion obtained in (a) a useful series for numerical calculations? (For example, $a = \frac{1}{2}, x = 10, (\Gamma(\frac{1}{2}, 10) = \sqrt{\pi} + 0(10^{-5}).)$

If x is large and positive, it seems better to consider the function

$$\Gamma(a, x) = \Gamma(a) - \gamma(a, x) = \int_x^{\infty} e^{-t} t^{a-1} dt, a \in \mathbb{R}.$$

c. Give an asymptotic expansion for $x \to \infty$ and an error estimate.
d. How suitable is the result, obtained in (c), for numerical calculations?
e. Put $x = \frac{1}{\varepsilon}, \Gamma(a, \frac{1}{\varepsilon}) = f(\varepsilon)$. Which asymptotic sequence is used in the expansion of $f(\varepsilon)$?

Exercise 3.2 Consider the error function

$$\text{erfc}(x) = \frac{2}{\sqrt{\pi}} \int_0^x e^{-t^2} dt.$$

Give an asymptotic expansion of $\text{erfc}(x)$ for $x \to \infty$, and give also an error estimate. (Remark: The error function is by transformation related to the incomplete gammafunction.)

Exercise 3.3 Consider the oscillatory integral

$$\begin{cases} \phi(x) = \int_\alpha^\beta e^{ixt} f(t) dt \\ \alpha, \beta \in \mathbb{R}, f(t) \text{ is } N \text{ times continuously differentiable} \end{cases}$$

for $x \to \infty$ (or $x \to -\infty$). Prove that:

$$\phi(x) = \sum_{n=0}^{N-1} (f^n(\alpha)e^{ix\alpha} - f^n(\beta)e^{ix\beta})(-ix)^{-n-1} + o(x^{-N})$$

for $x \to \infty$ (and $x \to -\infty$).
(Remark: Use Weierstrass' approximation theorem.)

Exercise 3.4 Consider the integral

$$f(w) = \int_0^{\infty} e^{-w(t-\ln(1+t))} dt.$$

The function $\phi(t) = t - \ln(1+t)$ is concave with minimum zero at $t = 0$.

a. Compute the Taylor expansion for $\phi(t)$ in a neighbourhood of $t = 0$:

$$\phi(t) = \sum_{n=0}^{\infty} a_n t^n.$$

b. Substitute this expansion in $\frac{1}{2}u^2 = \phi(t)$ and obtain

$$u = \sum_{n=0}^{\infty} b_n t^n.$$

c. Compute the inverse expansion of u:

$$t = \sum_{n=0}^{\infty} c_n t^n.$$

d. Differentiate this expansion and compute the first two terms of the asymptotic expansion of $f(w)$.

Remark:
$f(w) = \int_{-\infty}^{\infty} e^{-\frac{1}{2}wu^2} \frac{dt}{du} du.$
$f(w) = w! w^{-w-1} e^w.$

Exercise 3.5 Prove that

$$\int_0^{\infty} \cos\left(t + \frac{1}{2}\varepsilon t^2\right) dt = 0(1), \quad \text{for } \varepsilon \downarrow 0,$$

and

$$\int_0^{\infty} \cos\left(-t + \frac{1}{2}\varepsilon t^2\right) dt = 0\left(\frac{1}{\varepsilon}\right), \quad \text{for } \varepsilon \downarrow 0.$$

(Remark: Find a suitable transformation and use partial integration.)

4

Boundary Layer Behaviour

In this chapter, we take a closer look at boundary layer phenomena. The tools we shall develop for our analysis are local boundary layer variables and degenerations of operators.

4.1 Regular Expansions and Boundary Layers

In Chapter 1, we considered two examples of first-order differential equations. For the solutions, we substituted a formal expansion of the form

$$\phi_\varepsilon(x) = \sum_{n=0}^{\infty} \varepsilon^n \phi_n(x).$$

In the case of the problem

$$\phi' + \varepsilon\phi = \cos x, \phi(0) = 0,$$

this led to a consistent formal expansion. In the problem

$$\varepsilon\phi' + \phi = \cos x, \phi(0) = 0,$$

by comparing this with the exact solution, we have shown that the formal expansion is far too simple to represent the solution. The difficulty with the formal expansion arises when applying the boundary condition; a problem of this type is often called a *singular perturbation* or *boundary layer problem*. The asymptotic expansion in the second case looks like

$$\phi_\varepsilon(x) = \sum_{n=0}^{\infty} \delta_n(\varepsilon)\psi_n(x, \varepsilon).$$

The simpler expansion that we used in the first case will be called *regular*. Note that in the literature the term "regular" is used in many ways.

Definition
Consider the function $\phi_\varepsilon(x)$ defined on $D \subset \mathbb{R}^n$; an asymptotic expansion for $\phi_\varepsilon(x)$ will be called regular if it takes the form

$$\phi_\varepsilon(x) = \sum_{n=0}^{m} \delta_n(\varepsilon)\psi_n(x) + 0(\delta_{m+1})$$

with $\delta_n(\varepsilon), n = 0, 1, \cdots$ an asymptotic sequence and $\psi_n(x), n = 0, 1, \cdots$ functions on D.

It turns out that in studying a function $\phi_\varepsilon(x)$ on a domain D we have to take into account that regular expansions of $\phi_\varepsilon(x)$ often only exist on subdomains of D.

Example 4.1
Consider the function

$$\phi_\varepsilon(x) = e^{-x/\varepsilon} + e^{\varepsilon x}, x \in [0, 1].$$

On any subdomain $[d, 1]$ with $0 < d < 1$ a constant independent of ε, we have the regular expansion

$$\phi_\varepsilon(x) = \sum_{n=0}^{m} \varepsilon^n \frac{x^n}{n!} + 0(\varepsilon^{m+1}).$$

However, there exists no regular expansion in functions of x on $[0, d]$ or $[0, 1]$.

Example 4.2
The function $\phi_\varepsilon(x)$ is for $x \in [0, 1]$ defined as the solution of the two-point boundary value problem

$$\varepsilon\phi'' + \phi' = 0, \phi(0) = 1, \phi(1) = 0.$$

The solution is

$$\phi_\varepsilon(x) = \frac{e^{-x/\varepsilon}}{1 - e^{-1/\varepsilon}} - \frac{e^{-1/\varepsilon}}{1 - e^{-1/\varepsilon}}.$$

Ignoring this exact solution and looking for a regular expansion by substituting into the differential equation

$$\phi_\varepsilon(x) = \sum_{n=0}^{m} \varepsilon^n \phi_n(x),$$

we find

$$\phi_0' = 0,$$

$$\phi_0'' + \phi_1' = 0, \text{ etc.}$$

This leads to $\phi_0(x) = $ constant, $\phi_1(x) = $ constant, etc. There is no way to satisfy the boundary conditions with the regular expansion.

The exact solution can be written as

$$\phi_\varepsilon(x) = e^{-x/\varepsilon} + 0(e^{-1/\varepsilon}),$$

which shows that the behaviour of the solutions is different in two subdomains of $[0, 1]$: in a small region of size $0(\varepsilon)$ near $x = 0$, the solution decreases very rapidly from 1 towards 0; in the remaining part of $[0, 1]$, the solution is very near 0. The regular expansion is valid here with, rather trivially, $\phi_n(x) = 0, n = 0, 1, \cdots$. Note that even in the domain where the regular expansion is valid, the choice to expand with respect to order functions of the form ε^n is not a fortunate one as $e^{-1/\varepsilon} = o(\varepsilon^n), n = 1, 2, \cdots$.

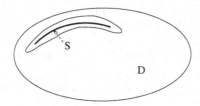

Fig. 4.1. Domain D with boundary layer near S.

The small region near $x = 0$ where no regular expansion exists in Example 4.2, is called a boundary layer. We now characterise such regions.

4.1.1 The Concept of a Boundary Layer

Consider the function $\phi_\varepsilon(x)$ defined on $D \subseteq \mathbb{R}^n$. Suppose there exists a connected subset $S \subset D$ of dimension $\leq n$, with the property that $\phi_\varepsilon(x)$ has no regular expansion in each subset of D containing points of S (see Fig. 4.1). Then a neighbourhood of S in D with a size to be determined, will be called a boundary layer of the function $\phi_\varepsilon(x)$.

In Examples 4.1 and 4.2 the domain is one-dimensional. A boundary layer, corresponding with the subset S, has been found near the boundary point $x = 0$ of the domain. Note that in applications we also may find a boundary layer in the interior of the domain and, if we have an evolution equation, the boundary layer may even be moving in time.

To study the behaviour of a function $\phi_\varepsilon(x)$ in a boundary layer, a fundamental technique is to use a *local analysis*. This is a technique that we shall meet again and again. In this chapter, we consider local analysis in a one-dimensional context. Later we shall meet higher-dimensional problems.

Suppose that near a point $x_0 \in S$ the boundary layer is characterised in size by an order function $\delta(\varepsilon)$. We "rescale" or "stretch" the variable x by introducing the local variable

$$\xi = \frac{x - x_0}{\delta(\varepsilon)}.$$

If $\delta(\varepsilon) = o(1)$, we call ξ a local (stretched or boundary layer) variable. The function $\phi_\varepsilon(x)$ transforms to

$$\phi_\varepsilon(x) = \phi_\varepsilon(x_0 + \delta(\varepsilon)\xi)$$
$$= \phi_\varepsilon^*(\xi).$$

It is then natural to continue the local analysis by expanding the function ϕ_ε^* with respect to the local variable ξ; we hope to find again a *regular* expansion. To be more precise, assume that

$$\phi_\varepsilon^*(\xi) = 0_s(1) \text{ near } \xi = 0.$$

We wish to find local approximations of ϕ_ε^* by a regular expansion of the form

$$\phi_\varepsilon^*(\xi) = \sum_n \delta_n^*(\varepsilon)\psi_n(\xi)$$

with $\delta_n^*(\varepsilon), n = 0, 1, 2, \cdots$ an asymptotic sequence.

When solving the problem of approximating a function $\phi_\varepsilon(x)$ in a domain D, our program will then be as follows.

1. Try to construct a regular expansion in the original variable x. This is possible outside the boundary layers (by definition), and this expansion is usually called the *outer expansion*.
2. Construct in the boundary layer(s) a local expansion in an appropriate local variable. Such a regular expansion is usually called the *inner expansion* or *boundary layer expansion*.
3. The inner and outer expansions should be matched to obtain a formal expansion for the whole domain D. As we shall see later, in a number of problems, techniques have been developed to combine the three stages, which makes the process more efficient. This formal expansion, which is valid in the whole domain, is sometimes called a uniform expansion. Note, however, that in the literature and also in this book, expressions that are called "uniformly valid expansion" are more often than not formal expansions. So we come to the next point.
4. Prove that formal expansions, obtained in the stages 1–3, represent valid asymptotic approximations of the function $\phi_\varepsilon(x)$ that we set out to study.

In many problems, the function $\phi_\varepsilon(x)$ has been implicitly defined as the solution of a system of differential equations with initial and/or boundary conditions. To study such a problem by perturbation theory, we have to be more precise in the use of the expressions "formal expansion" and "formal approximation". Suppose that we have to study the perturbation problem

$$L_\varepsilon \phi = f(x), x \in D + \text{ other conditions.}$$

L_ε is an operator containing a small parameter ε. For instance, in Example 4.2 we have

$$L_\varepsilon = \varepsilon \frac{d^2}{dx^2} + \frac{d}{dx}$$

$D = [0, 1]$, $f(x) = 0$ and boundary conditions $\phi(0) = 1, \phi(1) = 0$.

The function $\tilde{\phi}_\varepsilon(x)$ will be called a *formal approximation* or *formal expansion* of $\phi_\varepsilon(x)$ if $\tilde{\phi}$ satisfies the boundary conditions to a certain approximation and if

$$L_\varepsilon \tilde{\phi} = f(x) + o(1).$$

We shall see that to require $\tilde{\phi}$ to satisfy the boundary conditions in full is asking too much and in practice it suffices for $\tilde{\phi}$ to satisfy the boundary conditions to a certain approximation.

To prove that if $\tilde{\phi}$ is a formal approximation it also is an asymptotic approximation of ϕ is in general a difficult problem. Moreover, one can give simple and realistic examples in which this is not true. Remarkably enough, we shall later also meet cases where we have an asymptotic approximation that is not a formal approximation.

Example 4.3
(a formal approximation that is not asymptotic)
Consider the harmonic oscillator

$$\ddot{\phi} + (1 + \varepsilon)^2 \phi = 0, t \geq 0$$

with initial conditions $\phi(0) = 1, \dot{\phi}(0) = 0$. The solutions of the equation are bounded, and with these initial conditions we have

$$\phi(t) = \cos((1 + \varepsilon)t).$$

On the other hand, $\tilde{\phi}(t) = \cos t$ satisfies the initial conditions and moreover

$$\ddot{\tilde{\phi}} + (1 + \varepsilon)^2 \tilde{\phi} = 2\varepsilon \cos t + \varepsilon^2 \cos t = O(\varepsilon).$$

However, using the sup norm, it is easy to see that $\phi(t) - \tilde{\phi}(t) = 0_s(1), t \geq 0$.

4.2 A Two-Point Boundary Value Problem

We shall now illustrate the process of constructing a formal expansion for a simple boundary value problem.

Example 4.4

Consider the equation

$$\varepsilon \frac{d^2\phi}{dx^2} - \phi = f(x), x \in [0,1]$$

with boundary values $\phi_\varepsilon(0) = \phi_\varepsilon(1) = 0$. The function $f(x)$ is sufficiently smooth on $[0,1]$ to allow for the construction that follows. We note that the choice of boundary values equal to zero (homogeneous boundary values) is not a restriction: aif $\phi_\varepsilon(0) = \alpha, \phi_\varepsilon(1) = \beta$, we introduce $\psi_\varepsilon(x) = \phi_\varepsilon(x) - \alpha - (\beta - \alpha)x$, which produces zero boundary values for the problem in ψ_ε (while of course changing the right-hand side).

In the spirit of Section 3.1, we assume that in some subset D_0 of $[0,1]$ the solution has a regular expansion of the form

$$\phi_\varepsilon(x) = \sum_{n=0}^{m} \varepsilon^n \phi_n(x).$$

Substitution in the preceding equation produces successively

$$\phi_0(x) = -f(x),$$

$$\phi_n(x) = \frac{d^2\phi_{n-1}}{dx^2}, n = 1, 2, \cdots.$$

The expansion coefficients ϕ_n are determined completely by the recurrency relation so that we *cannot impose* the boundary conditions to the regular expansion. (Even if accidentally $f(0) = f(1) = 0$, the next order will change the boundary conditions again.) We conclude that for the regular expansion to make sense in D_0, this subset should not contain the boundary points $x = 0$ and $x = 1$ (see Fig. 4.2). Before analysing what is going on near the boundary

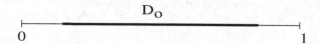

Fig. 4.2. Boundary layers near the end points.

points, we carry out the *subtraction trick*; this trick is of no fundamental importance but is computationally convenient, as it shifts the equation to a homogeneous one. For nonlinear equations, the subtraction trick can be less convenient. Introduce

$$\psi_\varepsilon(x) = \phi_\varepsilon(x) - \sum_{n=0}^{m} \varepsilon^n \phi_n(x).$$

The two-point boundary value problem for ψ_ε becomes

$$\varepsilon \frac{d^2\psi}{dx^2} - \psi = 0(\varepsilon^{m+1})$$

and

$$\psi_\varepsilon(0) = -\sum_{n=0}^{m} \varepsilon^n \phi_n(0), \, \psi_\varepsilon(1) = -\sum_{n=0}^{m} \varepsilon^n \phi_n(1).$$

Near $x = 0$, we introduce the local variable

$$\xi = \frac{x}{\delta(\varepsilon)}$$

with $\delta(\varepsilon) = o(1)$; at this point, we have no a priori knowledge of a suitable choice of $\delta(\varepsilon)$.

The equation with respect to this local variable becomes

$$\frac{\varepsilon}{\delta^2} \frac{d^2\psi^*}{d\xi^2} - \psi^* = 0(\varepsilon^{m+1}),$$

where $\psi^* = \psi_\varepsilon(\delta(\varepsilon)\xi)$. How to choose $\delta(\varepsilon)$ will be the subject of discussion later on; here we just remark that

$$\delta^2(\varepsilon) = \varepsilon \text{ or } \delta(\varepsilon) = \sqrt{\varepsilon}$$

seems to be a well-balanced choice. We now assume that there exists a regular expansion of ψ_ε^*, so

$$\lim_{\varepsilon \downarrow 0} \psi_\varepsilon^*(\xi) = \psi_0(\xi) \text{ exists}$$

and satisfies the formal limit equation

$$\frac{d^2\psi_0}{d\xi^2} - \psi_0 = 0.$$

ψ_0 would be the first term in a formal regular expansion in ξ near the point $x = 0$. Solving this limit equation, we find

$$\psi_0(\xi) = ae^{-\xi} + be^{+\xi}.$$

Imposing the boundary condition at $\xi = 0$, we have

$$a + b = -\phi_0(0) = f(0).$$

We have to find a second relation to determine a and b; for this we observe that the solution in the boundary layer near $x = 0$ should be matched with the regular expansion in D_0. Rewriting $\psi_0(\xi)$ in x, we note that the term $\exp(x/\sqrt{\varepsilon})$ becomes exponentially large with x in D_0 unless its coefficient b vanishes, so we propose the *matching relation* $\lim_{\xi \to \infty} \psi_0(\xi) = 0$ or $b = 0$ and we have

$$\psi_0(\xi) = f(0)e^{-\xi}.$$

We can proceed to calculate higher-order terms by assuming a regular expansion of the form

$$\psi_\varepsilon^*(\xi) = \sum_{n=0}^{m} \varepsilon^{n/2} \psi_n(\xi),$$

where for each term in the expansion we have again one boundary condition and one matching relation. The equations for $\psi_1, \psi_2, \cdots$ will become increasingly complicated. We can repeat this local analysis near $x = 1$ by introducing the local variable

$$\eta = \frac{1 - x}{\sqrt{\varepsilon}}.$$

The calculation runs along exactly the same lines and is left to the reader. Adding the terms from the outer expansion and the two local (boundary layer) expansions, we find to first order the formal uniform expansion

$$\tilde{\phi}_\varepsilon(x) = -f(x) + f(0)e^{-x/\sqrt{\varepsilon}} + f(1)e^{-(1-x)/\sqrt{\varepsilon}} + \cdots.$$

The dots are standing for terms such as $\varepsilon\phi_1(x)$, $\sqrt{\varepsilon}\psi_1(\xi)$, etc. It follows from the construction that $\tilde{\phi}_\varepsilon(x)$ satisfies the equation to a certain approximation and it is a formal approximation of the solution, as we allow for exponentially small deviations of the boundary conditions. Later we shall return to the question of whether $\tilde{\phi}$ is an asymptotic approximation of ϕ. At this stage, it is interesting to note that one easily finds an affirmative answer by analysing the exact solution, which can be found by variation of constants.

4.3 Limits of Equations and Operators

We consider differential operators L_ε parametrised by ε. We are interested in discussing the limit as $\varepsilon \to 0$ of L_ε while keeping in mind our experience of the preceding sections where in various subdomains different variables have played a part. Consider again Example 4.4 of Section 4.2,

$$L_\varepsilon\phi = f(x), \phi_\varepsilon(0) = \phi_\varepsilon(1) = 0,$$

with

$$L_\varepsilon = \varepsilon \frac{d^2}{dx^2} - 1.$$

Taking the formal limit of L_ε as $\varepsilon \to 0$, we obtain

$$L_0 = -1.$$

Introducing local variables of course changes the limiting behaviour of the operator. Consider for instance the boundary layer near $x = 0$ and introduce

$$\xi = \frac{x}{\delta(\varepsilon)} \text{ with } \delta(\varepsilon) = o(1).$$

We find

$$L^*_\varepsilon = \frac{\varepsilon}{\delta^2} \frac{d^2}{d\xi^2} - 1.$$

Our choice of $\delta(\varepsilon)$ determines which operator L^*_ε degenerates into as $\varepsilon \to 0$. For instance, if $\delta^2(\varepsilon) = \varepsilon$, then

$$L^*_0 = \frac{d^2}{d\xi^2} - 1.$$

If $\varepsilon = o(\delta^2(\varepsilon))$, and for instance $\delta = \varepsilon^{1/4}$, we have

$$L^*_0 = -1.$$

If $\delta^2(\varepsilon) = o(\varepsilon)$, the limit does not exist and it makes sense to rescale,

$$\frac{\delta^2}{\varepsilon} L^*_\varepsilon = L^{**}_\varepsilon.$$

We find the degeneration

$$L^{**}_0 = \frac{d^2}{d\xi^2}.$$

Making these calculations, we observe that, taking the formal limits, the degenerations of the operator in the cases $\varepsilon = o(\delta^2)$ and $\delta^2 = o(\varepsilon)$ are *contained* in the degeneration obtained on choosing $\delta^2 = \varepsilon$. We call L^*_0 in this case,

$$L^*_0 = \frac{d^2}{d\xi^2} - 1,$$

a *significant degeneration* of the operator L_ε near $x = 0$. Put in a different way, a significant degeneration implies a well-balanced choice of the local variable ξ such that the corresponding operator as $\varepsilon \to 0$ contains as much information as possible.

Note that a number of authors are using the term *distinguished limit* instead of significant degeneration.

Definition
Consider the operator L_ε, written in the variable x, near the boundary layer point $x = x_0$ and the operator L_ε rewritten in all possible local variables of the form $(x - x_0)/\delta(\varepsilon)$ near x_0. L^*_0 is called a significant degeneration of L_ε if L^*_0 is obtained by writing L in the local variable ξ and taking the formal limit as $\varepsilon \to 0$ (possibly after rescaling), whereas the corresponding degenerations in the other local variables are contained in L^*_0.

Remark
In practice, many operators can be analysed for significant degenerations by considering the set of order functions $\delta(\varepsilon) = \varepsilon^\nu, \nu > 0$. For instance, in our example

$$L_\varepsilon = \varepsilon \frac{d^2}{dx^2} - 1,$$

we have, introducing the local variable $\xi_\nu = x/\varepsilon^\nu$,

$$L_\varepsilon = \varepsilon^{1-2\nu} \frac{d^2}{d\xi_\nu^2} - 1,$$

$$\nu = \frac{1}{2}, \ L_0^* = \frac{d^2}{d\xi_{\frac{1}{2}}^2} - 1,$$

$$\nu > \frac{1}{2}, \ L_0^* = \frac{d^2}{d\xi_\nu^2},$$

$$\nu < \frac{1}{2}, \ L_0^* = -1.$$

Remark

This definition is too simple in some cases. For instance, it is possible that near a point $x = x_0$, more than one significant degeneration exists corresponding with a more complicated boundary layer structure. For these cases, we have to adjust the definition somewhat, but we omit this here.

By introducing the concept of significant degeneration, we have a formal justification of our choice of boundary layer variables in the problem of Example 3.3 in Section 3.2. The underlying assumption here and in the following is that on analysing the significant degenerations of an operator, we find the correct boundary layer variables and the corresponding expansions in these variables. It turns out that this asumption works very well in most problems.

Still we have to keep in mind the possibility of hidden pitfalls. One of the assumptions in the construction is that the problem obtained by taking the formal limit of the operator does have something to do with the original problem. That is, if we study the function $\phi_\varepsilon(x)$ in a domain D given by the equation

$$L_\varepsilon \phi = f_\varepsilon(x),$$

by taking the limit we have

$$L_0 \psi = f_0(x).$$

Does this mean that we have $\lim_{\varepsilon \downarrow 0} \phi_\varepsilon(x) = \psi(x)$ in some nontrivial subdomain $D_0 \subset D$?

In applications, the answer seems to be affirmative. The reason is that equations in practice have been obtained by modelling reality. In these equations, various distinct effects or forces (in mechanics) play a part. For a degeneration to make sense, we have that locally in space or time one or a few of these effects or forces is dominant. This is quite natural in applications. Mathematically, this is not a simple question, and we give an example, admittedly artificial, to show that we may have a serious problem here.

Example 4.5
(Eckhaus, 1979)
Consider the following initial value problem in a neighbourhood of $x = 0$:

$$L_\varepsilon \phi = f_\varepsilon(x), \phi_\varepsilon(0) = 1, \phi'_\varepsilon(0) = 0,$$

$$L_\varepsilon = \varepsilon^3 \cos(\frac{x}{\varepsilon^2})\frac{d^2}{dx^2} + \varepsilon \sin(\frac{x}{\varepsilon^2})\frac{d}{dx} - 1,$$

$$f_\varepsilon(x) = -\varepsilon(1 - \cos(\frac{x}{\varepsilon^2})).$$

Taking the formal limit $L_0\psi = f_0(x)$, we find $\psi(x) = 0$, but the solution of the problem is

$$\phi_\varepsilon(x) = 1 + \varepsilon - \varepsilon \cos\left(\frac{x}{\varepsilon^2}\right)$$

with, for all $x \geq 0$,

$$\lim_{\varepsilon \to 0} \phi_\varepsilon(x) = 1.$$

Note that $\psi(x) = 0$ is not a formal approximation since it does not satisfy the boundary conditions. However, more dramatically, there is also no subdomain where $\psi(x)$ represents an asymptotic approximation of ϕ. On the other hand, the function $\phi_0(x) = 1$ does represent an asymptotic approximation, but this function does not satisfy the limit equation!

Fortunately, this example is not typical for applications, but it is instructive to keep it in mind as a possible phenomenon. It also illustrates the importance of giving proofs of asymptotic validity.

4.4 Guide to the Literature

The idea of a boundary layer orginates from physics, in particular fluid mechanics; see Prandtl (1905) and Prandtl and Tietjens (1934). Degenerations of operators, significant degenerations in local variables, and matching techniques are more recent concepts. Both the terminology and the techniques may take rather different forms.

One of the approaches to validate matching is the assumption of the so-called *overlap hypothesis*. This assumes that if one has two neighbouring local expansions or a neighbouring local and a regular expansion, there exists a common subdomain where both expansions are valid. This provides a sufficient condition to match the expansions by using in this subdomain intermediate variables. Such variables were considered by Kaplun and Lagerstrom (1957). More discussion of matching rules is found in Section 6.2 and Section 15.4.

The foundations are further discussed by Van Dyke (1964), Fraenkel (1969), Lagerstrom and Casten (1972), Eckhaus (1979) and Kevorkian and Cole (1996).

A different approach to identify scales and layers is to use blow-up transformations; see Krupa and Szmolyan (2001) and Popović and Szmolyan (2004).

The theory of singular perturbations as far as boundary layer theory is concerned, is still largely a collection of inductive methods in which taste and inventiveness play an important part.

4.5 Exercises

Exercise 4.1 Compute a second-order approximation of Example 4.4 treated in Section 4.2. Discuss the asymptotic character of the approximation.

Exercise 4.2 Consider the boundary value problem

$$\begin{cases} \varepsilon y'' + y' + y = 0, \\ y(0) = a, y(1) = b. \end{cases}$$

Compute a first-order approximation of the solution $y_\varepsilon(x)$ of the boundary value problem and compare this approximation with the exact solution. Is this first-order approximation a formal approximation?

Exercise 4.3 Consider the operator

$$L_\varepsilon = \varepsilon(\varepsilon^2 + x - 1)\frac{d}{dx} + \varepsilon(\varepsilon + 1) + x - 1.$$

a. Compute the significant degenerations of the operator L_ε in a neighbourhood of $x = 0$ and $x = 1$. (Show that the other degenerations are contained in them.)

b. To illustrate the result of (a), we solve the initial value problem

$$L_\varepsilon y = 0, y(1) = 1.$$

Compute the solution of this problem and compare it with the result in (a).

Exercise 4.4 Consider the boundary value problem

$$\varepsilon L_1 y - y = 1, x \in (0,1),$$
$$y(0) = a, y(1) = b, a, b \neq 0,$$
$$L_1 = (1 + x^2)\frac{d^2}{dx^2} + \frac{d}{dx} - x.$$

Compute the first- and second-order terms of a formal approximation; show explicitly that this expansion is a formal approximation.

5

Two-Point Boundary Value Problems

In linear problems, much more qualitative information is available than for nonlinear systems. Still, in general we cannot solve linear equations with variable coefficients explicitly. Therefore it is very surprising that in the case of singularly perturbed linear systems, we can often obtain asymptotic expansions to any accuracy required.

In most of this chapter, we shall look at rather general, linear two-point boundary value problems. To start, consider a second-order equation of the form

$$\varepsilon L_1 \phi + L_0 \phi = f(x), \quad x \in (0, 1)$$

with boundary values $\phi_\varepsilon(0) = \alpha, \phi_\varepsilon(1) = \beta$. For the operator L_1, we write

$$L_1 = a_2(x)\frac{d^2}{dx^2} + a_1(x)\frac{d}{dx} + a_0(x)$$

with $a_0(x), a_1(x), a_2(x)$ continuous functions in $[0, 1]$; moreover,

$$a_2(x) > 0, \quad x \in [0, 1].$$

We exclude that $a_2(x)$ vanishes in $[0, 1]$, as this introduces singularities requiring a special approach. Taking the formal limit for $\varepsilon \to 0$, we retain $L_0 \phi = f(x)$; for the operator L_0, we have

$$L_0 = b_1(x)\frac{d}{dx} + b_0(x).$$

We assume $b_0(x)$ and $b_1(x)$ to be continuous in $[0, 1]$. It turns out that the analysis of the problem depends very much on the properties of b_0 and b_1.

5.1 Boundary Layers at the Two Endpoints

We first treat the case where $b_1(x)$ vanishes everywhere in $[0, 1]$, so we are considering the equation

$$\varepsilon L_1 \phi + b_0(x)\phi = f(x).$$

Assume that $b_0(x)$ does not change sign; for instance,

$$b_0(x) < 0, \quad x \in [0,1].$$

If $b_0(x)$ changes sign, the analysis is more complicated; we shall deal with such a case later on. As before, we suppose that in a subdomain of $[0,1]$ there exists a regular expansion of the solution of the form

$$\phi_\varepsilon(x) = \sum_{n=0}^{m} \varepsilon^n \phi_n(x) + O(\varepsilon^{m+1}).$$

On substituting this into the preceding equation, we find

$$\phi_0(x) = f(x)/b_0(x)$$

$$\phi_n = -L_1 \phi_{n-1}(x)/b_0(x), \quad n = 1, 2, \cdots.$$

It is clear that to perform this construction $b_0(x)$ and $f(x)$ have to be sufficiently differentiable. By these recurrency relations, the terms ϕ_n of the expansion are completely determined and, of course, in general they will not satisfy the boundary conditions. Therefore we make the assumption that this regular expansion will exist in the subdomain $[d, 1-d]$ with d a constant, $0 < d < \frac{1}{2}$. Near the boundary points $x = 0$ and $x = 1$, we expect the presence of boundary layers to make the transitions to the boundary values possible.

First we use the "subtraction trick" by putting

$$\phi_\varepsilon(x) = \sum_{n=0}^{m} \varepsilon^n \phi_n(x) + \psi_\varepsilon(x)$$

to obtain, using the relations for ϕ_n,

$$\varepsilon L_1 \psi + b_0(x)\psi = -\varepsilon^{m+1} L_1 \phi_m(x).$$

The subtraction trick is not essential, to use it is a matter of taste. Its advantage is that in the transformed problem the regular (outer) expansion has coefficients zero, which facilitates matching. The boundary conditions become

$$\psi_\varepsilon(0) = \alpha - \sum_{n=0}^{m} \varepsilon^n \phi_n(0),$$

$$\psi_\varepsilon(1) = \beta - \sum_{n=0}^{m} \varepsilon^n \phi_n(1).$$

We start with the analysis of the boundary layer near $x = 0$. In a neighbourhood of $x = 0$, we introduce the local variable

$$\xi = \frac{x}{\delta(\varepsilon)}, \quad \delta(\varepsilon) = o(1),$$

and write $\psi_\varepsilon(\delta\xi) = \psi_\varepsilon^*(\xi)$. The orderfunction $\delta(\varepsilon)$ and correspondingly the size of the boundary layer still have to be determined. The equation in the local variable ξ becomes

$$L^*\psi^* = \frac{\varepsilon}{\delta^2}a_2(\delta\xi)\frac{d^2\psi^*}{d\xi^2} + \frac{\varepsilon}{\delta}a_1(\delta\xi)\frac{d\psi^*}{d\xi} + \varepsilon a_0(\delta\xi)\psi^* + b_0(\delta\xi)\psi^* = O(\varepsilon^{m+1}).$$

A significant degeneration L_0^* arises if $\delta(\varepsilon) = \sqrt{\varepsilon}$, which yields

$$L_0^* = a_2(0)\frac{d^2}{d\xi^2} + b_0(0).$$

The equation with this local variable has terms with coefficients containing ε and $\sqrt{\varepsilon}$. This suggests that we look for a regular expansion in ξ of the form

$$\psi_\varepsilon^*(\xi) = \sum_{n=0}^{2m} \varepsilon^{n/2}\psi_n(\xi) + O(\varepsilon^{m+\frac{1}{2}}).$$

We find at the lowest order

$$L_0^*\psi_0 = a_2(0)\frac{d^2\psi_0}{d\xi^2} + b_0(0)\psi_0 = 0.$$

Assuming that the coefficients have Taylor expansions to sufficiently high order near $x = 0$, we can expand a_2, a_1, a_0, and b_0, rewritten in the local variable ξ, in the equation for ψ^*. So we have for instance

$$a_2(\sqrt{\varepsilon}\xi) = a_2(0) + \sqrt{\varepsilon}\xi a_2'(0) + \varepsilon\cdots.$$

Collecting terms of equal order of ε, we deduce equations for $\psi_1, \psi_2, \cdots$. Keeping terms to order $\sqrt{\varepsilon}$, this looks like

$$\left(a_2(0) + \sqrt{\varepsilon}\,\xi a_2'(0) + \cdots\right)\left(\frac{d^2\psi_0}{d\xi^2} + \sqrt{\varepsilon}\frac{d^2\psi_1}{d\xi^2} + \cdots\right)$$

$$+ \sqrt{\varepsilon}\,(a_1(0) + \cdots)\left(\frac{d\psi_0}{d\xi} + \cdots\right)$$

$$+(b_0(0) + \sqrt{\varepsilon}\,\xi b_0'(0) + \cdots)(\psi_0 + \sqrt{\varepsilon}\psi_1 + \cdots) = O(\varepsilon^{m+1}),$$

where the dots stand for $O(\varepsilon)$ terms. For ψ_1, we find the equation

$$L_0^*\psi_1 = a_2(0)\frac{d^2\psi_1}{d\xi^2} + b_0(0)\psi_1 = -\xi a_2'(0)\frac{d^2\psi_0}{d\xi^2} - a_1(0)\frac{d\psi_0}{d\xi} - \xi b_0'(0)\psi_0.$$

In fact, it is easy to see that the coefficients ψ_n all satisfy the same type of differential equation (with different right-hand sides) of the form

$$L_0^*\psi_n = F_n(\psi_0(\xi), ..., \psi_{n-1}(\xi), \xi).$$

The boundary values at $x = 0$ are

$$\psi_0 = \alpha - \phi_0(0),$$

$$\psi_1(0) = 0,$$

$$\psi_2(0) = -\phi_1(0), \quad \text{etc.}$$

So we have for ψ_n, $n = 0, 1, \ldots$ a second-order equation and one boundary condition at $\xi = 0$, which is not enough to determine ψ_n completely. As in Chapter 3, we determine the functions ψ_n by requiring the boundary layer functions $\psi_n(\xi)$ to vanish outside the boundary layer. This means that we will add the matching relation

$$\lim_{\xi \to \infty} \psi_n(\xi) = 0.$$

That we have to match towards zero is a result of the subtraction trick. Of course, for $\xi \to \infty$ the variable ξ leaves the domain we are considering. However, if we let x tend to 1, ξ tends to $1/\sqrt{\varepsilon}$, which is very large, and we use $+\infty$ instead.

The solution of the equation for ψ_0 is

$$\psi_0(\xi) = Ae^{-\omega_0 \xi} + Be^{\omega_0 \xi},$$

where we have abbreviated $\omega_0 = (-b_0(0)/a_2(0))^{\frac{1}{2}}$; note that we assumed $b_0(0) < 0$. From the boundary condition, we find

$$A + B = \alpha - \phi_0(0).$$

The matching condition yields $B = 0$, so ψ_0 is now determined completely. The determination of ψ_1 runs along the same lines. We find

$$\psi_1(\xi) = A_1(\xi)e^{-\omega_0 \xi} + B_1(\xi)e^{\omega_0 \xi},$$

where A_1 and B_1 are polynomial functions of ξ that are determined completely by the boundary condition and the matching relation.

To satisfy the boundary conditions of the original problem, we have to repeat this analysis near the boundary at $x = 1$; the calculations mirror the preceding analysis. We give the results; the details are left as an exercise for the reader.

A suitable local variable is

$$\eta = \frac{1 - x}{\sqrt{\varepsilon}}.$$

The operator degenerates into

$$a_2(1)\frac{d^2}{d\eta^2} + b_0(1).$$

We propose a regular expansion of the form

$$\sum_{n=0}^{2m} \varepsilon^{n/2} \bar{\psi}_n(\eta),$$

which yields for the first term

$$\bar{\psi}_0(\eta) = \bar{A} e^{-\omega_1 \eta} + \bar{B} e^{\omega_1 \eta}$$

with $\omega_1 = (-b_0(1)/a_2(1))^{\frac{1}{2}}$. The boundary condition at $x = 1$ produces

$$\bar{A} + \bar{B} = \beta - \phi_0(1).$$

The matching relation becomes

$$\lim_{\eta \to \infty} \bar{\psi}_0(\eta) = 0,$$

so $\bar{B} = 0$. Collecting the first terms of the expansion with respect to x, ξ, and η in the three domains, we find

$$\phi_\varepsilon(x) = \frac{f(x)}{b_0(x)} + \left(\alpha - \frac{f(0)}{b_0(0)} \right) e^{-\omega_0 x/\sqrt{\varepsilon}} + \left(\beta - \frac{f(1)}{b_0(1)} \right) e^{-\omega_1(1-x)/\sqrt{\varepsilon}} + O(\sqrt{\varepsilon}).$$

For an illustration see Fig. 5.1.

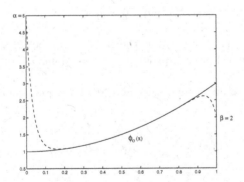

Fig. 5.1. Matching at two endpoints for the equation $\varepsilon d^2\phi/dx^2 + \phi = 1 + 2x^2$, $\phi(0) = 5$, $\phi(1) = 2$, $\varepsilon = 0.001$.

Remark
Omitting the $O(\sqrt{\varepsilon})$ term in the preceding expression and substituting the result in the differential equation, after checking the boundary conditions, we find that we have obtained a formal approximation of the solution of the boundary value problem. Because of the construction, this seems like a natural

result; in the next section, we shall see, however, that this is due to the absence of the term $b_1(x)$.

Once we knew how to proceed, the calculation itself was as simple as in Section 4.2 even though the problem is much more general. We shall now consider other boundary value problems and shall encounter some new phenomena.

5.2 A Boundary Layer at One Endpoint

We consider again the boundary value problem formulated at the beginning of this chapter,

$$\varepsilon L_1 \phi + L_0 \phi = f(x), \ \phi_\varepsilon(0) = \alpha, \phi_\varepsilon(1) = \beta,$$

but now with

$$L_0 = b_1(x)\frac{d}{dx} + b_0(x).$$

As before, we have

$$L_1 = a_2(x)\frac{d^2}{dx^2} + a_1(x)\frac{d}{dx} + a_0(x)$$

with $a_2(x) > 0$, $x \in [0,1]$; all coefficients are assumed to be sufficiently differentiable. Suppose, moreover, that $b_1(x)$ does not change sign, say

$$b_1(x) < 0, \quad x \in [0,1].$$

The case in which $b_1(x)$ vanishes in the interior of the interval is called a turning-point problem. For such problems, see Sections 5.4 and 5.5 and some of the exercises.

Here we assume again the existence of a regular expansion in a subdomain of $[0,1]$ of the form

$$\phi_\varepsilon(x) = \sum_{n=0}^{m} \varepsilon^n \phi_n(x) + O(\varepsilon^{m+1}).$$

After substitution of this expansion into the equation, we find

$$L_0 \phi_0 = f(x),$$
$$L_0 \phi_n = -L_1 \phi_{n-1}(x), \quad n = 1, 2, \cdots.$$

For $\phi_0(x)$, we find the first-order equation

$$b_1(x)\frac{d\phi_0}{dx} + b_0(x)\phi_0 = f(x),$$

which can be solved by variation of constants. We find, abbreviating,

$$g(x) = \int_0^x \frac{b_0(t)}{b_1(t)}\, dt,$$

$$\phi_0(x) = Ae^{-g(x)} + e^{-g(x)} \int_0^x e^{g(t)} f(t)\, dt.$$

In contrast with the preceding boundary value problem, we have one free constant in the expression for $\phi_0(x)$; we can verify that the same holds for $\phi_1, \phi_2, \cdots$. This means that the regular expansion in the variable x can be made to satisfy the boundary condition at $x = 0$ or at $x = 1$. Which one do we choose?

Suppose that we choose to satisfy the boundary condition at $x = 1$ and that we expect the existence of a boundary layer near $x = 0$. Does this lead to a consistent construction of a formal expansion? First we perform the subtraction trick

$$\phi_\varepsilon(x) = \sum_{n=0}^{m} \varepsilon^n \phi_n(x) + \psi_\varepsilon(x).$$

The equation for $\psi_\varepsilon(x)$ becomes

$$\varepsilon L_1 \psi + L_0 \psi = O(\varepsilon^{m+1})$$

with the boundary conditions

$$\psi_\varepsilon(0) = \alpha - \sum_{n=0}^{m} \varepsilon^n \phi_n(0),$$
$$\psi_\varepsilon(1) = \beta - \sum_{n=0}^{m} \varepsilon^n \phi_n(1).$$

With our assumption that the regular expansion satisfies the boundary condition at $x = 1$, we have $\phi_0(1) = \beta$ so $\psi_\varepsilon(1) = 0$. Introduce the local variable

$$\xi = \frac{x}{\delta(\varepsilon)}.$$

The differential operator written in the variable ξ takes the form

$$L^* = \frac{\varepsilon}{\delta^2} a_2(\delta\xi) \frac{d^2}{d\xi^2} + \frac{\varepsilon}{\delta} a_1(\delta\xi) \frac{d}{d\xi} + \varepsilon a_0(\delta\xi) + \frac{1}{\delta} b_1(\delta\xi) \frac{d}{d\xi} + b_0(\delta\xi).$$

Looking for a significant degeneration, we find $\delta(\varepsilon) = \varepsilon$ and the degeneration

$$L_0^* = a_2(0) \frac{d^2}{d\xi^2} + b_1(0) \frac{d}{d\xi}.$$

Expanding

$$\psi_\varepsilon(\varepsilon\xi) = \sum_{n=0}^{m} \varepsilon^n \psi_n(\xi) + O(\varepsilon^{m+1}),$$

we find

$$L_0^* \psi_0 = a_2(0) \frac{d^2 \psi_0}{d\xi^2} + b_1(0) \frac{d\psi_0}{d\xi} = 0.$$

The solution is

$$\psi_0(\xi) = B + Ce^{-\frac{b_1(0)}{a_2(0)}\xi}.$$

Because of the subtraction trick, the matching relation will again be

$$\lim_{\xi \to \infty} \psi_0(\xi) = 0.$$

As we assumed $b_1(0)/a_2(0)$ to be negative, this implies

$$B = C = 0.$$

A similar result holds for $\psi_1, \psi_2, \cdots$, namely all boundary layer terms (functions of ξ) vanish from the expansion. We conclude that the assumption of the existence of a boundary layer near $x = 0$ is not correct.

We consider now the other possibility, which is assuming that the regular expansion in the variable x satisfies the boundary condition at $x = 0$ that produces $\psi_\varepsilon(0) = 0$; we then expect the existence of a boundary layer near $x = 1$. Introduce the local variable

$$\eta = \frac{1 - x}{\delta(\varepsilon)}.$$

Looking for a significant degeneration of the operator written in the variable η, we find $\delta(\varepsilon) = \varepsilon$. Expanding

$$\psi_\varepsilon(1 - \varepsilon\eta) = \sum_{n=0}^{m} \varepsilon^n \bar{\psi}_n(\eta) + O(\varepsilon^{m+1}),$$

we find

$$\bar{L}_0^* \bar{\psi}_0 = a_2(1)\frac{d^2 \bar{\psi}_0}{d\eta^2} - b_1(1)\frac{d\bar{\psi}_0}{d\eta} = 0,$$

$$\bar{L}_0^* \bar{\psi}_n = F_n(\bar{\psi}_0, ..., \bar{\psi}_{n-1}, \eta), \quad n = 1, 2, \cdots.$$

Putting $\omega = -b_1(1)/a_2(1)$, we have

$$\bar{\psi}_0(\eta) = B + Ce^{-\omega\eta}.$$

The matching relation is

$$\lim_{\eta \to \infty} \bar{\psi}_0(\eta) = 0$$

so that $B = 0$. (Note that $\omega > 0$.) The boundary condition yields

$$C = \beta - \phi_0(1).$$

We compose an expansion from regular expansions in two subdomains, in the variables x and η, respectively, to obtain

$$\phi_\varepsilon(x) = \phi_0(x) + (\beta - \phi_0(1))e^{-\omega(1-x)/\varepsilon} + O(\varepsilon),$$

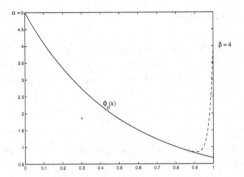

Fig. 5.2. Matching at one endpoint for the equation $\varepsilon d^2\phi/dx^2 - d\phi/dx - 2\phi = 0, \phi(0) = 5, \phi(1) = 4, \varepsilon = 0.02.$

where

$$\phi_0(x) = \alpha e^{-g(x)} + e^{-g(x)} \int_0^x e^{g(t)} f(t)\, dt,$$
$$g(x) = \int_0^x \frac{b_0(t)}{b_1(t)}\, dt.$$

For an illustration, see Fig. 5.2.

Remark

When omitting the $O(\varepsilon)$ terms, do we have a formal approximation of the solution? The boundary condition at $x = 0$ is satisfied with an exponentially small error. However, on calculating $(\varepsilon L_1 + L_0)\bar{\psi}_0$, we find a result that is $O_s(1)$, so we have not obtained a formal approximation. It is easy to see that to obtain a formal approximation we have to include the $O(\varepsilon)$ terms of the expansion. This is in contrast with the calculation in the preceding section.

On the other hand, it can be proved (see Section 5.6 for references) that on omitting the $O(\varepsilon)$ terms we have an *asymptotic* approximation of the solution! This looks like a paradox, but one should realise that a second-order linear ODE is characterised by a two-dimensional solution space. In the case of a scalar equation this is a space spanned by the solution and its derivative. In the problem at hand, omitting the $O(\varepsilon)$ terms produces an asymptotic approximation of the solution but not of the derivative.

Remark

Note that the location of the boundary layer is determined by the sign of $b_1(x)/a_2(x)$. If we were to choose $b_1(x) > 0, \quad x \in [0,1]$, the boundary layer would be located near $x = 0$, while the regular expansion in the variable x would extend to the boundary $x = 1$. Note, however, that it is not necessary to know this a priori, as the location of the boundary layer is determined while constructing the formal approximation.

5.3 The WKBJ Method

The method ascribed to Wentzel, Kramers, Brillouin, and Jeffreys plays a part in theoretical physics, in particular in quantum mechanics. One of the simplest examples is the analysis of the one-dimensional Schrödinger equation

$$\frac{d^2\psi}{dx^2} + (k^2 - U(x))\psi = 0.$$

$U(x)$ is the potential associated with the problem and k the wave number ($k/2\pi = 1/\lambda$ with λ the wavelength). We are looking for solutions of the Schrödinger equation with short wavelength (i.e., k is large). Over a few (short) wavelengths, $U(x)$ will not vary considerably, so it seems reasonable to introduce an effective wave number $q(x)$ by

$$q(x) = \sqrt{k^2 - U(x)}$$

and propose as a first approximation of the Schrödinger equation

$$\tilde{\psi} = e^{\pm i \int q(x)\, dx}.$$

One expects that this type of formal approximation may break down if, after all, $U(x)$ changes very quickly ($dU/dx \gg 1$) or if $k^2 - U(x)$ has zeros. Both situations occur in practice; in the case of zeros of $k^2 - U(x)$, one usually refers to turning points. For a discussion of a number of applications of the WKBJ method in physics, the reader may consult Morse and Feshbach (1953, Vol. II, Chapter 9.3). Here we shall explore the method from the point of view of asymptotic analysis for one-dimensional problems. Consider the two-point boundary value problem

$$\varepsilon\frac{d^2\phi}{dx^2} - w(x)\phi = 0, \quad 0 < x < 1.$$

$w(x)$ is sufficiently smooth and positive in $[0,1]$ with boundary values $\phi(0) = \alpha, \quad \phi(1) = \beta$. We analysed this problem in this chapter to find boundary layers near $x = 0$ and $x = 1$. We propose to interpret the WKBJ method as a regularising transformation in the following sense. We try to find solutions in the form

$$\exp.(Q(x)/\delta(\varepsilon)).$$

The regularisation assumption implies that we expect Q to have a regular expansion that is valid in the whole domain. In the case of the two-point boundary value problem, we substitute

$$\phi = \exp.(Q/\sqrt{\varepsilon})$$

to find

$$\sqrt{\varepsilon}Q'' + (Q')^2 - w(x) = 0.$$

This does not look like an equation with a regular expansion. However, if it has one, and because of $\sqrt{\varepsilon}$ in the equation, we expect such a regular expansion to take the form

$$Q_\varepsilon(x) = \sum_{n=0}^{2m} \varepsilon^{n/2} q_n(x) + O(\varepsilon^{m+\frac{1}{2}}).$$

We find after substitution

$$(q_0')^2 = w(x),$$
$$2q_0' q_1' = -q_0'', \quad \text{etc.}$$

with solutions

$$q_0(x) = \pm \int_0^x \sqrt{w(t)}\, dt + C_0,$$
$$q_1(x) = -\ln w^{\frac{1}{4}}(x) + C_1.$$

The original differential equation is linear and has two independent solutions. Using the first two terms q_0, q_1 to determine Q, we find from the calculation up to now two expressions that we propose to use as approximations for the independent solutions:

$$\psi_1(x) = \frac{1}{w^{\frac{1}{4}}(x)} e^{-\frac{1}{\sqrt{\varepsilon}} \int_0^x \sqrt{w(t)}\, dt},$$
$$\psi_2(x) = \frac{1}{w^{\frac{1}{4}}(x)} e^{-\frac{1}{\sqrt{\varepsilon}} \int_x^1 \sqrt{w(t)}\, dt}.$$

Note that $\psi_1(1)$ and $\psi_2(0)$ are exponentially small. A linear combination of ψ_1 and ψ_2 should represent a formal approximation of the boundary value problem; we put

$$\tilde{\phi}_\varepsilon(x) = A\psi_1(x) + B\psi_2(x).$$

Imposing the boundary values, we have

$$A = \alpha w^{\frac{1}{4}}(0) + O(e^{-\Omega/\sqrt{\varepsilon}}),$$
$$B = \beta w^{\frac{1}{4}}(1) + O(e^{-\Omega/\sqrt{\varepsilon}}),$$

with $\Omega = \int_0^1 w(t)\, dt$. In Section 5.1, we found a formal approximation $\tilde{\phi}_\varepsilon(x)$ of the boundary value problem with two boundary layers and a regular expansion identically zero. To compare the results, we expand $\tilde{\phi}_\varepsilon(x)$ with respect to x and $1 - x$:

$$\tilde{\phi}_\varepsilon(x) = \alpha e^{-\sqrt{w(0)}x/\sqrt{\varepsilon}} + \beta e^{-\sqrt{w(1)}(1-x)/\sqrt{\varepsilon}} + o(1), \quad x \in [0,1].$$

So, to a first approximation, the results of the boundary layer method in Section 5.1 (which can be proved to be asymptotically valid) agree with the results of the WKBJ method. The quantitative difference between the methods can be understood in terms of relative and absolute errors. One can show that in this case one can write for the independent solutions ϕ_1 and ϕ_2 of the original differential equation solutions ϕ_1 and ϕ_2 of the original differential equation

$$\phi_1 = \psi_1(1 + O(\sqrt{\varepsilon})),$$
$$\phi_2 = \psi_2(1 + O(\sqrt{\varepsilon})), \quad x \in [0,1].$$

The error outside the boundary layers is in the case of our boundary layer expansion an *absolute* one of order $\sqrt{\varepsilon}$; in the case of the WKBJ expansion, we have a *relative* error $O(\sqrt{\varepsilon})$, and as ψ_1, ψ_2 are exponentially decreasing, this is a much better result. However, this advantage of the WKBJ method is lost in slightly more general perturbation problems, as it rests on the regular expansion being identically zero. As soon as we find nontrivial regular expansions, the corresponding errors destroy the nice exponential estimates. Finally, we note that proofs of asymptotic validity involving WKBJ expansions are still restricted to relatively simple cases.

5.4 A Curious Indeterminacy

If we omit some of the assumptions of the preceding sections, the expansion and matching techniques that we have introduced may fail to determine the approximation. We shall demonstrate this for an example where we have an exact solution. The phenomenon itself is interesting but, even more importantly, it induced Grasman and Matkowsky (1977) to develop a new method to resolve the indeterminacy. We shall discuss this at the end of this section.

Consider the boundary value problem (see also Kevorkian and Cole, 1996, Section 2.3.4)

$$\varepsilon \frac{d^2\phi}{dx^2} - x\frac{d\phi}{dx} + \phi = 0, -1 < x < +1,$$
$$\phi(-1) = \alpha, \phi(+1) = \beta.$$

Assuming the existence of a regular expansion of the form

$$\psi_\varepsilon(x) = \sum_{n=0}^{m} \varepsilon^n \phi_n(x) + O(\varepsilon^{m+1}),$$

we find

$$-x\frac{d\phi_0}{dx} + \phi_0 = 0$$

so that $\phi_0(x) = c_0 x$ and actually to any order $\phi_n(x) = c_n x$ with $c_n, n = 0, \cdots, m$ arbitrary constants. We can satisfy one of the boundary conditions by setting either $-c_0 = \alpha$ or $+c_0 = \beta$.

In fact, if $\alpha = -\beta$, $\phi_0(x)$ solves the boundary value problem exactly. In the following we assume that $\alpha \neq -\beta$ with the presence of a boundary layer near $x = -1$ or $x = +1$. We expect that one of the choices will lead to an obstruction when trying to match the boundary layer solution to the regular expansion. Subtraction of the regular expansion by

$$\phi_\varepsilon(x) = \sum_{n=0}^{m} \varepsilon^n c_n x + \psi_\varepsilon(x)$$

leads to

$$\varepsilon\frac{d^2\psi}{dx^2} - x\frac{d\psi}{dx} + \psi = O(\varepsilon^{m+1}), \psi(-1) = \alpha + c_0 - \varepsilon \cdots, \psi(+1) = \beta - c_0 + \varepsilon \cdots.$$

Suppose we have a boundary layer near $x = -1$ with local variable

$$\xi = \frac{x+1}{\varepsilon^\nu}.$$

We find

$$\varepsilon^{1-2\nu}\frac{d^2\psi}{d\xi^2} - \frac{\varepsilon^\nu\xi - 1}{\varepsilon^\nu}\frac{d\psi}{d\xi} + \psi = O(\varepsilon^{m+1}),$$

with a significant degeneration for $\nu = 1$; expanding ψ_ε produces to first order

$$\frac{d^2\psi_0}{d\xi^2} + \frac{d\psi_0}{d\xi} = 0$$

with solution

$$\psi_0(\xi) = A_1 + A_2 e^{-\xi}$$

with A_1, A_2 constants. The matching relation will be

$$\lim_{\xi\to\infty} \psi_0(\xi) = 0$$

so that $A_1 = 0$; the boundary condition yields $A_2 = \alpha + c_0$.

We expect no boundary layer near $x = +1$; let's check this. Introduce the local variable

$$\eta = \frac{1-x}{\varepsilon^\nu}$$

so that we have locally

$$\varepsilon^{1-2\nu}\frac{d^2\bar\psi}{d\eta^2} + \frac{1 - \varepsilon^\nu\eta}{\varepsilon^\nu}\frac{d\bar\psi}{d\eta} + \bar\psi = O(\varepsilon^{m+1})$$

with a significant degeneration for $\nu = 1$. To first order, we find

$$\frac{d^2\bar\psi_0}{d\eta^2} + \frac{d\bar\psi_0}{d\eta} = 0$$

with solution

$$\bar\psi_0(\eta) = B_1 + B_2 e^{-\eta}$$

with B_1, B_2 constants. The matching relation

$$\lim_{\eta\to\infty} \bar\psi_0(\eta) = 0$$

produces $B_1 = 0$, and the boundary condition yields $B_2 = \beta - c_0$.

It turns out there is no obstruction to the presence of boundary layers near $x = -1$ *and* near $x = +1$. The approximation obtained until now takes the form

$$\phi_\varepsilon(x) = c_0 x + (\alpha + c_0)e^{-\frac{x+1}{\varepsilon}} + (\beta - c_0)e^{-\frac{1-x}{\varepsilon}} + O(\varepsilon)$$

with undetermined constant c_0. It can easily be checked that introducing higher-order approximations does not resolve the indeterminacy.

This is an unsatisfactory situation. In what follows we analyse the exact solution, which luckily we have in this case. In general, this is not a possible option and we shall discuss a general method that enables us to resolve the indeterminacy.

As $\phi_\varepsilon(x) = x$ solves the equation, we can construct a second independent solution to obtain the general solution

$$\phi_\varepsilon(x) = C_1 x + C_2\left(e^{\frac{x^2}{2\varepsilon}} - \frac{x}{\varepsilon}\int_{-1}^{x} e^{\frac{t^2}{2\varepsilon}}\,dt\right).$$

We assume again $\alpha \neq -\beta$ to avoid this simple case. C_1 and C_2 are determined by the boundary conditions. Analysing the exact solution is quite an effort; see Exercise 5.6 or Kevorkian and Cole (1996). The conclusion is that indeed near $x = -1$ and $x = +1$ a boundary layer of size $O(\varepsilon)$ exists. In the interior of the interval, there exists a regular expansion with first-order term $c_0 x$, $c_0 = (\beta - \alpha)/2$.

The method developed by Grasman and Matkowsky (1977) to resolve the indeterminacy is based on variational principles. The solution to our boundary value problem can be viewed as the element of the set

$$V = \{C^2(-1,+1)|y(-1) = \alpha, y(+1) = \beta\}$$

that extremalises the functional

$$I_\varepsilon = \int_{-1}^{+1} L(x, \phi, \phi'; \varepsilon)\,dx,$$

where L is a suitable Lagrangian function. Extremalisation of the functional leads to the Euler-Lagrange equation

$$\frac{d}{dx}\left(\frac{\partial L}{\partial \phi'}\right) - \frac{\partial L}{\partial \phi} = 0.$$

For a general reference to variational principles, see Stakgold (2000). In the case of the equation

$$\varepsilon\phi'' - x\phi' + \phi = 0,$$

a suitable Lagrangian function is

$$L = \frac{1}{2}(\varepsilon\phi'^2 - \phi^2)e^{-\frac{x^2}{2\varepsilon}}.$$

The approximation that we derived can be seen as a one-parameter family of functions, a subset of V, parameterised by c_0. We can look for a member of this family that extremalises the functional I_ε by substituting the expression and looking for an extremal value by satisfying the condition

$$\frac{dI_\varepsilon}{dc_0} = 0.$$

Keeping the terms of L to leading order in ε, we find again $c_0 = (\beta - \alpha)/2$. This result has been obtained without any explicit knowledge of the exact solution.

We can apply this elegant method to many other boundary value problems where a combination of boundary layer and variational methods is fruitful.

5.5 Higher Order: The Suspension Bridge Problem

Following Von Kármán and Biot (1940), we consider a model for a suspension bridge consisting of a beam supported at the endpoints and by hangers attached to a cable. Without a so-called live load on the bridge, the cable assumes a certain shape while bearing the beam that forms the bridge (the dead weight position). Adding a live load, and upon linearising, we obtain an equation describing the deflection $w(x)$ from the dead weight position of the cable,

$$\varepsilon \frac{d^4 w}{dx^4} - \frac{d^2 w}{dx^2} = p(x).$$

On deriving the equation, we have assumed that the beam and cable axes are lined up with the x-axis and that the total tension in the cable is large relative to the flexural rigidity (Young's elasticity modulus times the inertial moment). Also, we have rescaled such that $0 \le x \le 1$; $p(x)$ represents the result of a dead weight and live load. Natural boundary conditions are clamped supports at the endpoints which means

$$w(0) = w(1) = 0; \quad w'(0) = w'(1) = 0.$$

We assume that in a subdomain of $[0, 1]$ a regular expansion exists of the form

$$w(x) = \sum_{n=0}^{m} \varepsilon^n w_n(x) + O(\varepsilon^{m+1}).$$

We find after substitution

$$-\frac{d^2 w_0}{dx^2} = p(x), \quad \frac{d^2 w_n}{dx^2} = \frac{d^4 w_{n-1}}{dx^4}, \quad n = 1, 2, \cdots.$$

The second derivative $d^2 w/dx^2$ is inversely proportional to the curvature and so proportional to the bending moment of the cable. In the domain of the regular expansion, tension induced by the load $p(x)$ dominates the elasticity effects. Solving the lowest-order equation, we have

$$w_0(x) = -\int_0^x \int_0^s p(t)dtds + ax + b$$

with a and b constants. On assuming $p(x)$ to be sufficiently differentiable, we obtain higher-order terms of the same form.

We cannot apply the four boundary conditions, but a good choice turns out to be

$$w_0(0) = w_0(1) = 0.$$

We could also leave this decision until matching conditions have to be applied, but we shall run ahead of this. We find

$$a = \int_0^1 \int_0^s p(t)dtds, b = 0.$$

Subtracting the regular expansion

$$\psi(x) = w(x) - \sum_{n=0}^m \varepsilon^n w_n(x)$$

produces

$$\varepsilon \frac{d^4\psi}{dx^4} - \frac{d^2\psi}{dx^2} = O(\varepsilon^{m+1})$$

with boundary conditions

$$\psi(0) = \psi(1) = 0, \quad \psi\prime(0) = -a, \psi\prime(1) = \int_0^1 p(s)ds - a.$$

Expecting boundary layers at $x = 0$ and $x = 1$, we analyse what happens near $x = 0$; near $x = 1$ the analysis is similar. Introduce the local variable

$$\xi = \frac{x}{\varepsilon^\nu},$$

and the equation becomes

$$\varepsilon^{1-4\nu} \frac{d^4\psi^*}{d\xi^4} - \varepsilon^{-2\nu} \frac{d^2\psi^*}{d\xi^2} = O(\varepsilon^{m+1}).$$

A significant degeneration arises if $1 - 4\nu = -2\nu$ or $\nu = \frac{1}{2}$. Expanding $\psi^* = \psi_0(\xi) + \varepsilon^{1/2}\psi_1(\xi) + \cdots$, we have

$$\frac{d^4\psi_0}{d\xi^4} - \frac{d^2\psi_0}{d\xi^2} = 0$$

with boundary conditions

$$\psi_0(0) = \frac{\psi_0(0)}{d\xi} = 0$$

and general solution

$$\psi_0(\xi) = c_0 + c_1\xi + c_2 e^{-\xi} + c_3 e^{+\xi}.$$

Applying the matching condition

$$\lim_{\xi \to +\infty} \psi_0(\xi) = 0,$$

we have $c_0 = c_1 = c_3 = 0$; the boundary conditions yield $c_2 = 0$ so we have to go to the next order to find a nontrivial boundary layer contribution. Note that this is not unnatural because of the clamping conditions of the cable ($w(0) = w\prime(0) = 0$). For ψ_1 (and to any order) we have the same equation; the boundary conditions are

$$\psi_1(0) = 0, \frac{d\psi_1(0)}{d\xi} = -a.$$

However, applying the matching condition, we find again $c_0 = c_1 = c_3 = 0$, and we cannot apply both boundary conditions.

What is wrong with our assumptions and construction? At this point, we have to realise that the matching rule

$$\lim_{\xi \to +\infty} \psi(\xi) = 0$$

is matching in its elementary form. What we expect of matching is what this terminology expresses: the boundary layer expansion should be smoothly fitted to the regular (outer) expansion. If the boundary layer expansion is growing exponentially as $\exp(+\xi)$, there is no way to fit this behaviour with a regular expansion. Polynomial growth, however, is a different matter; a term such as $\varepsilon^{1/2}\xi$ behaves as x outside the boundary layer and poses no problem for incorporation in the regular expansion.

To allow for polynomial growth, we have to devise slightly more general matching rules. This will not be a subject of this chapter, but it is important to realise that one may encounter these problems. See for more details Section 6.2 and Section 15.4.

To illustrate this here and to conclude the discussion, one can compute the exact solution of the problem by variation of constants and by applying the boundary conditions. It is easier to look at the equation for $\psi(x)$ obtained by the subtraction trick. Its general solution is

$$\psi(x) = c_0 + c_1 x + c_2 e^{-x/\sqrt{\varepsilon}} + c_3 e^{(x-1)/\sqrt{\varepsilon}}.$$

Applying the boundary conditions, one finds that in general $c_1 = O_s(\sqrt{\varepsilon})$.

5.6 Guide to the Literature

Linear two-point boundary value problems have been studied by a number of authors. An elegant technique to prove asymptotic validity is the use of maximum principles. It was introduced by Eckhaus and De Jager (1966) to study elliptic problems. Such problems will be considered in Chapter 7; we give an example of a proof in Section 15.6. The technique of using maximum principles was applied extensively by Dorr, Parter, and Shampine (1973). Other general references for two-point boundary value problems, including turning-point problems, are Wasow (1965), Eckhaus (1979), Smith (1985), O'Malley (1991), and De Jager and Jiang Furu (1996). A number of basic aspects were analysed by Ward (1992, 1999).

An interesting phenomenon involving turning points is called Ackerberg-O'Malley resonance and has inspired a large number of authors. De Groen (1977, 1980) clarified the relation of this resonance with spectral properties of the related differential operator; for other references, see his papers.

Higher-dimensional linear boundary value problems can present themselves in various shapes. One type of problem is the linear scalar equation

$$\varepsilon \phi_\varepsilon^{(n)} + L_{n-1}\phi_\varepsilon = f(x)$$

with L_{n-1} a linear operator of order $(n-1)$ and appropriate boundary values. Another formulation is for systems of first-order equations of the form

$$\dot{x} = a(t)x + b(t)y,$$
$$\varepsilon \dot{y} = c(t)x + d(t)y,$$

with x an n-dimensional vector, y an m-dimensional vector, a, b, c and d matrices, and appropriate boundary values added. The vector form is also relevant for control problems. For a more systematic treatment, see O'Malley (1991).

More details about the WKBJ method can be found in Eckhaus (1979), Vainberg (1989), O'Malley (1991) and Holmes (1998). For turning points, see Wasow (1984), Smith (1985), and De Jager and Jiang Furu (1996).

5.7 Exercises

Exercise 5.1 We consider the following boundary problem on $[0, 1]$:

$$\varepsilon \left(\frac{d^2\phi}{dx^2} + \arctan(x)\frac{d\phi}{dx} - e^{x^2}\cos(x)\phi \right) - \cos(x)\phi = x^2,$$

$$\phi(0) = \alpha, \phi(1) = \beta.$$

Compute a first-order approximation. Is this a formal approximation?

Exercise 5.2 Consider the following boundary value problem on $[0, 1]$:

$$\varepsilon \frac{d^2\phi}{dx^2} + \frac{d\phi}{dx} + \cos x\phi = \cos x,$$

$$\phi(0) = \alpha, \phi(1) = \beta.$$

Compute a first-order approximation. Is the approximation formal?

Exercise 5.3 Consider the boundary value problem

$$\varepsilon \frac{d^2 y}{dx^2} + (1 + 2x)\frac{dy}{dx} - 2y = 0, \quad x \in (0, 1),$$
$$y(0) = \alpha, \quad y(1) = \beta.$$

a. Compute a first-order approximation of $y(x)$ using Section 5.2.
b. Compute a first-order approximation of $y(x)$ by the WKBJ method and compare the results of (a) and (b).

Note that for a one-parameter set of boundary values no boundary layer is present.

Exercise 5.4 Consider the so-called "turning-point problem":

$$L_\varepsilon y = (\varepsilon L_1 + L_0)(y) = 0, x \in [0, 1],$$
$$L_1 = a_2(x)\frac{d^2}{dx^2} + a_1(x)\frac{d}{dx} + a_0(x), a_2(x) \neq 0,$$
$$L_0 = b_1(x)\frac{d}{dx} + b_0(x).$$

Suppose $b_1(x)$ has a simple zero $x_0 \in (0, 1)$ and $b_0(x) \neq 0$ in $[0, 1]$. This is usually called a turning-point problem.

a. Compute the significant degenerations of L_ε in a neighbourhood of $x = x_0$. Take for instance $b_1(x) = \beta_0(x - x_0) + \cdots$, where β_0 is a nonzero constant and the dots indicate the higher-order terms in $(x - x_0)$ so that $b_1(x)$ has a simple zero.
b. Does a significant degeneration arise if $b_1 \equiv 0$ in $[0, 1]$ and b_0 has a simple zero?

Exercise 5.5 To recognise some of the difficulties arising with singular differential equations mentioned in the introduction to this chapter, we consider the Euler equation

$$\varepsilon(x^2 y'' + 3xy') - y = 0, x \in (0, 1),$$

$$y(0) = \alpha, y(1) = \beta, \alpha^2 + \beta^2 \neq 0, p \in \mathbb{R}.$$

a. Try to find a suitable local variable near $x = 0$.
b. Show that the boundary value problem has no solution.

Exercise 5.6 In Section 5.4 we obtained an exact solution,

$$\phi_\varepsilon(x) = C_1 x + C_2 \left(e^{\frac{x^2}{2\varepsilon}} - \frac{x}{\varepsilon} \int_{-1}^{x} e^{\frac{t^2}{2\varepsilon}} dt \right),$$

which has to satisfy the boundary conditions $\phi_\varepsilon(-1) = \alpha$, $\phi_\varepsilon(-1) = \beta$. We wish to determine the first-order term of the regular expansion in the interior of $[-1, +1]$. We also want to show that there exist boundary layers near $x = -1$ and $x = +1$. The calculation closely follows Kevorkian and Cole (1996), Section 2.3.4.

a. Apply the boundary conditions to find

$$C_1 = \frac{(\beta - \alpha)e^{1/2\varepsilon} + \alpha A(\varepsilon)}{2e^{1/2\varepsilon} - A(\varepsilon)}, C_2 = \frac{\beta + \alpha}{2e^{1/2\varepsilon} - A(\varepsilon)},$$

with

$$A(\varepsilon) = \frac{1}{\varepsilon} \int_{-1}^{+1} e^{t^2/2\varepsilon} dt.$$

In the following, we assume $\beta + \alpha \neq 0$. Putting $\beta + \alpha = 0$ eliminates the boundary layers and produces the exact solution $\phi_\varepsilon(x) = \beta x$.

b. Use Laplace's method (Chapter 3) to evaluate

$$e^{-1/2\varepsilon} A(\varepsilon) = 2(1 + \varepsilon + 3\varepsilon^2) + O(\varepsilon^3),$$

$$C_1 = -\frac{\beta + \alpha}{2\varepsilon} + \frac{3\beta + \alpha}{2} + O(\varepsilon), C_2 = (\beta + \alpha)e^{-1/2\varepsilon} \left(-\frac{1}{2\varepsilon} + \frac{3}{2} + O(\varepsilon) \right).$$

c. To expand $\phi_\varepsilon(x)$ in the interior of $[-1, +1]$, note that, away from the boundary layers, $C_2 e^{x^2/2\varepsilon}$ is exponentially small.

d. Again with Laplace's method show that in the interior of $[-1, +1]$ with $x \neq 0$

$$\int_{-1}^{x} e^{\frac{t^2}{2\varepsilon}} dt = \varepsilon e^{1/2\varepsilon} (1 + \varepsilon + O(\varepsilon^2)) + \frac{2\varepsilon}{x} \left(1 + \frac{\varepsilon}{x^2} + O(\varepsilon^2) \right) e^{x^2/2\varepsilon}.$$

e. Conclude that in the interior of $[-1, +1]$

$$\phi_\varepsilon(x) = \frac{\beta - \alpha}{2} x + O(\varepsilon).$$

f. Introduce local variables to analyse the exact solution near the endpoints.

6

Nonlinear Boundary Value Problems

For nonlinear problems a number of ideas have been developed to obtain formal approximations and proofs of asymptotic validity; however, there are so many complications and interesting phenomena that the theory is far from complete. To start with it is not a priori clear that a nonlinear boundary value problem has a solution. There may be various reasons for this.

Example 6.1
Consider the problem

$$\varepsilon\frac{d^2\phi}{dx^2} = \left(\frac{d\phi}{dx}\right)^2 + 1, \ \phi_\varepsilon(0) = \alpha, \phi_\varepsilon(1) = \beta.$$

Note that we cannot obtain a regular expansion, as putting $\varepsilon = 0$ produces no solution. Integration of the equation is possible by putting $v = d\phi/dx$ and separation of variables. We find

$$\frac{d\phi}{dx} = \tan\left(\frac{x}{\varepsilon} + a\right)$$

with a a constant of integration. We observe that, whatever the value of a, within an $O(\varepsilon)$-neighbourhood of any point in the interval, the derivative of $\phi_\varepsilon(x)$ becomes unbounded. There exists no solution of the boundary value problem. Van Harten (1975) gives a more general formulation for "explosion" of solutions to occur; see Section 15.5.

Another cause of unsolvability may be the presence of restrictions on the boundary conditions: the solutions may not be malleable enough to admit boundary layers. In a paper on the existence of solutions of nonlinear boundary value problems, Coddington and Levinson (1952) give an example; see also Chang and Howes (1984, Section 6.1).

Example 6.2
Consider

$$\varepsilon\frac{d^2\phi}{dx^2} + \left(\frac{d\phi}{dx}\right) + \left(\frac{d\phi}{dx}\right)^3 = 0, \ \phi_\varepsilon(0) = \alpha, \phi_\varepsilon(1) = \beta.$$

By putting $v = d\phi/dx$ and by separation of variables, we can obtain the solution of the equation

$$\phi_\varepsilon(x) = \pm\varepsilon \arcsin(e^{(x+a)/\varepsilon}) + b$$

with a and b constants of integration. As the arcsin function is bounded, $\phi_\varepsilon(x)$ can only produce $O(\varepsilon)$ variations around a constant. If $|\alpha - \beta| = O_s(1)$, the problem has no solution.

As we shall show in this chapter, there are examples that can be studied along the same lines as in the preceding chapter but also cases where new phenomena take place.

6.1 Successful Use of Standard Techniques

We start with some simple-looking equations that can be associated with conservative problems. For a related analysis, see O'Malley (1991).

6.1.1 The Equation $\varepsilon\phi'' = \phi^3$

We consider the equation with special boundary values in the following example.

Example 6.3

$$\varepsilon\frac{d^2\phi}{dx^2} = \phi^3, \ \phi_\varepsilon(0) = 1, \phi_\varepsilon(1) = 0.$$

Starting with a regular expansion, one finds that to any order the terms vanish. This suggests that the regular expansion extends to $x = 1$ and that a boundary layer exists at $x = 0$. Introducing the boundary layer variable

$$\xi = \frac{x}{\varepsilon^\nu},$$

we find a significant degeneration for $\nu = 1/2$. Transforming $\phi(x)$ to $\psi(\xi)$, we find

$$\frac{d^2\psi}{d\xi^2} = \psi^3,$$

which is the original equation in a slightly different form. Fortunately, we can find an integral by multiplying with $d\psi/d\xi$ and integrating. We find

$$\frac{1}{2}\left(\frac{d\psi}{d\xi}\right)^2 = \frac{1}{4}\psi^4 + a$$

with a a constant of integration. As the solution has to decrease from the value 1 to 0, we choose the minus sign:

$$\frac{d\psi}{d\xi} = -\sqrt{\frac{1}{2}\psi^4 + 2a}.$$

In the limit for $\xi \to +\infty$, ψ will vanish to any order, so matching will require $a = 0$. Integrating once more, applying the boundary condition at $x = 0$, and transforming back to x, we find

$$\tilde{\phi}_\varepsilon(x) = \frac{1}{x/\sqrt{2\varepsilon} + 1}.$$

This turns out to be an exact solution of the equation and an asymptotic approximation of the boundary value problem.

The problem becomes less simple if we consider more general boundary conditions, $\phi_\varepsilon(0) = \alpha, \phi_\varepsilon(1) = \beta$. However, the equation will have the integral formulated above, which facilitates the study of the solutions; see Exercise 6.1.

6.1.2 The Equation $\varepsilon\phi'' = \phi^2$

For this equation, there are some interesting aspects of existence or non-existence of solutions. Consider the equation with Dirichlet boundary values in the following example.

Example 6.4

$$\varepsilon\frac{d^2\phi}{dx^2} = \phi^2, \ \phi_\varepsilon(0) = \alpha, \phi_\varepsilon(1) = \beta.$$

As in Example 6.3, the regular expansion vanishes to any order. So, if $\alpha\beta \neq 0$, we expect boundary layers at both endpoints. As before, we find $O(\sqrt{\varepsilon})$ boundary layers. Near $x = 0$, the boundary layer equation is, with $\xi = x/\sqrt{\varepsilon}$,

$$\frac{d^2\psi}{d\xi^2} = \psi^2,$$

which can be integrated to

$$\frac{1}{2}\left(\frac{d\psi}{d\xi}\right)^2 = \frac{1}{3}\psi^3 + a$$

with a a constant of integration. As before, matching yields $a = 0$ and we have

$$\frac{d\psi}{d\xi} = \pm\sqrt{\frac{2}{3}}\psi^{\frac{3}{2}}.$$

If $\alpha < 0$, we choose the $+$ sign; integration produces

$$\psi^{-\frac{1}{2}} = \frac{\xi}{\sqrt{6}} + b$$

with b a constant. However, we cannot satisfy the condition $\psi(0) = \alpha$.

A similar conclusion holds when calculating the boundary layer behaviour near $x = 1$ for $\beta < 0$. We conclude that the boundary value problem has no solutions if α or β is negative.

Chang and Howes (1984, Chapter 3) concluded this in a different way. It is clear that near $x = 0$ and $x = 1$, where $\phi(x)$ does not vanish, we have $\phi'' > 0$. So the graph of $\phi(x)$ has to be convex. This makes it impossible to connect negative boundary values with the trivial regular solution.

If $\alpha, \beta > 0$, there is no such problem. The reader may verify that we find as a first-order approximation

$$\tilde{\phi}_\varepsilon(x) = \frac{\alpha}{(1 + \sqrt{\frac{\alpha}{6\varepsilon}}x)^2} + \frac{\beta}{\left(1 + \sqrt{\frac{\beta}{6\varepsilon}}(1 - x)\right)^2}.$$

The asymptotic validity of this approximation can be shown by analysing the exact solution in terms of elliptic functions.

6.1.3 A More General Equation

The function ϕ is for $x \in [0, 1]$ the solution of

$$L_\varepsilon \phi = \varepsilon \left(\frac{d^2}{dx^2} + a_1(x)\frac{d}{dx} + a_0(x) \right) \phi - g(x, \phi) = 0$$

with boundary values $\phi(0) = \alpha > 0$, $\phi(1) = 0$; $a_0(x)$, $a_1(x)$, and $g(x, \phi)$ are sufficiently smooth functions. Furthermore,

$$g(x, 0) = 0.$$

Suppose that a regular expansion exists in a subdomain of $[0, 1]$:

$$\phi_\varepsilon(x) = \sum_{n=0}^{m} \varepsilon^n \phi_n(x) + O(\varepsilon^{m+1}).$$

Substitution produces

$$g(x, \phi_0) = 0.$$

$\phi_0(x) = 0$ is a solution of this equation, and we suppose this solution to be unique by assuming $g_z(x, z) \neq 0$. It is easy to see that we have also $\phi_n(x) = 0$, $n = 1, ..., m$. Note that the regular expansion satisfies the boundary condition at $x = 1$. We expect a boundary layer near $x = 0$ and we introduce the local variable

$$\xi = \frac{x}{\varepsilon^\nu}.$$

The equation becomes, after transformation,

$$L_\varepsilon^* \phi^* = \varepsilon^{1-2\nu} \frac{d^2 \phi^*}{d\xi^2} + \varepsilon^{1-\nu} a_1(\varepsilon^\nu \xi) \frac{d\phi^*}{d\xi} + \varepsilon a_0(\varepsilon^\nu \xi) \phi^* - g(\varepsilon^\nu \xi, \phi^*) = 0.$$

A significant degeneration arises if $\nu = \frac{1}{2}$. Assuming that

$$\lim_{\varepsilon \to 0} \phi_\varepsilon^*(\xi) = \psi_0(\xi),$$

we find

$$\frac{d^2 \psi_0}{d\xi^2} - g(0, \psi_0) = 0.$$

We add the boundary value $\psi_0(0) = \alpha$ and the matching relation

$$\lim_{\xi \to \infty} \psi_0(\xi) = 0.$$

To solve this problem for ψ_0, we have to be more explicit about g. Suppose for instance that $g_z(x, z) > 0$, $x \in [0, 1]$, $z \in \mathbb{R}$, and introduce

$$G(\psi_0) = \int_0^{\psi_0} g(0, t)\, dt.$$

The equation for ψ_0 can be integrated to give

$$\frac{1}{2} \left(\frac{d\psi_0}{d\xi} \right)^2 - G(\psi_0) = a$$

with a a constant. We have

$$\frac{d\psi_0}{d\xi} = \pm (2G(\psi_0) + 2a)^{\frac{1}{2}}.$$

As $\alpha > 0$, the matching relation requires us to take the minus sign, corresponding with a decreasing solution (the plus sign if $\alpha < 0$). Matching produces $a = 0$. The solution for ψ_0 is given implicitly by

$$\xi = - \int_\alpha^{\psi_0} \frac{d\eta}{(2G(\eta))^{\frac{1}{2}}}.$$

Note that higher-order terms satisfy *linear* equations that we can solve in terms of ψ_0. In the boundary value problem, a boundary layer near $x = 1$ would arise if we were to put $\phi_\varepsilon(1) = \beta$, $\beta \neq 0$. The analysis of the behaviour of the solution near $x = 1$ in this case runs along exactly the same lines.

6.2 An Intermezzo on Matching

In the examples of the preceding section, we have deliberately introduced conditions such that the regular expansion vanishes to all orders. This facilitates matching, which obeys for instance a rule such as

$$\lim_{\xi \to +\infty} \psi_0(\xi) = 0,$$

where ξ is a local variable at the left endpoint of the interval.

In the linear problems of the preceding chapters, this type of matching is easy to achieve by applying the subtraction trick. In the case of nonlinear problems, this trick can also be used but it may cause technical complications. Also, in more difficult problems, we need a matching principle that can be generalised.

In Section 4.4, we noted that in the justification of matching rules the *overlap hypothesis* plays a part. To fix the idea, consider an interval with left endpoint $x = 0$. Suppose that in the interior of the interval we have a regular expansion of the form $\phi_0(x) + \varepsilon\phi_1(x) + \cdots$. Near $x = 0$, we have a boundary layer expansion $\psi_0(\xi) + \kappa(\varepsilon)\psi_1(\xi) + \cdots$, $\kappa(\varepsilon) = o(1)$, in the local variable

$$\xi = \frac{x}{\delta(\varepsilon)}, \quad \delta(\varepsilon) = o(1).$$

The overlap hypothesis assumes that there exists a subdomain near $x = 0$ where both the regular and the boundary layer expansions are valid and where they can be matched. A local variable ξ_0 corresponding with such a subdomain can be of the form

$$\xi_0 = \frac{x}{\delta_0(\varepsilon)}, \quad \delta_0(\varepsilon) = o(1), \delta(\varepsilon) = o(\delta_0(\varepsilon)).$$

In the overlap domain, the first terms of the regular and the boundary layer expansions are transformed:

$$\psi_0(\xi) \to \psi_0\left(\frac{\delta_0(\varepsilon)}{\delta(\varepsilon)}\xi_0\right), \quad \phi_0(x) \to \phi_0(\delta_0(\varepsilon)\xi_0).$$

For $\psi_0(\xi)$ and $\phi_0(x)$ to be matched in the overlap domain as $\varepsilon \to 0$, we have, upon expanding from the equality of the leading terms,

$$\lim_{\xi \to +\infty} \psi_0(\xi) = \lim_{x \to 0} \phi_0(x).$$

This is a slightly more general matching rule than the one we derived in Section 4.2. We illustrate this as follows.

Example 6.5

$$\varepsilon \frac{d^2\phi}{dx^2} + 2\frac{d\phi}{dx} + e^\phi = 0, \quad \phi_\varepsilon(0) = \phi_\varepsilon(1) = 0.$$

The first-order term of a regular expansion $\phi_0(x) + \varepsilon \dots$ will be a solution of

$$2\frac{d\phi_0}{dx} + e^{\phi_0} = 0$$

so that

$$\phi_0(x) = \ln\left(\frac{2}{x+a}\right)$$

with a a constant of integration. Assume that we have a boundary layer at $x = 0$ and that $\phi_0(1) = 0$; in this case $a = 1$. Near $x = 0$, we introduce the boundary layer variable

$$\xi = \frac{x}{\varepsilon^\nu},$$

which produces the equation

$$\varepsilon^{1-2\nu}\frac{d^2\psi}{d\xi^2} + 2\varepsilon^{-\nu}\frac{d\psi}{d\xi} + e^\psi = 0.$$

A significant degeneration arises for $\nu = 1$; expanding $\psi(\xi) = \psi_0(\xi) + \varepsilon \cdots$ yields for $\psi_0(\xi)$ the equation

$$\frac{d^2\psi_0}{d\xi^2} + 2\frac{d\psi_0}{d\xi} = 0$$

with solution

$$\psi_0(\xi) = A + Be^{-2\xi}.$$

From the boundary condition $\psi_0(0) = 0$, we have $A + B = 0$. Matching according to our new rule produces

$$A = \ln 2$$

and

$$\psi_0(\xi) = \ln 2(1 - e^{-2\xi}).$$

In composing a first-order approximation, we cannot merely add $\phi_0(x)$ and $\psi_0(\xi)$; we have to subtract the common part, which is in this case $\ln 2$. The approximation takes the form

$$\tilde{\phi}_\varepsilon(x) = \ln\frac{2}{x+1} - \ln 2e^{-2x/\varepsilon}.$$

Remark

The matching rule that we have formulated here is slightly more general than the rule formulated in Section 4.2. In applications, we may still encounter cases where this rule does not enable us to match different local expansions. One of the problems lies in the requirement that we restrict ourselves to leading terms. A more general matching rule, intermediate matching, is discussed in Section 15.4.

6.3 A Nonlinearity with Unexpected Behaviour

We consider now a problem with new, rather unexpected phenomena.

Example 6.6
The function ϕ is for $x \in [0,1]$ the solution of

$$\varepsilon \frac{d^2\phi}{dx^2} - \frac{d\phi}{dx}\left(\frac{d\phi}{dx} + 1\right) = 0$$

with boundary values $\phi_\varepsilon(0) = \alpha$, $\phi_\varepsilon(1) = 0$; we take $0 < \alpha < 1$. The problem was discussed by Van Harten (1975). The assumption of the existence of a regular expansion in x suggests the expansion

$$\phi_\varepsilon(x) = \sum_{n=0}^{m} \varepsilon^n \phi_n(x) + O(\varepsilon^{m+1}).$$

We find for ϕ_0 the equation

$$\frac{d\phi_0}{dx}\left(\frac{d\phi_0}{dx} + 1\right) = 0$$

with solutions

$$\phi_{01}(x) = c_1 \quad \text{and} \quad \phi_{02}(x) = -x + c_2.$$

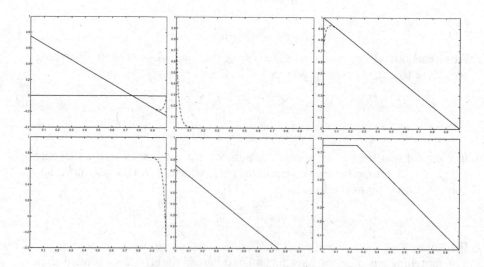

Fig. 6.1. Various possibilities in Example 6.6

From this point on, there are several possibilities. We could for instance assume that one of the solutions can be used in a subdomain of $(0,1)$ and the

other should be discarded; there are many other choices possible, see Fig. 6.1.
It turns out, however, that the following construction leads to correct results:
assume that $\phi_{01}(x)$ is a valid first approximation in a subdomain that has
$x = 0$ as a boundary, so we take $\phi_{01}(x) = \alpha$; $\phi_{02}(x)$ is valid in a subdomain
that has $x = 1$ as a boundary, so take $\phi_{02}(x) = 1 - x$. The two local approximations intersect at $x = 1 - \alpha$; in this case we find a boundary layer at
$x = 1 - \alpha$, the location of which clearly depends on the boundary values!

To obtain more insight into what is going on, we analyse the exact solution.
The solution for $\alpha \in (0, 1)$ is of the form

$$-\varepsilon \ln \left(A e^{\frac{x}{\varepsilon}} + B \right)$$

with A and B constants. Applying the boundary conditions, we can write for
the solution

$$\phi_\varepsilon(x) = \alpha - \varepsilon \ln \left(\frac{1 + e^{\frac{x-1+\alpha}{\varepsilon}} - e^{\frac{x-1}{\varepsilon}} - e^{-\frac{1-\alpha}{\varepsilon}}}{1 - e^{-\frac{1}{\varepsilon}}} \right),$$

which on $[0, 1]$ simplifies to

$$\phi_\varepsilon(x) = \alpha - \varepsilon \ln \left(1 + e^{\frac{x-1+\alpha}{\varepsilon}} - e^{\frac{x-1}{\varepsilon}} \right) + o(\varepsilon).$$

Inspection of this expression confirms the validity of the two regular expansions:

$$\phi_\varepsilon(x) = \alpha + O(\varepsilon), \quad 0 \leq x \leq 1 - \alpha - d, \quad d > 0,$$
$$\phi_\varepsilon(x) = 1 - x + O(\varepsilon), \quad 1 - \alpha + d \leq x \leq 1.$$

To describe the behaviour in the free boundary layer, we introduce the local
variable

$$\xi = \frac{x - (1 - \alpha)}{\varepsilon}$$

and we find

$$\phi_\varepsilon^*(\xi) = \alpha - \varepsilon \ln \left(1 - e^{-\frac{\alpha}{\varepsilon} + \xi} + e^\xi \right) + o(\varepsilon)$$
$$= \alpha - \varepsilon \ln(1 + e^\xi) + o(\varepsilon).$$

What happens if we take $\alpha > 1$?

The equation that we studied here is an example of an equation with
quadratic nonlinearities of the form

$$\varepsilon \frac{d^2\phi}{dx^2} = a(x, \phi) \left(\frac{d\phi}{dx} \right)^2 + b(x, \phi) \frac{d\phi}{dx} + c(x, \phi).$$

For more details, see Section 15.5.

6.4 More Unexpected Behaviour: Spikes

As we have seen, relatively simple differential equations can have solutions with striking behaviour. Another type of equation, producing spike-like behaviour, has attracted a lot of attention. A transparent demonstration of such behaviour for autonomous, conservative equations was given by O'Malley (1976); see also O'Malley (1991) and De Jager and Jiang Furu (1996).
We sketch the ideas by considering the problem in Example 6.7.

Example 6.7

$$\varepsilon \frac{d^2\phi}{dx^2} + \phi - \phi^2 = 0, \ \phi_\varepsilon(0) = \alpha, \phi_\varepsilon(1) = \beta.$$

In the $\phi, d\phi/dx$ phase-plane (see Fig. 6.2), there are two critical points, $(\phi, d\phi/dx) = (0,0)$, which is a centre, and $(\phi, d\phi/dx) = (1,0)$, which is a saddle. The phase orbits are given by the integral of the equation

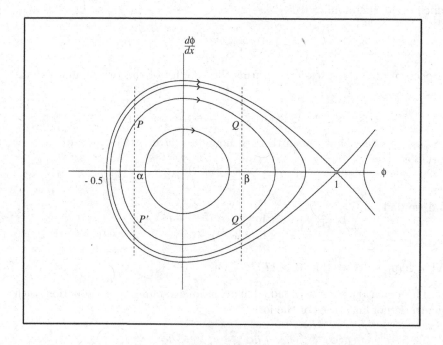

Fig. 6.2. The phase-plane of Example 6.7. Phase orbits such as PQQ' or $PQQ' \cdots PQ$ correspond with spiked solutions with boundary conditions $\phi_\varepsilon(0) = \alpha, \phi_\varepsilon(1) = \beta$.

$$\frac{1}{2}\varepsilon\left(\frac{d\phi}{dx}\right)^2 + \frac{1}{2}\phi^2 - \frac{1}{3}\phi^3 = E_0$$

for all possible values of the constant E_0. The closed orbits around $(0,0)$ and within the saddle loop, given by

$$\varepsilon\left(\frac{d\phi}{dx}\right)^2 + \phi^2 - \frac{2}{3}\phi^3 = \frac{1}{3},$$

correspond with periodic solutions (i.e., periodic in x when disregarding the boundedness of the interval). Near $(0,0)$, the frequency approaches $1/\sqrt{\varepsilon}$, which means very fast oscillations.

To apply the boundary values, we have to select a part of a phase orbit corresponding with length 1 in x, starting in $\phi = \alpha$ and ending in $\phi = \beta$. Suppose that $-1/2 < \alpha, \beta < 1$ and that α and β are not close to either $-1/2$ or 1 as $\varepsilon \to 0$. It is always possible to find a suitable part of a phase orbit inside the saddle loop as, moving out from the centre point to the saddle, the period increases monotonically from 0 (as $\varepsilon \to 0$) to ∞.

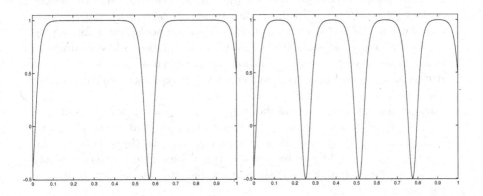

Fig. 6.3. Spikes in Example 6.7 with $\alpha = -0.4$, $\beta = 0.5$, $\varepsilon = 0.00025$.

To use a phase orbit such as PQQ', we have to pass fairly close to the saddle to obtain a "time interval" of length 1 in x. The motion near the saddle will be slow. (The solution ϕ stays for a long "time" near the value 1.) Outside a neighbourhood of the saddle, the motion will be fast. This solution corresponds with boundary layer behaviour near $x = 0$ and $x = 1$, connecting α and β with regular behaviour near $\phi = 1$.

Moving to a phase orbit slightly more inwards, we can construct a solution $PQQ' \cdots PQ$ corresponding with a chosen number of rotations of the phase orbit. Each time the phase orbit moves from a neighbourhood of the saddle through $Q'P'PQ$ back to this neighbourhood, we observe a spike (see Fig. 6.3). The sizes of the spikes are bounded by the size of the saddle loop, and they will hardly increase as $\varepsilon \to 0$. So remarkably enough, we can construct as many solutions with as many spikes as we want. For a more detailed analysis and proofs we refer to the literature cited.

6.4.1 Discussion

1. In our example, the analysis is made easier by the a priori knowledge that the period function, when going from the centre to the saddle, increases *monotonically* from $O(\sqrt{\varepsilon})$ to ∞. We know for equations of the form

$$\varepsilon \frac{d^2 \phi}{dx^2} + f(\phi) = 0$$

with $f(\phi)$ quadratic that the period function is a monotonic function (as in the case of the pendulum equation), see Chow and Sanders (1986). If $f(\phi)$ is cubic, the period function can already be nonmonotonic. This opens the possibility of more complications.

2. In the analysis of a boundary value problem for $\varepsilon \phi'' + f(\phi) = 0$, the critical points $(\phi, d\phi/dx) = (x_0, 0)$ with $f(x_0) = 0$ play an essential part. Excluding the degenerate case $f'(x_0) = 0$, all critical points are either centres (when $f'(x_0) > 0$) or saddles (when $f'(x_0) < 0$). Near the centres, we have high-frequency oscillations that are not suitable for solving boundary value problems. Near the saddles, the motion is slowed down as much as we want to produce solutions as discussed above. For a general background on conservative equations of this type, see Verhulst (2000, Chapter 2).

3. Various generalisations of the boundary value problem for the equation $\varepsilon \phi'' + f(\phi) = 0$ are possible. A natural one is to add linear damping. The qualitative analysis remains the same as a saddle (hyperbolic critical point) remains a saddle under perturbation. The centre becomes a slowly attracting focus, but the time-scale of circulation remains asymptotically the same.

6.5 Guide to the Literature

Nonlinear boundary value problems are extensively discussed in Chang and Howes (1984), O'Malley (1991), and De Jager and Jiang Furu (1996). Eckhaus (1979) uses the examples of van Harten (1975), discussed in Section 6.3 (see also Section 15.5), to analyse significant degenerations of operators and matching problems.

   In applied mathematics, difficult examples often play an important and stimulating part. Lagerstrom and Cole (see Kevorkian and Cole, 1996) introduced the problem

$$\varepsilon \frac{d^2 \phi}{dx^2} + \phi \frac{d\phi}{dx} - \phi = 0, \ \ \phi_\varepsilon(0) = \alpha, \phi_\varepsilon(1) = \beta,$$

which involves all kind of problems such as shocks, the location of boundary layers, and matching of expansions. Of the many studies see Dorr, Parter, and Shampine (1973), Howes (1978) and Kevorkian and Cole (1996). For another Lagerstrom model problem studied by blow-up transformation, see Popović and Szmolyan (2004).

Carrier and Pearson (1968) formulated the problem

$$\varepsilon \frac{d^2\phi}{dx^2} + 2(\phi - \phi^3) = 0, \ \phi_\varepsilon(-1) = \phi_\varepsilon(1) = 0,$$

which displays shock-layer phenomena and spikes. This has triggered a lot of research, of which Section 6.4 is only a very small part. Carrier and Pearson noticed that the perturbation scheme produces "spurious solutions" that do not represent actual solutions. It is not easy to find a selection-rejection criterion to handle these spurious solutions. The phase-plane method for autonomous problems (O'Malley, 1976) solves this by adding a new, geometric element to the discussion.

A number of authors have tried to solve this selection-rejection problem *within* the theory of singular perturbations. We mention Lange (1983), who uses exponentially small terms in his analysis. In the same spirit, MacGillivray (1997) also uses transcendentally small terms in the matching process.

A completely different analysis of these problems has been proposed by Ward (1992); for a survey and introduction, see Ward (1999). He associates the Carrier-Pearson problem with a linear eigenvalue problem that is ill-conditioned (i.e., there are exponentially small eigenvalues). By using an asymptotic projection on the space spanned by the eigenfunctions, the asymptotic approximation can be identified.

Another very different approach is to use nonstandard analysis. For a readable account, one should consult Lutz and Goze (1981). In this type of analysis, quantitative information is more difficult to obtain without returning to standard analysis of singular perturbations.

A large and important field that we do not discuss in this book is the application of singular perturbations to control theory. Some of its topics can be found in O'Malley (1991); for a systematic study and references, see Kokotović, Khalil, and O'Reilly (1999).

6.6 Exercises

Exercise 6.1 Consider the problem

$$\varepsilon \frac{d^2\phi}{dx^2} = \phi^3, \ \phi_\varepsilon(0) = \alpha, \phi_\varepsilon(1) = \beta,$$

with $\alpha < 0, \beta > 0$.
a. Note that the regular expansion vanishes to each order.
b. Compute a first-order approximation of the solution.
c. Can you say anything about the validity of the approximation?

Exercise 6.2 Consider the nonlinear boundary value problem

$$\varepsilon \frac{d^2y}{dx^2} - \frac{dy}{dx} + g(y) = 0,$$
$$y(0) = \alpha, y(1) = \beta.$$

The function g is sufficiently differentiable and is chosen in such a way that the reduced equation ($\varepsilon = 0$), for any given $y(x_0) = y_0$, $x_0 \in [0, 1]$, has a unique solution $y(x)$ for $x \in [0, 1]$.
a. Localise the possible boundary layer(s).
b. Give a first-order approximation.

Exercise 6.3 Consider the nonlinear boundary value problem

$$\varepsilon \frac{dx}{dt} = y - x^2, \quad x(0) = 0,$$
$$\frac{dy}{dt} = x, \qquad y(1) = 1, \ x, y \geq 0.$$

Compute a first-order approximation.

Exercise 6.4 Consider the boundary value problem on $[0, 1]$:

$$\varepsilon \frac{d^2y}{dx^2} = \left(\frac{dy}{dx}\right)^2, \ y(0) = \alpha, y(1) = \beta.$$

We exclude the trivial case $\alpha = \beta$. This example will show that the location of the boundary layer depends on whether $\alpha > \beta$ or $\alpha < \beta$.

a. Consider the various possibilities for constructing a regular expansion. Is it possible to locate boundary layers? Can you find a significant degeneration?
b. Compute the exact solution and show that there is one boundary layer location.
c. Identify a boundary layer variable in the case $\alpha > \beta$.

Exercise 6.5 Consider again the problem

$$\varepsilon \frac{d^2\phi}{dx^2} + \phi - \phi^2 = 0, \ \phi_\varepsilon(0) = \alpha, \phi_\varepsilon(1) = \beta.$$

a. Does the boundary value problem have solutions if $\alpha, \beta < -1/2$?
b. The same question as in (a) if $\alpha < -1/2, \beta > 1$.
c. Sketch the solutions for the orbits PQQ' and $PQQ'P'PQ$.

Exercise 6.6 Consider the solvability of Dirichlet boundary value problems on $[0, 1]$ for the equations $\phi'' - \phi + \phi^3 = 0$ and $\phi'' + \phi - \phi^3 = 0$.

7

Elliptic Boundary Value Problems

We now turn our attention to partial differential equations and, in particular, to elliptic boundary value problems in the plane. At the end of this chapter, we shall sketch some developments for nonconvex domains and for domains in $\mathbb{R}^3$.

Consider a bounded domain $D \subset \mathbb{R}^2$ with boundary Γ. We shall study a function of two variables $\phi_\varepsilon(x, y)$, that in D is given implicitly as the solution of a differential equation; on the boundary Γ, ϕ will take values prescribed by the function θ. This type of prescription of boundary values is usually called a Dirichlet problem.

We shall restrict ourselves to Dirichlet problems to keep the presentation simple but note that other boundary value problems can be treated by a similar analysis. We start with what appears to be a direct extension of the boundary value problem of Section 5.1.

7.1 The Problem $\varepsilon\Delta\phi - \phi = f$ for the Circle

Consider the domain $D = \{x, y | x^2 + y^2 < 1\}$ with the circle $\Gamma = \{x, y | x^2 + y^2 = 1\}$ as boundary. The problem is to find $\phi_\varepsilon(x, y)$ as the solution of

$$\varepsilon\Delta\phi - \phi = f(x, y)$$

with boundary condition $\phi|_\Gamma = \theta(x, y)$. The functions f and θ are defined in D and Γ, respectively, and are sufficiently differentiable; Δ is the Laplace operator.

Note that choosing the boundary condition to be homogeneous, $\phi|_\Gamma = 0$, we have no restriction of generality.

We assume that, in a subdomain of D, a regular expansion of the solution exists:

$$\phi_\varepsilon(x, y) = \sum_{n=0}^{m} \varepsilon^n \phi_n(x, y) + O(\varepsilon^{m+1}).$$

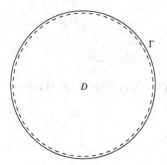

Fig. 7.1. A boundary layer along a circle.

Substitution in the equation produces

$$\phi_0(x, y) = -f(x, y),$$
$$\phi_n(x, y) = \Delta\phi_{n-1}(x, y), \qquad n = 1, \cdots.$$

In general, this expansion will not satisfy the boundary conditions, so that we expect the existence of a boundary layer along Γ. As in the preceding chapters, it is convenient to subtract the regular expansion; we put

$$\psi(x, y) = \phi(x, y) - \sum_{n=0}^{m} \varepsilon^n \phi_n(x, y)$$

and calculate

$$\begin{aligned}
\varepsilon\Delta\psi &= \varepsilon\Delta\phi - \sum_{n=0}^{m} \varepsilon^{n+1}\Delta\phi_n \\
&= \phi + f - \sum_{n=0}^{m} \varepsilon^{n+1}\phi_{n+1} \\
&= \psi + \sum_{n=0}^{m} \varepsilon^n\phi_n + f - \sum_{n=0}^{m} \varepsilon^{n+1}\phi_{n+1} \\
&= \psi - \varepsilon^{m+1}\phi_{m+1}.
\end{aligned}$$

So we have

$$\varepsilon\Delta\psi - \psi = O(\varepsilon^{m+1})$$

with boundary value $\psi|_\Gamma = \theta(x, y) - \sum_{n=0}^{m} \varepsilon^n\phi_n|_\Gamma$. To study the behaviour of the solution along the boundary, it is natural to introduce the radius-angle coordinates ρ, α by

$$x = (1 - \rho)\cos\alpha,$$
$$y = (1 - \rho)\sin\alpha.$$

The equation for ψ becomes

$$L\psi = \varepsilon\left(\frac{\partial^2}{\partial\rho^2} - \frac{1}{1-\rho}\frac{\partial}{\partial\rho} + \frac{1}{(1-\rho)^2}\frac{\partial^2}{\partial\alpha^2}\right)\psi - \psi = O(\varepsilon^{m+1}).$$

The boundary corresponds with $\rho = 0$, and we put

$$\psi|_{\rho=0} = \sum_{n=0}^{m} \varepsilon^n \theta_n(\alpha),$$

where θ_0 is obtained by transforming $x, y \to \rho, \alpha$ in $\theta(x, y) - \phi_0(x, y)$ and putting $\rho = 0$; θ_1 is obtained from $-\phi_1(x, y)$ etc.

We introduce a boundary layer variable along Γ by

$$\xi = \frac{\rho}{\varepsilon^\nu}.$$

The equation becomes

$$L^*\psi^* = \varepsilon^{1-2\nu}\frac{\partial^2\psi^*}{\partial\xi^2} - \frac{\varepsilon^{1-\nu}}{1 - \varepsilon^\nu\xi}\frac{d\psi^*}{d\xi} + \frac{\varepsilon}{(1 - \varepsilon^\nu\xi)^2}\frac{\partial^2\psi^*}{\partial\alpha^2} - \psi^* = 0(\varepsilon^{m+1}).$$

The differential operator has a significant degeneration L_0^* if $\nu = \frac{1}{2}$ so that $\xi = \rho/\varepsilon^{\frac{1}{2}}$ and then

$$L_0^* = \frac{\partial^2}{\partial\xi^2} - 1.$$

The equation has terms containing ε and $\varepsilon^{\frac{1}{2}}$, so we suppose that near the boundary a regular expansion exists of the form

$$\psi_\varepsilon^*(\xi, \alpha) = \sum_{n=0}^{2m} \varepsilon^{n/2}\psi_n(\xi, \alpha) + O\left(\varepsilon^{m+\frac{1}{2}}\right).$$

Substitution of the expansion and comparing terms of the same order in ε produces

$$L_0^*\psi_0 = 0, \quad \psi_0|_\Gamma = \theta_0(\alpha),$$
$$L_0^*\psi_1 = \frac{\partial\psi_0}{\partial\xi}, \quad \psi_1|_\Gamma = 0,$$
$$L_0^*\psi_2 = \frac{\partial\psi_1}{\partial\xi} + \xi\frac{\partial\psi_0}{\partial\xi} - \frac{\partial^2\psi_0}{\partial\alpha^2}, \quad \psi_2|_\Gamma = \theta_1(\alpha), \quad \text{etc.}$$

Note that the equations for $\psi_0, \psi_1, \ldots$ are all ordinary differential equations. For this reason, we call the boundary layer along Γ an *ordinary boundary layer* (see Fig. 7.1). The equations are easy to solve; we find for ψ_0

$$\psi_0(\xi, \alpha) = A(\alpha)e^{-\xi} + B(\alpha)e^{+\xi}.$$

Again we introduce a matching relation, and it takes the form

$$\lim_{\xi\to\infty} \psi_n(\xi, \alpha) = 0, \quad n = 0, \ldots, m.$$

For ψ_0 this implies that $B(\alpha) = 0$. We find successively

$$\psi_0(\xi, \alpha) = \theta_0(\alpha)e^{-\xi},$$
$$\psi_1(\xi, \alpha) = -\frac{1}{2}\theta_0(\alpha)\xi e^{-\xi}.$$

The formal approximation of the original boundary value problem becomes, to $O(\varepsilon)$,

$$\phi_\varepsilon = -f(x,y) + (\theta(x,y) + f(x,y))|_\Gamma \left(1 - \frac{1}{2}\rho\right) e^{-\rho/\sqrt{\varepsilon}} + \varepsilon \cdots,$$

in which $\rho = 1 - \sqrt{x^2 + y^2}$. It would appear that in the expansion only terms that are $O_s(1)$ arise; the reader may verify, however, that we have

$$\rho e^{-\rho/\sqrt{\varepsilon}} = O(\sqrt{\varepsilon}), \quad 0 \le \rho \le 1.$$

7.2 The Problem $\varepsilon\Delta\phi - \frac{\partial\phi}{\partial y} = f$ for the Circle

Consider again the domain $D = \{x, y | x^2 + y^2 < 1\}$ with boundary the circle $\Gamma = \{x, y | x^2 + y^2 = 1\}$. The equation is

$$\varepsilon\Delta\phi - \frac{\partial\phi}{\partial y} = f(x,y)$$

with boundary condition

$$\phi|_\Gamma = \theta(x,y).$$

The equation with boundary condition models the laminar flow of a conducting, incompressible fluid in the presence of a magnetic field; see Van Harten (1979) and references therein.

We start again in the usual way: assume that in a subset of D a regular expansion exists,

$$\phi_\varepsilon(x,y) = \sum_{n=0}^{m} \varepsilon^n \phi_n(x,y) + O(\varepsilon^{m+1}).$$

Substitution in the differential equation produces

$$\frac{\partial\phi_0}{\partial y} = -f(x,y),$$
$$\frac{\partial\phi_n}{\partial y} = \Delta\phi_{n-1}, \quad n = 1, 2, \cdots,$$

which are first-order equations. For ϕ_0, we find

$$\phi_0(x,y) = -\int^y f(x,t)\,dt + A(x).$$

We are still free to choose the function $A(x)$. On the boundary, we can eliminate y; we separate Γ into an upper and a lower half,

$$\Gamma^+ = \{x, y | x^2 + y^2 = 1, \quad y \ge 0\},$$
$$\Gamma^- = \{x, y | x^2 + y^2 = 1, \quad y \le 0\},$$

and accordingly the boundary conditions become

$$\Gamma^+ : \theta^+(x) = \theta\left(x, \sqrt{1-x^2}\right),$$
$$\Gamma^- : \theta^-(x) = \theta\left(x, -\sqrt{1-x^2}\right).$$

It is clear that we can choose $A(x)$ such that the boundary condition on either Γ^+ or Γ^- is satisfied. A good guess is then that imposing the boundary condition on one half will evoke the existence of a boundary layer along the other half of the boundary. We have to analyse the boundary layer equations before making this choice. We subtract the regular expansion

$$\psi(x,y) = \phi(x,y) - \sum_{n=0}^{m} \varepsilon^n \phi_n(x,y)$$

to find, in an analogous way as in Example 7.1,

$$\varepsilon\Delta\psi - \frac{\partial\psi}{\partial y} = O\left(\varepsilon^{m+1}\right).$$

The boundary conditions must also be transformed, but we postpone this calculation until we know the location of the boundary layer. Again we introduce the radius-angle coordinates to study what happens along the boundary

$$x = (1-\rho)\cos\alpha,$$
$$y = (1-\rho)\sin\alpha.$$

The equation becomes

$$\varepsilon\left(\frac{\partial^2\psi}{\partial\rho^2} - \frac{1}{1-\rho}\frac{\partial\psi}{\partial\rho} + \frac{1}{(1-\rho)^2}\frac{\partial^2\psi}{\partial\alpha^2}\right) + \sin\alpha\frac{\partial\psi}{\partial\rho} - \frac{\cos\alpha}{1-\rho}\frac{\partial\psi}{\partial\alpha} = O(\varepsilon^{m+1}).$$

The boundary corresponds with $\rho = 0$, and so we introduce the local variable

$$\xi = \frac{\rho}{\varepsilon^\nu}.$$

Transforming $x, y \to \rho$, $\alpha \to \xi, \alpha$ and replacing $\psi(x,y)$ by $\psi^*(\xi,\alpha)$, we find the equation

$$L^*\psi^* = \varepsilon^{1-2\nu}\frac{\partial^2\psi^*}{\partial\xi^2} - \frac{\varepsilon^{1-\nu}}{1-\varepsilon^\nu\xi}\frac{\partial\psi^*}{\partial\xi} + \frac{\varepsilon}{(1-\varepsilon^\nu\xi)^2}\frac{\partial^2\psi^*}{\partial\alpha^2}$$
$$+\varepsilon^{-\nu}\sin\alpha\frac{\partial\psi^*}{\partial\xi} - \frac{\cos\alpha}{1-\varepsilon^\nu\xi}\frac{\partial\psi^*}{\partial\alpha} = O(\varepsilon^{m+1}).$$

The operator L^* has a significant degeneration if $\nu = 1$. After rescaling, the degeneration takes the form

$$L_0^* = \frac{\partial^2}{\partial\xi^2} + \sin\alpha\frac{\partial}{\partial\xi}.$$

Note, however, that we may have to exclude the points $\alpha = 0$ and $\alpha = \pi$ where the second term vanishes and which corresponds with the points where Γ^- and Γ^+ are connected. As we have only entire powers of ε in the equation, it is natural to postulate the regular expansion

$$\psi_\varepsilon^*(\xi, \alpha) = \sum_{n=0}^{m} \varepsilon^n \psi_n(\xi, \alpha) + O(\varepsilon^{m+1}).$$

We find

$$L_0^* \psi_0 = 0$$

with the solution

$$\psi_0(\xi, \alpha) = B(\alpha) + C(\alpha)e^{-\xi \sin \alpha}.$$

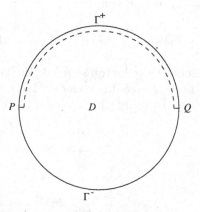

Fig. 7.2. Boundary layer at one side of a circular domain.

If we move into the interior of the domain, the boundary layer solution has to match the regular expansion in x, y. So we have the matching relation

$$\lim_{\xi \to \infty} \psi_n(\xi, \alpha) = 0.$$

Along Γ^-, $\pi < \alpha < 2\pi$, the exponent of ψ_0 is positive so that matching produces $B = C = 0$. Clearly, we then have no boundary layer along Γ^-, so we propose the regular expansion in x, y to satisfy the boundary condition on Γ^- and we expect a boundary layer along Γ^+ (see Fig. 7.2). This determines $A(x)$ (and the lower constant of integration) as

$$\theta^-(x) = - \int^{-\sqrt{1-x^2}} f(x,t)dt + A(x).$$

Furthermore, we have the boundary condition

$$\psi_{\varepsilon}|_{\Gamma^+} = \theta^+(x) - \sum_{n=0}^{m} \varepsilon^n \phi_n|_{\Gamma^+}.$$

We find with this condition and the matching relation

$$\psi_0(\xi,\alpha) = (\theta^+(\cos\alpha) - \phi_0(\cos\alpha, \sin\alpha))e^{-\xi\sin\alpha}$$

with $0 < \alpha < \pi$. The next term, $\psi_1(\xi,\alpha)$, follows from the boundary condition, the matching relation, and the equation

$$L_0^* \psi_1 = +\frac{\partial\psi_0}{\partial\xi} + \cos\alpha\frac{\partial\psi_0}{\partial\alpha}.$$

In our analysis, we excluded the points where $\alpha = 0, \pi$; let us call them P and Q. There is another reason to exclude a neighbourhood of these points. The regular expansion coefficients $\phi_n(x,y)$ will in general be singular at P and Q. To illustrate this, take for instance $\theta^-(x) = 1$, $f(x,y) = 1$. We find

$$A(x) = 1 - \sqrt{1-x^2},$$
$$\phi_0(x,y) = -y + 1 - \sqrt{1-x^2},$$

which has branch points at $x = \pm 1$. On calculating ϕ_1, one finds expressions that contain singularities at $x = \pm 1$. It is not difficult to see that this example illustrates the general situation. We conclude that our first-order calculation is complete except for a neighbourhood of the points P and Q. To complete the analysis, we have to introduce local variables near these points. It turns out that boundary layer regions arise near P and Q that form a transition between the boundary layer along Γ^+ and the boundary Γ^- where no boundary layer exists. The analysis becomes much more sophisticated, and we do not perform it here.

Remarks

1. On adding the analysis for a neighbourhood of the points P and Q, one still has not obtained a formal approximation, although the result can be proved to be an asymptotic approximation. This is analogous to the discussion in Section 5.2.
2. In the equation for ϕ, the term $\partial\phi/\partial y$ has coefficient -1. Replacing -1 by $+1$, we can repeat the analysis and find that the parts played by Γ^- and Γ^+ with respect to the boundary layer structure are interchanged.
3. A difference between the results of Section 7.1 and the present one is that we have in the second case a part of the boundary where no boundary layer exists and special problems in the transition regions near P and Q. In general, such problems arise if we have an equation such as

$$\varepsilon L_1 \phi + L_0 \phi = f$$

in which L_1 is an elliptic second-order operator and L_0 is a first-order operator. Special boundary layers, as above near P and Q, arise whenever the characteristics of L_0 are tangent to the boundary. This is in agreement with the theory of boundary value problems in partial differential equations and the part played by characteristics.

7.3 The Problem $\varepsilon \Delta \phi - \frac{\partial \phi}{\partial y} = f$ for the Rectangle

We have seen in the preceding section that the presence of characteristics in the reduced equation ($\varepsilon = 0$) causes special problems at the points where these characteristics are tangent to the boundary. In this way, the geometry of the domain plays an important part.

We shall now consider a problem that is typical for the case in which part of the boundary coincides with a characteristic (see Fig. 7.3). The equation is again

$$\varepsilon \Delta \phi - \frac{\partial \phi}{\partial y} = f(x, y),$$

which we consider on the rectangular domain D given by

$$D = \{x, y | a < x < b, \ c < y < d\}.$$

On the boundary Γ of D, we have the condition

$$\phi|_\Gamma = \theta(x, y).$$

We assume again that in a subdomain of D a regular expansion exists of the form

$$\phi_\varepsilon(x, y) = \sum_{n=0}^{m} \varepsilon^n \phi_n(x, y) + O(\varepsilon^{m+1}).$$

We find after substitution

$$\frac{\partial \phi_0}{\partial y} = -f(x, y),$$
$$\frac{\partial \phi_n}{\partial y} = \Delta \phi_{n-1}, \quad n = 1, 2, \cdots.$$

Inspired by the problem of the preceding section, we suppose that the regular expansion satisfies the boundary condition for $y = c$. We find

$$\phi_0(x, y) = -\int_c^y f(x, t)dt + \theta(x, c).$$

We subtract the regular expansion

$$\psi(x, y) = \phi(x, y) - \sum_{n=0}^{m} \varepsilon^n \phi_n(x, y)$$

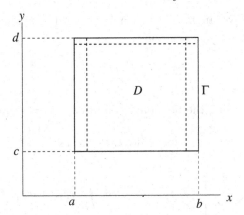

Fig. 7.3. Boundary layers in the case where characteristics coincide with part of the boundary.

to find, as before,

$$\varepsilon \Delta \psi - \frac{\partial \psi}{\partial y} = O(\varepsilon^{m+1}).$$

The boundary conditions for ψ are

$$\psi(x, c) = 0,$$

$$\psi(x, d) = \theta(x, d) - \sum_{n=0}^{m} \varepsilon^n \phi_n(x, d),$$

$$\psi(a, y) = \theta(a, y) - \sum_{n=0}^{m} \varepsilon^n \phi_n(a, y),$$

$$\psi(b, y) = \theta(b, y) - \sum_{n=0}^{m} \varepsilon^n \phi_n(b, y).$$

We expect the presence of boundary layers along the part of Γ where $x = a, b$ and $y = d$. We shall compute the first terms of the boundary layer expansions. Near $y = d$, we propose the local variable

$$\xi_d = \frac{d - y}{\varepsilon^\nu}.$$

The equation becomes (we omit *)

$$\varepsilon \frac{\partial^2 \psi}{\partial x^2} + \varepsilon^{1-2\nu} \frac{\partial^2 \psi}{\partial \xi_d^2} + \varepsilon^{-\nu} \frac{\partial \psi}{\partial \xi_d} = O(\varepsilon^{m+1}).$$

The significant degeneration of the operator arises if $\nu = 1$ and becomes

$$\frac{\partial^2}{\partial \xi_d^2} + \frac{\partial}{\partial \xi_d}.$$

Near $y = d$, we have an *ordinary* boundary layer. Expanding here $\psi_\varepsilon = \sum_{n=0}^{m} \varepsilon^n \psi_n^d(x, \xi_d) + O(\varepsilon^{m+1})$, we find

$$\frac{\partial^2 \psi_0^d}{\partial \xi_d^2} + \frac{\partial \psi_0^d}{\partial \xi_d} = 0,$$

which we can solve using the boundary condition at $y = d$ and the matching condition

$$\lim_{\xi_d \to \infty} \psi_0^d(x, \xi_d) = 0.$$

We find

$$\psi_0^d(x, \xi_d) = \left[\theta(x, d) - \theta(x, c) + \int_c^d f(x, t)\, dt \right] e^{-\xi_d}.$$

Near $x = a$, we propose the local variable

$$\xi_a = \frac{x - a}{\varepsilon^\nu}.$$

The equation becomes

$$\varepsilon^{1-2\nu} \frac{\partial^2 \psi}{\partial \xi_a^2} + \varepsilon \frac{\partial^2 \psi}{\partial y^2} - \frac{\partial \psi}{\partial y} = O(\varepsilon^{m+1}).$$

The significant degeneration arises if $\nu = \frac{1}{2}$ and becomes

$$L_0^* = \frac{\partial^2}{\partial \xi_a^2} - \frac{\partial}{\partial y}.$$

This is a parabolic differential operator, and therefore we call the boundary layer near $x = a$ *parabolic*. The equation contains only integer powers of ε, so we propose an expansion of the form

$$\psi_\varepsilon = \sum_{n=0}^{m} \varepsilon^n \psi_n^a(\xi_a, y) + O(\varepsilon^{m+1}).$$

To first order, we have

$$L_0^* \psi_0^a = 0$$

with boundary condition

$$\psi_0^a(0, y) = \theta(a, y) - \theta(a, c) + \int_c^y f(a, t)\, dt = g(y)$$

and the matching condition

$$\lim_{\xi_a \to \infty} \psi_0^a(\xi_a, y) = 0.$$

The solution of the equation is well-known, and we find

$$\psi_0^a(\xi_a, y) = \sqrt{\frac{2}{\pi}} \int_{\xi_a/2\sqrt{y}}^{\infty} e^{-\frac{1}{2}t^2} g\left(y - \frac{\xi_a^2}{2t^2}\right) dt.$$

It is not difficult to calculate higher-order terms of the expansion, and it can be shown that in this calculation difficulties arise near the boundary points (a, c) and (a, d). At (a, c) there is a transition between a parabolic boundary layer and the lower boundary where no boundary layer exists; at point (a, d) there is a transition between a parabolic and an ordinary boundary layer. In the analysis given here, we must exclude a neighbourhood of these points. It will be clear that the discussion of the formal expansion near the boundary $x = b$ runs along exactly the same lines. We find the boundary layer variable

$$\xi_b = \frac{b - x}{\sqrt{\varepsilon}}$$

and a parabolic boundary layer. In this case, we have to exclude a neighbourhood of the points (b, c) and (b, d).

7.4 Nonconvex Domains

In applications, nonconvex domains arise naturally, for instance in ocean and seabasin dynamics. The analysis of such problems shows some interesting new features. We demonstrate this by looking at an example (see Fig. 7.4).

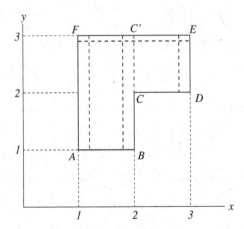

Fig. 7.4. Boundary layers in the case of a nonconvex domain; a free boundary layer arises.

Consider again the Dirichlet problem for the elliptic equation

$$\varepsilon \Delta \phi - \frac{\partial \phi}{\partial y} = 0$$

on the interior of the domain $ABCDEFA$. The boundary conditions are specified as follows. On the boundary $\Gamma, \phi_\varepsilon(x)$ is given by

$$AB : 1, DE : y,$$
$$BC : y, EF : 3,$$
$$CD : 2, FA : y.$$

We start again with the regular expansion

$$\phi_\varepsilon(x, y) = \sum_{n=0}^{m} \varepsilon^n \phi_n(x, y) + O(\varepsilon^{m+1}).$$

From the analysis in Section 7.3, we expect:

- $\phi_0(x, y)$ satisfies the boundary condition on AB and CD. Explicitly, $\phi_0(x, y) = 1$ on the rectangle $ABC'F, \phi_0(x, y) = 2$ on $CDEC'$.
- An ordinary boundary layer of size $O(\varepsilon)$ exists along EF and parabolic boundary layers of size $O(\sqrt{\varepsilon})$ exist along BC, DE, and FA.
- Transition boundary layers are present at all the corner points.

The verification runs along the same lines as in Section 7.3 and is left to the reader. What concerns us here is that, because of the nonconvexity of the domain, $\phi_0(x, y)$ has a nonsmooth transition on CC', where it equals 1 to the left and 2 to the right.

To describe this transition by our approximation scheme, we have to assume the presence of a *free boundary layer* along CC' and an additional transition layer at C and C'. Note that this free boundary layer is located along a characteristic of the reduced ($\varepsilon = 0$) equation. Introducing the local variable

$$\xi = \frac{2 - x}{\varepsilon^\nu},$$

the equation becomes

$$\varepsilon^{1-2\nu} \frac{\partial^2 \psi}{\partial \xi^2} + \varepsilon \frac{\partial^2 \psi}{\partial y^2} - \frac{\partial \psi}{\partial y} = O.$$

A significant degeneration arises if $\nu = 1/2$ with leading operator

$$\frac{\partial^2}{\partial \xi^2} - \frac{\partial}{\partial y},$$

so the free boundary layer is parabolic. Expanding $\psi_\varepsilon(\xi)$ gives

$$\frac{\partial^2 \psi_0}{\partial \xi^2} - \frac{\partial \psi_0}{\partial y} = 0.$$

If $\xi \to +\infty, x$ has to decrease, so we move into $ABC'F$. This produces the matching condition

$$\lim_{\xi \to +\infty} \psi_0(\xi, y) = 1.$$

Moving into $CDEC'$ we must have

$$\lim_{\xi \to -\infty} \psi_0(\xi, y) = 2.$$

The smoothness that produces these matching conditions is a global effect induced by the smoothness of the boundary conditions of the problem. The complete analysis is rather technical; for details see Mauss (1969, 1970).

7.5 The Equation $\varepsilon \Delta \phi - \frac{\partial \phi}{\partial y} = f$ on a Cube

The analysis until now was for second-order elliptic problems in one and two dimensions. The transition from one- to two-dimensional domains produced a number of new phenomena, and the same holds for the transition to three-dimensional problems.

Explicit calculations for three dimensions are less common; an example can be found in Van Harten (1975, Chapter 3.7) and in Lelikova (1978). One considers the analogue of the problem treated in Section 7.3: the interior Dirichlet problem for

$$\varepsilon \Delta \phi - \frac{\partial \phi}{\partial z} = f(x, y, z)$$

on the cube in 3-space. In this case, one finds from a formal analysis the following phenomena:

- a regular expansion in the variables x, y, z in the interior of the cube that satisfies the boundary condition on the base plane;
- an ordinary boundary layer near the upper plane of size $O(\varepsilon)$;
- parabolic boundary layers along the side planes of size $O(\sqrt{\varepsilon})$;
- parabolic boundary layers along the vertical edges of size $O(\sqrt{\varepsilon})$;
- elliptic boundary layers along the edges in the base plane of size $O(\varepsilon)$;
- elliptic boundary layers at the corner points of size $O(\varepsilon)$.

It is interesting to note that although there are more complications as the dimension of the domain is increased, the calculation scheme can still be carried out as explained before.

7.6 Guide to the Literature

The subject of elliptic equations was pioneered by Levinson (1950) and Vishik and Liusternik (1957); there is a survey paper by Trenogin (1970). Regarding proofs to establish the asymptotic character of approximations in elliptic problems, the first study introducing maximum principles was the paper by Eckhaus and De Jager (1966); see also Dorr, Parter, and Shampine (1973) and

Section 15.6 of this book. Since then, the theory has been extended considerably to involve other a priori estimates and Hilbert space techniques. Good references are Lions (1973), Huet (1977), and Il'in (1999); see also Van Harten (1978).

Constructions as in Section 7.2, where a characteristic of L_0 is tangent to the boundary (points P and Q), were handled by Grasman (1971, 1974). The phenomenon is called "birth of boundary layers". Applications to exit problems, groundwater flow, and pollution are described by Grasman and van Herwaarden (1999). For a general introduction, extensions to quasilinear and higher-order problems and for more references, we refer to the books by Eckhaus (1979, Chapter 7) and De Jager and Jiang Furu (1996).

We did not discuss singular perturbations of eigenvalue problems. Sanchez Hubert and Sanchez Palencia (1989) discuss spectral singular perturbations, for instance in the case of clamped plate vibrations; more references can be found in their monograph.

7.7 Exercises

Exercise 7.1 Consider a pipe with a circular cross section in the x, y-plane and a flow in the z direction of an incompressible, electrically charged fluid. The wall of the pipe is insulating. The velocity field $\phi(x, y)$ of the flow is described by the problem

$$\begin{cases} \varepsilon \Delta \phi - \frac{d\phi}{dy} = 1, \\ \phi(x, y) = 0, & \text{if } (x, y) \in \Gamma = \{(x, y) \in \mathbb{R}^2 | x^2 + y^2 = 1\}, \\ \varepsilon = \frac{1}{M}, & \text{(where } M \text{ is the Hartmann-number)} \end{cases}$$

Compute an approximation of ϕ up to and including the second-order term with the possible exception of some point(s) with their neighbourhood(s).

Exercise 7.2 Consider the elliptic problem

$$\begin{cases} \varepsilon \Delta \phi - \frac{\partial \phi}{\partial y} + \phi = 0 & \text{with} \\ \phi|_\Gamma = 1, & \text{where } \Gamma \text{ is the boundary of the rectangle,} \\ R = \{(x, y) \in \mathbb{R}^2 | \ 0 \leq x \leq a, \ 0 \leq y \leq b.\} \end{cases}$$

a. Localise the boundary layer(s).
b. Compute the regular expansion.
c. Formulate the boundary layer equations to third order for the horizontal boundary layer. Solve the first of these equations.

Exercise 7.3 Consider again the interior problem for the circle described in this chapter but now for the equations

$$\varepsilon \Delta \phi - g(x, y)\phi = f(x, y),$$

$$\varepsilon \Delta \phi - g(x, y)\phi_y = f(x, y).$$

Assume that $g(x, y)$ is sufficiently smooth and positive for $x^2 + y^2 \leq 1$. Show, without giving all detailed calculations, that the same perturbation scheme as in Section 7.1 resp. Section 7.2 carries over.

Exercise 7.4 We assume that $\phi_\varepsilon(x, y)$ is the solution of the equation

$$\varepsilon \Delta \phi - \phi_y = f(x, y), \qquad 1 < x^2 + y^2 < 4,$$
$$\phi|_{x^2+y^2=1} = \theta_1, \qquad \phi|_{x^2+y^2=4} = \theta_2,$$

$f(x, y)$ is sufficiently smooth, and θ_1 and θ_2 are functions describing the behaviour of ϕ on the boundary. Construct an expansion of the solution by giving a regular expansion, locating the boundary layers and giving the first terms of the boundary layer expansions.

Exercise 7.5 Assume that $\phi_\varepsilon(x, y)$ is the solution of the equation

$$\varepsilon \Delta \phi - \phi = x \quad \text{in the rectangle} \quad 0 < x < a, \quad 0 < y < b.$$

On the boundary Γ, we have the Neumann condition

$$\left.\frac{\partial \phi}{\partial n}\right|\Gamma = 0.$$

Construct an expansion of the solution by giving regular and boundary layer expansions while ignoring corner boundary layers.

8

Boundary Layers in Time

In the preceding chapters, we have analysed problems in which fast changes in the solutions take place near the spatial boundary of the domain. In this chapter, we consider initial value problems and restrict ourselves to ordinary differential equations. PDEs will be studied later.

As usual, the independent variable will be $t \in \mathbb{R}$. It is clear from the outset that there exist various types of initial value problems. Consider for instance the harmonic oscillator

$$\varepsilon \ddot{\phi} + \phi = 0$$

with initial values. The solutions consist of high-frequency oscillations of the forms $\sin(t/\sqrt{\varepsilon})$ and $\cos(t/\sqrt{\varepsilon})$; no boundary layer can be found. Or consider the problem

$$\varepsilon \ddot{\phi} + a\dot{\phi} = 0, \quad \phi(0) = \alpha, \quad \dot{\phi}(0) = \beta,$$

where the constant a is positive. The solution is

$$\phi_\varepsilon(t) = \alpha + \varepsilon \frac{\beta}{a} - \varepsilon \frac{\beta}{a} e^{-at/\varepsilon}.$$

We have boundary layer behaviour of the solution near $t = 0$. If a were negative, this would not be the case.

Analysing this problem using the formal calculations of the preceding chapters, we would assume the existence of a regular or outer expansion of the form

$$\sum_n \varepsilon^n \phi_n(t),$$

and we would have found $\dot{\phi}_0 = 0$ or $\phi_0(t) = $ constant. The condition that a is positive guarantees that such an outer expansion exists. Conditions such as this, in a more general form, will arise naturally in initial value problems; we shall formulate theorems later.

We start with two examples to show that these assumptions arise while constructing expansions.

8.1 Two Linear Second-Order Problems

In the first example, our construction is rather straightforward. In the second example, because of its slightly different character, decisions have to be made in constructing the expansion. However, it is interesting that for a consistent construction of asymptotic expansions, we need not know the Tikhonov theorem to be formulated in the next section. Of course, the knowledge of that theorem will clarify the background of these initial value problems enormously.

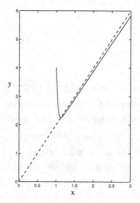

Fig. 8.1. Solution in Example 8.1 starting in $(x, y) = (1, 4)$. The solution is to $O(\varepsilon)$ attracted to the approximate slow manifold $y = 2x$ (dashed), which corresponds with the regular expansion $x_0(t), y_0(t), \varepsilon = 0.1$.

Example 8.1
Consider the two-dimensional system

$$\dot{x} = x + y, \quad x(0) = 1,$$
$$\varepsilon \dot{y} = 2x - y, \quad y(0) = 4.$$

As usual, we assume the existence of a regular expansion

$$x(t) = x_0(t) + \varepsilon x_1(t) + \varepsilon^2 \cdots,$$
$$y(t) = y_0(t) + \varepsilon y_1(t) + \varepsilon^2 \cdots,$$

which yields

$$\dot{x}_0 = x_0 + y_0, \quad \dot{x}_n = x_n + y_n,$$
$$y_0 = 2x_0, \quad y_n = 2x_n - \dot{y}_{n-1}, \quad n = 1, 2, \cdots.$$

At lowest order, we find $x_0(t) = Ae^{3t}, y_0(t) = 2Ae^{3t}$. We can apply the initial condition for x so we take $A = 1$. At the next order, we have

$$\dot{x}_1 = x_1 + y_1, \quad y_1 = 2x_1 - 6e^{3t},$$

so that

$$\dot{x}_1 = 3x_1 - 6e^{3t}$$

and

$$x_1(t) = Be^{3t} - 6te^{3t}, \quad y_1(t) = (2B - 6)e^{3t} - 12te^{3t}.$$

We could take $x_1(0) = 0$ so $B = 0$, but then we will be in trouble with the boundary layer expansion later on. So for the moment we leave the decision on the value of B. Summarising, we have the regular expansion (see Fig. 8.1)

$$x: \quad e^{3t} + \varepsilon(B - 6t)e^{3t} + \varepsilon^2 \cdots,$$
$$y: \quad 2e^{3t} + \varepsilon(2B - 6 - 12t)e^{3t} + \varepsilon^2 \cdots.$$

Note that in the regular expansion $y(0) = 2 + \varepsilon(2B - 6) + \varepsilon^2 \cdots$ so we expect a boundary layer jump near $t = 0$ as initially y equals 4. Subtracting the regular expansion, we have

$$X(t) = x(t) - (x_0(t) + \varepsilon x_1(t) + \varepsilon^2 \cdots),$$
$$Y(t) = y(t) - (y_0(t) + \varepsilon y_1(t) + \varepsilon^2 \cdots),$$

and because of the linearity, the same equations with different initial values

$$\dot{X} = X + Y, \quad X(0) = -\varepsilon B + \varepsilon^2 \cdots,$$
$$\varepsilon\dot{Y} = 2X - Y, \quad Y(0) = 2 + \varepsilon(6 - 2B) + \varepsilon^2 \cdots.$$

Introducing the local time-like variable

$$\tau = \frac{t}{\varepsilon^\nu},$$

we find the significant degeneration $\nu = 1$ and the equations

$$\frac{dX}{d\tau} = \varepsilon X + \varepsilon Y,$$

$$\frac{dY}{d\tau} = 2X - Y.$$

For X and Y, we propose the expansions

$$X(\tau) = \alpha_0(\tau) + \varepsilon\alpha_1(\tau) + \varepsilon^2 \cdots, \quad Y(\tau) = \beta_0(\tau) + \varepsilon\beta_1(\tau) + \varepsilon^2 \cdots,$$

to find at lowest order

$$\frac{d\alpha_0}{d\tau} = 0, \quad \frac{d\beta_0}{d\tau} = 2\alpha_0 - \beta_0, \quad \alpha_0(0) = 0, \beta_0(0) = 2,$$

so that $\alpha_0(\tau) = \text{constant}$. The matching rules

$$\lim_{\tau \to +\infty} \alpha_n(\tau), \beta_n(\tau) = 0, \quad n = 0, 1, 2, \cdots,$$

produce $\alpha_0(\tau) = 0$; this fits with the initial condition. Solving the equation for β_0, we find exponential decay, which agrees with the matching rule, and after applying the initial condition

$$\beta_0(\tau) = 2e^{-\tau}.$$

At the next order, we have

$$\frac{d\alpha_1}{d\tau} = \alpha_0 + \beta_0, \quad \frac{d\beta_1}{d\tau} = 2\alpha_1 - \beta_1, \quad \alpha_1(0) = -B, \quad \beta_1(0) = 6 - 2B.$$

Solving the equations, applying matching and the initial conditions, we find $B = 2$ and

$$\alpha_1(\tau) = -2e^{-\tau}, \beta_1(\tau) = -4\tau e^{-\tau} + 2e^{-\tau}.$$

As an expansion of the solution of the original initial-value problem, we then propose

$$x(t) = e^{3t} + \varepsilon(2e^{3t} - 2e^{-\frac{t}{\varepsilon}} - 6te^{3t}) + \varepsilon^2 \cdots,$$
$$y(t) = 2e^{-\frac{t}{\varepsilon}} + 2e^{3t} + \varepsilon(-\frac{4t}{\varepsilon}e^{-\frac{t}{\varepsilon}} + 2e^{-\frac{t}{\varepsilon}} - (2 + 12t)e^{3t}) + \varepsilon^2 \cdots.$$

The term $\beta_0(\tau)$ represents the major part of the boundary layer jump from the initial value to the regular expansion.

Example 8.2
Consider the linear second-order equation

$$L_\varepsilon\phi = \varepsilon L_1\phi + L_0\phi = f(t), \quad t \geq 0,$$

with

$$L_1 = \frac{d^2}{dt^2} + a_1(t)\frac{d}{dt} + a_0(t),$$
$$L_0 = b_1(t)\frac{d}{dt} + b_0(t);$$

$a_0, a_1, b_0, b_1, f(t)$ are sufficiently smooth functions. The initial values are $\phi_\varepsilon(0) = \alpha$, $\dot{\phi}_\varepsilon(0) = \beta$. As usual, it is not clear a priori what kind of expansion we should expect; we proceed, however, as before.

Suppose that a regular expansion exists in a subdomain of $\mathbb{R}^+$,

$$\phi_\varepsilon(t) = \sum_{n=0}^{m} \varepsilon^n \phi_n(t) + O(\varepsilon^{m+1}).$$

Substitution produces

$$L_0\phi_0 = f(t),$$
$$L_0\phi_n = -L_1\phi_{n-1}, \quad n = 1, 2, \cdots.$$

We avoid singularities in the regular expansion by assuming $b_1(t) \neq 0$, $t \geq 0$. Introduce

$$B(t) = \int_0^t \frac{b_0(s)}{b_1(s)} \, ds$$

to find

$$\phi_0(t) = C_0 e^{-B(t)} + e^{-B(t)} \int_0^t e^{B(s)} f(s) \, ds$$

and similar expressions for $\phi_n(t)$. There is one free constant at each step, so we can impose one of the initial values. We leave this decision until later. In any case, we expect a boundary layer near $t = 0$. Subtract the regular expansion

$$\psi(t) = \phi(t) - \sum_{n=0}^{m} \varepsilon^n \phi_n(t)$$

to find an equation for ψ with initial conditions

$$L_\varepsilon \psi = O(\varepsilon^{m+1}),$$

$$\psi(0) = \alpha - \sum_{n=0}^{m} \varepsilon^n \phi_n(0),$$

$$\dot{\psi}(0) = \beta - \sum_{n=0}^{m} \varepsilon^n \dot{\phi}_n(0).$$

Introduce the local variable $\tau = \frac{t}{\varepsilon^\nu}$, and then

$$L_\varepsilon^* \psi^* = \varepsilon^{1-2\nu} \frac{d^2 \psi^*}{d\tau^2} + \varepsilon^{1-\nu} a_1(\varepsilon^\nu \tau) \frac{d\psi^*}{d\tau} + \varepsilon a_0(\varepsilon^\nu \tau) \psi^*$$
$$+ \varepsilon^{-\nu} b_1(\varepsilon^\nu \tau) \frac{d\psi^*}{d\tau} + b_0(\varepsilon^\nu \tau) \psi^* = O(\varepsilon^{m+1}).$$

A significant degeneration arises if $\nu = 1$; we have

$$L_0^* = \frac{d^2}{d\tau^2} + b_1(0) \frac{d}{d\tau}.$$

For the solution ψ, we assume the existence of a regular expansion of the form

$$\psi_\varepsilon^* = \sum_{n=0}^{m} \varepsilon^n \psi_n(\tau) + O(\varepsilon^{m+1}).$$

The equations for the coefficients are obtained after expanding the coefficients a_1, a_0, b_1, b_0 and collecting terms of the same order of ε,

$$L_0^* \psi_0 = 0,$$

$$L_0^* \psi_1 = -a_1(0) \frac{d\psi_0}{d\tau} - b_0(0) \psi_0 - \tau b_1'(0) \frac{d\psi_0}{d\tau},$$

with initial values derived from

$$\psi_\varepsilon^*(0) = \alpha - \sum_{n=0}^{m} \varepsilon^n \phi_n(0),$$

$$\frac{d\psi_\varepsilon^*}{d\tau}(0) = \varepsilon\beta - \sum_{n=0}^{m} \varepsilon^{n+1} \dot{\phi}_n(0).$$

For ψ_0, we find

$$\psi_0(\tau) = A_1 + A_2 e^{-b_1(0)\tau}.$$

We expect $\psi_0, \psi_1, \ldots$ to vanish outside the boundary layer so that we have the matching relation

$$\lim_{\tau \to \infty} \psi_0(\tau) = 0.$$

It is clear that we have to require $b_1(0) > 0$; if not, our formal construction breaks down completely. Assuming $b_1(0) > 0$, we also deduce from the matching relation that

$$A_1 = 0.$$

The initial values $\psi_0(0) = \alpha - \phi_0(0)$, $\frac{d\psi_0}{d\tau}(0) = 0$ imply also that $A_2 = 0$ and we have to choose

$$(C_0 =) \; \phi_0(0) = \alpha,$$

which determines $\phi_0(t)$. It looks like we lost our boundary layer, but to show that this construction still involves boundary layer behaviour near $t = 0$, we must find the next order of the formal approximation. We have

$$\phi_1(t) = c_1 e^{-B(t)} - e^{-B(t)} \int_0^t e^{B(s)} L_1 \phi_0(s) \, ds.$$

$\psi_1(\tau)$ follows from the equation $L_0^* \psi_1 = 0$ with initial values $\psi_1(0) = -\phi_1(0)$, $\frac{d\psi_1}{d\tau}(0) = \beta - \dot{\phi}_0(0)$. This produces nontrivial terms for $\psi_1(\tau)$:

$$\psi_1(\tau) = \frac{\beta - \dot{\phi}_0(0)}{b_1(0)} + \frac{\dot{\phi}_0(0) - \beta}{b_1(0)} e^{-b_1(0)\tau} - \phi_1(0).$$

The matching relation $\lim_{\tau \to \infty} \psi_1(\tau) = 0$ produces

$$(C_1 =) \; \phi_1(0) = \frac{\beta - \dot{\phi}_0(0)}{b_1(0)},$$

so $\psi_1(\tau)$ and $\phi_1(t)$ have been determined completely. The formal approximation satisfies the initial values and is of the form

$$\phi_\varepsilon(t) = \sum_{n=0}^{m} \varepsilon^n \phi_n(t) + \sum_{n=1}^{m} \varepsilon^n \psi_n\left(\frac{t}{\varepsilon}\right) + O(\varepsilon^{m+1}).$$

We remark finally that the requirement $b_1(0) > 0$ corresponds with a kind of *attraction property* of the regular (outer) solution. Initially the solution moves very fast (initial layer), after which it settles, (at least for some time) in a state described by the regular expansion, which corresponds with relatively slow behaviour.

8.2 Attraction of the Outer Expansion

As we have seen, certain attraction properties of the regular (outer) expansion play an essential part in the construction of the formal approximation. This is also true for the analysis of nonlinear initial value problems; in the constructions, the following theorem provides a basic boundary layer property of the solution.

Theorem 8.1
(Tikhonov, 1952)
Consider the initial value problem

$$\dot{x} = f(x, y, t) + \varepsilon \cdots, \quad x(0) = x_0, \quad x \in D \subset \mathbb{R}^n, t \geq 0,$$
$$\varepsilon \dot{y} = g(x, y, t) + \varepsilon \cdots, \quad y(0) = y_0, \quad y \in G \subset \mathbb{R}^m.$$

For f and g, we take sufficiently smooth vector functions in x, y, and t; the dots represent (smooth) higher-order terms in ε.

a. We assume that a unique solution of the initial value problem exists and suppose this holds also for the reduced problem

$$\dot{x} = f(x, y, t), \quad x(0) = x_0,$$
$$0 = g(x, y, t),$$

with solutions $\bar{x}(t)$, $\bar{y}(t)$.
b. Suppose that $0 = g(x, y, t)$ is solved by $\bar{y} = \phi(x, t)$, where $\phi(x, t)$ is a continuous function and an isolated root. Also suppose that $\bar{y} = \phi(x, t)$ is an asymptotically stable solution of the equation

$$\frac{dy}{d\tau} = g(x, y, t)$$

that is uniform in the parameters $x \in D$ and $t \in \mathbb{R}^+$.
c. $y(0)$ is contained in an interior subset of the domain of attraction of $\bar{y} = \phi(x, t)$ in the case of the parameter values $x = x(0)$, $t = 0$.

Then we have
$$\lim_{\varepsilon \to 0} x_\varepsilon(t) = \bar{x}(t), \quad 0 \leq t \leq L,$$

$$\lim_{\varepsilon \to 0} y_\varepsilon(t) = \bar{y}(t), \quad 0 < d \leq t \leq L$$

with d and L constants independent of ε.

An *interior subset* of a domain is a subset of which all the points have a positive distance to the boundary of the domain, which is independent of ε. The necessity of this condition is illustrated by Example 8.6 at the end of this section.

In assumption (b), t and x are parameters and not variables. The idea is that during the fast motion of the variable y, the small variations of these parameters are negligible as long as the stability holds for values of the parameters $x \in D$ and $t \in \mathbb{R}^+$.

Example 8.3

We can apply Tikhonov's theorem to the second example of Section 8.1. Writing the equation in vector form by transforming $\phi, \dot{\phi} \to \phi_1, \phi_2$, we have,

$$\dot{\phi}_1 = \phi_2,$$
$$\varepsilon\dot{\phi}_2 = -b_1(t)\phi_2 - b_0(t)\phi_1 + f(t) - \varepsilon a_1(t)\phi_2 - \varepsilon a_0(t)\phi_1,$$
$$\phi_1(0) = \alpha, \phi_2(0) = \beta.$$

The reduced problem is

$$\dot{\phi}_1 = \phi_2, \phi_1(0) = \alpha,$$
$$0 = -b_1(t)\phi_2 - b_0(t)\phi_1 + f(t).$$

If for $0 \leq t \leq L$, $b_1(t) > 0$, then the root

$$\phi_2 = \frac{-b_0(t)}{b_1(t)}\phi_1 + \frac{f(t)}{b_1(t)}$$

satisfies the conditions of the theorem. In particular, this root corresponds with an asymptotically stable solution of

$$\frac{d\phi_2}{d\tau} = -b_1(t)\phi_2 - b_0(t)\phi_1 + f(t),$$

where t and ϕ_1 are parameters.

Example 8.4

(restriction of the time-interval)
Consider the initial value problem

$$\dot{x} = f(x, y), \quad x(0) = x_0,$$
$$\varepsilon\dot{y} = (2t - 1)yg(x), \quad y(0) = y_0$$

with $x, y \in \mathbb{R}, t \geq 0$; f and g are sufficiently smooth, and $g(x) > 0$. On the interval $0 \leq t \leq d$ with $0 < d < \frac{1}{2}$ (with d a constant independent of ε), $y = 0$ is an asymptotically stable solution of the equation

$$\frac{dy}{d\tau} = (2t - 1)yg(x)$$

with t, x as parameters. We expect a boundary layer in time near $t = 0$ after which the solution will settle near the solution of

$$\dot{x} = f(x, 0), y = 0, x(0) = x_0$$

as long as $0 \leq t \leq d$.

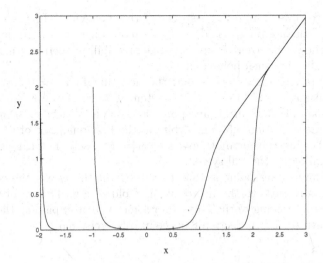

Fig. 8.2. Solutions in Example 8.5 starting in $(x, y) = (-2, 2)$ and $(-1, 2)$, $\varepsilon = 0.1$. If $x < 0$, they are attracted to the slow manifold $y = 0$; if $x > 0$, the attraction is to the approximate slow manifold $y = x$. Near the x-axis, the solutions get exponentially close and the closest one, starting in $(-2, 2)$, sticks longer to the x-axis after passing a neighbourhood of the origin; see Section 8.6.

Example 8.5

Consider the two-dimensional autonomous system

$$\dot{x} = 1, \quad x(0) = x_0,$$
$$\varepsilon \dot{y} = xy - y^2, \quad y(0) = y_0,$$

with $x, y \in \mathbb{R}, t \geq 0$.

In the case of autonomous (dynamical) systems as in this example, the attracting outer expansions correspond with manifolds in phase-space and so have a clear geometrical meaning. In this case, there are two roots corresponding with "critical points":

$$y = 0, \quad y = x.$$

If $x < 0, y = 0$ is stable and the second one unstable.
If $x > 0$, the second one is stable and $y = 0$ is unstable.

It follows that for $x < 0$ we expect a boundary layer jump towards the x-axis ($y = 0$) and for $x > 0$ we expect a boundary layer jump towards the line $y = x$. Note, however, the domains of attraction by sketching the phase-plane (see Fig. 8.2); solutions may also move off to infinity. The manifolds corresponding with the outer expansions, the lines $y = 0$ and $y = x$, are called *slow manifolds* to indicate the difference from the fast behaviour in the boundary layers.

There are two other interesting aspects of this behaviour. In the example, the solutions starting on the left-hand side move on to the right-hand side. On passing the y-axis, there is an exchange of stability; there are interesting applications showing such behaviour.

A second interesting aspect is that the slow manifold $y = 0$ corresponds with an invariant manifold of the full system; $y = x$, however (excluding a neighbourhood of the origin), is an approximation of a nearby existing invariant manifold; see Section 8.5. The orbits in the first quadrant of the phase-plane are following the manifold given by $y = x$ closely, but this is not an invariant manifold of the full system.

Another interesting point is that to the left of the y-axis the solutions get exponentially close to the invariant manifold $y = 0$. This results in the phenomenon of "sticking to the x-axis" for some time after passing the origin. We shall return to this in Section 8.6; see also Exercise 8.6.

Example 8.6

Condition c of the Tikhonov theorem requires us to choose the initial condition $y(0)$ in an *interior* subset of the domain of attraction. We show by an example that this is a necessary condition. Consider the equation

$$\varepsilon \dot{y} = -y + y^2 + 2\varepsilon y^2$$

with initial condition $y(0)$. Putting $\varepsilon = 0$, we find the critical points $y = 0$, which is asymptotically stable, and $y = 1$, which is unstable. Choose $y(0) = 1 - \varepsilon$, which is in the domain of attraction of $y = 0$.

For the full equation ($\varepsilon > 0$), the critical points are $y = 0, y = 1 - 2\varepsilon + \varepsilon^2 \cdots$, and our choice of $y(0)$ puts it outside the domain of attraction. The solution runs off to $+\infty$.

It is possible to weaken the interior subset condition. For instance, in this example, we may take $y(0) = 1 - \delta(\varepsilon)$ with $\delta(\varepsilon)$ an order function such that $\varepsilon = o(\delta(\varepsilon))$.

8.3 The O'Malley-Vasil'eva Expansion

How do we use Tikhonov's theorem to obtain approximations of solutions of nonlinear initial value problems? The theorem does not state anything about the size of the boundary layer (the parameter d in the theorem) or the timescales involved to describe the initial behaviour and the relative slow behaviour later on.

Asymptotic expansions are often based on Tikhonov's theorem *and* the additional assumption

$$\text{Re Sp } g_y(x, \bar{y}, t) \leq -\mu < 0, x \in D, 0 \leq t \leq L,$$

with Re Sp the real part of the spectrum (the collection of parameter-dependent eigenvalues of the matrix). This condition guarantees that the root

$\bar{y} = \phi(x,t)$ is asymptotically *stable in the linear approximation*. Following O'Malley (1974), we start with the system

$$\dot{x} = f(x,y,t,\varepsilon), \quad x(0) = x_0, \quad x \in D \subset \mathbb{R}^n, \quad t \geq 0,$$
$$\varepsilon \dot{y} = g(x,y,t,\varepsilon), \quad y(0) = y_0, \quad y \in G \subset \mathbb{R}^m, \quad t \geq 0,$$

in which f and g can be expanded in entire powers of ε. First, we substitute the regular expansion

$$x = \sum_{n=0}^{m} \varepsilon^n a_n(t) + O(\varepsilon^{m+1})$$
$$y = \sum_{n=0}^{m} \varepsilon^n b_n(t) + O(\varepsilon^{m+1}).$$

We write down the first equations for a_n, b_n:

$$\dot{a}_0 + \varepsilon \dot{a}_1 + \dots = f(a_0 + \varepsilon a_1 + \dots, b_0 + \varepsilon b_1 + \dots, t, \varepsilon),$$
$$\varepsilon \dot{b}_0 + \varepsilon^2 \dot{b}_1 + \dots = g(a_0 + \varepsilon a_1 + \dots, b_0 + \varepsilon b_1 + \dots, t, \varepsilon).$$

For a_0 and b_0, we find, on adding one initial condition,

$$\dot{a}_0 = f(a_0, \phi(a_0, t), t, 0), \quad a_0(0) = x_0,$$
$$0 = g(a_0, \phi(a_0, t), t, 0),$$

with $b_0(t) = \phi(a_0, t)$. The next order produces

$$\dot{a}_1 = f_x(a_0, b_0, t, 0)a_1 + f_y(a_0, b_0, t, 0)b_1 + f_\varepsilon(a_0, b_0, t, 0),$$
$$\dot{b}_0 = g_x(a_0, b_0, t, 0)a_1 + g_y(a_0, b_0, t, 0)b_1 + g_\varepsilon(a_0, b_0, t, 0).$$

The last equation enables us to express b_1 in terms of a_1 and known functions of t. The equation for a_1, after elimination of b_1, is *linear*; we add the initial condition $a_1(0) = 0$. It is easy to see that in higher order the equations are linear again.

In general, the regular expansion will not satisfy the initial condition for y. We now postulate the following expansion, which has to satisfy the equations and the initial conditions for $t \in [0, L]$:

$$x_\varepsilon(t) = \sum_{n=0}^{m} \varepsilon^n a_n(t) + \sum_{n=1}^{m} \varepsilon^n \alpha_n \left(\frac{t}{\varepsilon}\right) + \dots,$$
$$y_\varepsilon(t) = \sum_{n=0}^{m} \varepsilon^n b_n(t) + \sum_{n=0}^{m} \varepsilon^n \beta_n \left(\frac{t}{\varepsilon}\right) + \dots .$$

As we have two timescales, t and t/ε, in the expansion, this is called a *multiple timescales expansion*. Note that this also agrees with application of the subtraction trick. We substitute the expansion in the equations, finding to $O(\varepsilon)$

$$\dot{a}_0 + \varepsilon\dot{a}_1 + \varepsilon\dot{\alpha}_1 = f(a_0 + \varepsilon a_1 + \varepsilon\alpha_1, b_0 + \varepsilon b_1 + \beta_0 + \varepsilon\beta_1, t, \varepsilon) + O(\varepsilon^2),$$
$$\varepsilon\dot{b}_0 + \varepsilon\dot{\beta}_0 = g(a_0 + \varepsilon a_1 + \varepsilon\alpha_1, b_0 + \varepsilon b_1 + \beta_0 + \varepsilon\beta_1, t, \varepsilon) + O(\varepsilon^2),$$

where a_0, a_1, b_0, b_1 are known functions of t; a dot still denotes derivation with respect to t. Putting $t = \varepsilon\tau$, using the equation for a_0, and expanding, we have

$$\frac{d\alpha_1}{d\tau} = f(a_0(0), b_0(0) + \beta_0(\tau), 0, 0) - f(a_0(0), b_0(0), 0, 0),$$
$$\frac{d\beta_0}{d\tau} = g(a_0(0), b_0(0) + \beta_0(\tau), 0, 0).$$

Together with the initial condition (boundary layer jump)

$$\beta_0(0) = y_0 - b_0(0),$$

the last equation determines $\beta_0(\tau)$. The equation for α_1 can be integrated directly; as we took $a_1(0) = 0$, we have to put $\alpha_1(0) = 0$. The equations for α_n, β_n at higher order are linear and can be derived in the usual way. The initial conditions are

$$\alpha_n(0) = 0, \beta_n(0) = -b_n(0), \quad n = 1, 2, \cdots.$$

The construction given here follows the treatment given by O'Malley; see Section 8.8 for references. A computationally different construction has been given by A.B. Vasil'eva (1963) with asymptotically equivalent results; both authors prove the asymptotic validity of the expansions for which O'Malley uses an integral equation method based on a contraction argument. We summarise as follows:

Theorem 8.2
(O'Malley-Vasil'eva)
Consider the initial value problem in $\mathbb{R}^n \times \mathbb{R}^m \times \mathbb{R}^+$

$$\dot{x} = f(x, y, t, \varepsilon), \quad x(0) = x_0, \quad x \in D \subset \mathbb{R}^n, \ t \geq 0,$$
$$\varepsilon\dot{y} = g(x, y, t, \varepsilon), \quad y(0) = y_0, \quad y \in G \subset \mathbb{R}^m,$$

where f and g can be expanded in entire powers of ε. Suppose that the requirements of Tikhonov's theorem have been satisfied and moreover that for the solution of the reduced equation $0 = g(x, \bar{y}, t, 0)$, $\bar{y} = \phi(x, t)$, we have

$$\mathrm{Re} \ \mathrm{Sp} \ g_y(x, \bar{y}, t) \leq -\mu < 0, x \in D, 0 \leq t \leq L.$$

Then, for $t \in [0, L], x \in D, y \in G$, the formal approximation described above leads to asymptotic expansions of the form

$$x_\varepsilon(t) = \sum_{n=0}^{m} \varepsilon^n a_n(t) + \sum_{n=1}^{m} \varepsilon^n \alpha_n \left(\frac{t}{\varepsilon}\right) + O(\varepsilon^{m+1}),$$
$$y_\varepsilon(t) = \sum_{n=0}^{m} \varepsilon^n b_n(t) + \sum_{n=0}^{m} \varepsilon^n \beta_n \left(\frac{t}{\varepsilon}\right) + O(\varepsilon^{m+1}).$$

The constant L that bounds the domain of validity in time is in general an $O(1)$ quantity determined by the vector fields f and g. We shall return to this later on.

8.4 The Two-Body Problem with Variable Mass

In astrophysics, a model is being used that describes a binary system from which fast, spherical ejection of mass takes place. The equations of motion are

$$\ddot{r} = -G\frac{m(t)}{r^2} + \frac{c^2}{r^3},$$

$$c = r^2\dot{\theta},$$

in which r, θ are polar coordinates, and c and G are positive constants. We add initial values, rescale such that $c = G = 1$ and take for the first initial mass $m(0) = 1$. For the function describing quick loss of mass, we take

$$m(t) = m_r + (1 - m_r)e^{-t/\varepsilon},$$

where m_r is the rest mass, $0 \le m_r < 1$. For a discussion of the model and calculations in the case of other choices of $m(t)$, see Verhulst (1975). Putting $\rho = 1/r$, the problem can be transformed to the initial value problem

$$\frac{d^2\rho}{d\theta^2} + \rho = m_r + u, \quad \rho(0) = 1, \quad \frac{d\rho}{d\theta}(0) = \alpha,$$

$$\varepsilon\frac{du}{d\theta} = -\frac{u}{\rho^2}, \quad u(0) = 1 - m_r.$$

The first equation can be written in vector form by putting

$$\rho = \rho_1, \quad \frac{d\rho_1}{d\theta} = \rho_2$$

$$\frac{d\rho_2}{d\theta} = -\rho_1 + m_r + u.$$

Ejected shells transport momentum out of the system, and one of the questions of interest is whether orbits that are elliptic at $t = 0$ can become hyperbolic as a result of the process of loss of mass. The energy of the system is given by the expression

$$E = \frac{1}{2}\left(\frac{d\rho}{d\theta}\right)^2 + \frac{1}{2}\rho^2 - m_r\rho - u\rho.$$

Initially $E(0) = \frac{1}{2}(\alpha^2 - 1)$, so that starting with an elliptic orbit we have $0 \le \alpha^2 < 1$. It is not difficult to see that the requirements of the Vasil'eva - O'Malley theorem have been satisfied, and we propose to apply the O'Malley expansion method. We start with the regular expansion

$$\rho(\theta) = a_0(\theta) + \varepsilon a_1(\theta) + \varepsilon^2 \cdots,$$
$$\frac{d\rho}{d\theta}(\theta) = b_0(\theta) + \varepsilon b_1(\theta) + \varepsilon^2 \cdots,$$

$$u(\theta) = c_0(\theta) + \varepsilon c_1(\theta) + \varepsilon^2 \cdots.$$

We substitute these expansions in the equations to find $c_0 = c_1 = \ldots = 0$ and

$$\frac{da_0}{d\theta} = b_o, \qquad \frac{db_0}{d\theta} = -a_0 + m_r,$$
$$\frac{da_n}{d\theta} = b_n, \qquad \frac{db_n}{d\theta} = -a_n, \quad n = 1, 2, \cdots.$$

Satisfying the initial conditions for a_0, b_0, we have

$$a_0(\theta) = m_r + (1 - m_r) \cos\theta + \alpha \sin\theta,$$
$$b_0(\theta) = -(1 - m_r) \sin\theta + \alpha \cos\theta.$$

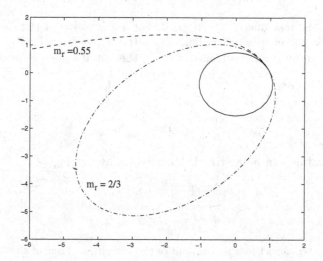

Fig. 8.3. Asymptotic approximations (dashed) of Kepler orbits after fast ejection of mass. The initial values for the original, unperturbed orbit (solid line) are $\rho(0) = 1, d\rho/d\theta(0) = 0.35, m_0 = 1$. If one third of the mass is shed ($m_r = 2/3$), the orbit remains elliptic; if 0.45 of the mass is shed ($m_r = 0.55$) the system becomes unbounded. For disruption of the two-body system, the limit value is $m_r = 0.57$.

Note that the regular expansion corresponds with a state of the two-body problem where the process of mass ejection has been finished. Following O'Malley, the complete expansion will be of the form

$$\rho_1 = \rho(\theta) = a_0(\theta) + \varepsilon a_1(\theta) + \varepsilon \alpha_1\left(\frac{\theta}{\varepsilon}\right) + \varepsilon^2 \cdots,$$

$$\rho_2 = \frac{d\rho}{d\theta}(\theta) = b_0(\theta) + \varepsilon b_1(\theta) + \varepsilon \beta_1\left(\frac{\theta}{\varepsilon}\right) + \varepsilon^2 \cdots,$$

$$u(\theta) = \gamma_0\left(\frac{\theta}{\varepsilon}\right) + \varepsilon \gamma_1\left(\frac{\theta}{\varepsilon}\right) + \varepsilon^2 \cdots.$$

The initial conditions for a_1, b_1, etc., will be derived from the boundary layer terms as

$$a_1(0) = -\alpha_1(0), b_1(0) = -\beta_1(0), \quad \text{etc.}$$

We have

$$a_n(\theta) = -\alpha_n(0) \cos\theta - \beta_n(0) \sin\theta,$$
$$b_n(\theta) = \alpha_n(0) \sin\theta - \beta_n(0) \cos\theta, \quad n = 1, 2, \cdots.$$

Substituting the complete expansion into the equations in vector form, we obtain with $\nu = \theta/\varepsilon$ and a dot for derivation with respect to θ

$$\dot{a}_0 + \varepsilon \dot{a}_1 + \frac{d\alpha_1}{d\nu} + \varepsilon^2 \cdots = b_0 + \varepsilon b_1 + \varepsilon \beta_1 + \varepsilon^2 \cdots,$$

$$\dot{b}_0 + \varepsilon \dot{b}_1 + \frac{d\beta_1}{d\nu} + \varepsilon^2 \cdots = -a_0 - \varepsilon a_1 - \varepsilon \alpha_1 + m_r + \gamma_0 + \varepsilon \gamma_1,$$

$$\frac{d\gamma_0}{d\nu} + \varepsilon \frac{d\gamma_1}{d\nu} + \varepsilon^2 \cdots = -(\gamma_0 + \varepsilon \gamma_1)(a_0 + \varepsilon a_1 + \varepsilon \alpha_1 + \varepsilon^2 \cdots)^{-2}.$$

For the first boundary layer terms, we find

$$\frac{d\alpha_1}{d\nu} = 0, \quad \frac{d\beta_1}{d\nu} = \gamma_0, \quad \frac{d\gamma_0}{d\nu} = -\gamma_0 a_0(0)^{-2}.$$

Note that we developed the outer expansion terms $a_0(\theta)$, etc., with respect to ε after putting $\theta = \varepsilon\nu$; $a_0(0) = 1$. As $c_0 = 0$, the boundary layer jump for γ_0 corresponds with the initial condition for u. Applying the matching conditions

$$\lim_{\nu \to \infty} \alpha_1(\nu) = \lim_{\nu \to \infty} \beta_1(\nu) = 0,$$

we find

$$\alpha_1(\nu) = 0, \beta_1(\nu) = -(1 - m_r)e^{-\nu}, \gamma_0(\nu) = (1 - m_r)e^{-\nu},$$

so $\alpha_1(0) = 0$, $\beta_1(0) = -(1 - m_r)$, which determines a_1 and b_1. Summarising our results, we have

$$\rho(\theta) = m_r + (1 - m_r)\cos\theta + \alpha\sin\theta + \varepsilon(1 - m_r)\sin\theta + O(\varepsilon^2),$$

$$\frac{d\rho}{d\theta}(\theta) = -(1 - m_r)\sin\theta + \alpha\cos\theta + \varepsilon(1 - m_r)\cos\theta$$

$$-\varepsilon(1 - m_r)e^{-\theta/\varepsilon} + O(\varepsilon^2),$$

$$u(\theta) = (1 - m_r)e^{-\theta/\varepsilon} + O(\varepsilon).$$

For an illustration, see Fig. 8.3. If we were to continue the expansion, we would find $\gamma_1(\nu) = 0$, so the last $O(\varepsilon)$ estimate can be improved to $O(\varepsilon^2)$. We can use these expressions to obtain an asymptotic approximation for the energy E; outside the boundary layer, when most of the ejected mass is shed, we omit the terms $\exp.(-\theta/\varepsilon)$ to find

$$E = E(0) + (1 - m_r) + \varepsilon\alpha(1 - m_r) + O(\varepsilon^2).$$

The orbits will become hyperbolic if $E > 0$ or

$$m_r < \frac{E(0) + 1 + \varepsilon\alpha}{1 + \varepsilon\alpha}.$$

Starting in a circular orbit, the radial velocity is zero, or $\alpha = 0$. In that case, we have simply

$$m_r < \frac{1}{2}.$$

So, for orbits starting in a circular orbit to become hyperbolic, more than half of the mass has to be shed; note that this statement is based on expansions that are valid to $O(\varepsilon)$. To find the dependence of m_r on ε, the rate of mass loss, we have to go to $O(\varepsilon^2)$ approximations in the case of orbits that are initially circular; see Hut and Verhulst (1981).

8.5 The Slow Manifold: Fenichel's Results

Tikhonov's theorem (Section 8.2) is concerned with the attraction, at least for some time, to the regular expansion that corresponds with a stable critical point of the boundary layer equation. The theory is quite general and deals with nonautonomous equations.

In the case of autonomous equations, it is possible to associate with the regular expansion a manifold in phase-space and to consider the attraction properties of the flow near this manifold. This raises the question of whether these manifolds really exist or whether they are just a phantom phenomenon. Such questions were addressed and answered in a number of papers by Fenichel (1971, 1974, 1977, 1979), Hirsch, Pugh, and Shub (1977), and other authors; the reader is referred to the survey papers by Jones (1994), Kaper (1999), and Kaper and Jones (2001).

Consider the autonomous system

$$\dot{x} = f(x, y) + \varepsilon \cdots, \quad x \in D \subset \mathbb{R}^n,$$
$$\varepsilon\dot{y} = g(x, y) + \varepsilon \cdots, \quad y \in G \subset \mathbb{R}^m.$$

In this context, one often transforms $t \to \tau = t/\varepsilon$ so that

$$x' = \varepsilon f(x, y) + \varepsilon^2 \cdots, \quad x \in D \subset \mathbb{R}^n,$$
$$y' = g(x, y) + \varepsilon \cdots, \quad y \in G \subset \mathbb{R}^m,$$

where the prime denotes differentiation with respect to τ.

As before, y is called the fast variable and x the slow variable. The zero set of $g(x, y)$ is given again by $y = \phi(x)$, which in this autonomous case represents a first-order approximation M_0 of the n-dimensional (slow) manifold M_ε. The flow on M_ε is to a first approximation described by $\dot{x} = f(x, \phi(x))$.

In Tikhonov's theorem, we assumed asymptotic stability of the approximate slow manifold; in the constructions of Section 8.3, we assumed a little bit more:

$$\text{Re Sp } g_y(x, \phi(x)) \leq -\mu < 0, x \in D.$$

That is, the eigenvalues of the linearised flow near M_0, derived from the equation for y, have negative real parts only.

In geometric singular perturbation theory, for which Fenichel's results are basic, we only assume that all real parts of the eigenvalues are nonzero. In this case, the slow manifold M_ε is called *normally hyperbolic*. A manifold is called hyperbolic if the local linearisation is structurally stable (real parts of eigenvalues all nonzero), and it is normally hyperbolic if in addition the expansion or contraction near the manifold in the transversal direction is larger than in the tangential direction (the slow drift along the slow manifold).

Note that, although this generalisation is not consistent with the constructions in Section 8.3 where all the real parts of the eigenvalues have to be negative, it allows for interesting phenomena. One might approach M_ε for instance by a stable branch, stay for some time near M_ε, and then leave again a neighbourhood of the slow manifold by an unstable branch. This produces solutions indicated as "pulse-like", "multibump solutions", etc. This type of exchanges of the flow near M_ε is what one often looks for in geometric singular perturbation theory.

8.5.1 Existence of the Slow Manifold

The question of whether the slow manifold M_ε approximated by $\bar{y} = \phi(x)$ persists for $\varepsilon > 0$ was answered by Fenichel. The main result is as follows. If M_0 is a compact manifold that is normally hyperbolic, it persists for $\varepsilon > 0$ (i.e., there exists for sufficiently small, positive ε a smooth manifold M_ε close to M_0). Corresponding with the signs of the real parts of the eigenvalues, there exist stable and unstable manifolds of M_ε, smooth continuations of the corresponding manifolds of M_0, on which the flow is fast.

There are some differences between the cases where M_0 has a boundary or not. For details, see Jones (1994), Kaper (1999) and the original papers by Fenichel.

8.5.2 The Compactness Property

Note that the assumption of compactness of D and G is essential for the existence and uniqueness of the slow manifold. In many examples and applications, M_0, the approximation of the slow manifold obtained from the fast equation, is not bounded. This can be remedied, admittedly in an artificial way, by applying a suitable cutoff of the vector field far away from the domain of interest. In this way, compact domains arise that coincide locally with D and G. However, this may cause some problems with the uniqueness of the slow manifold. Consider for instance the following example.

Example 8.7
Consider the system

$$\dot{x} = x^2, x(0) = x_0 > 0,$$
$$\varepsilon\dot{y} = -y, y(0) = y_0.$$

Putting $\varepsilon = 0$ produces $y = 0$, which corresponds with M_0. We can obtain a compact domain for x by putting $l \le x \le L$ with l and L positive constants independent of ε. However, the limiting behaviour of the solutions depends on the initial condition. Integration of the phase-plane equation yields

$$y(x) = y_0 \exp\left(\frac{1}{\varepsilon x} - \frac{1}{\varepsilon x_0}\right).$$

As $x(t)$ increases (for $t = 1/x_0$, $x(t)$ becomes infinite), the solution for $y(t)$ tends to

$$y_0 \exp\left(-\frac{1}{\varepsilon x_0}\right),$$

so the solutions are for $x(0) > 0$ and after an initial fast transition all exponentially close to $y = 0$. There are, however, an infinite number of slow manifolds dependent on x_0, all tunnelling into this exponentially small neighbourhood of M_0 given by $y = 0$.

A variation of this example is the system

$$\dot{x} = 1, x(0) = x_0 > 0,$$
$$\varepsilon\dot{y} = -\frac{y}{x^2}, y(0) = y_0.$$

To avoid the singularity at $x = 0$, we take $x_0 > 0$. We have the same phase-plane solutions but now we can take t arbitrarily large. The conclusions are similar.

One might wonder about the practical use of exponential closeness as such solutions cannot be distinguished numerically. The phenomenon is important and of practical use when there is a change of stability, a bifurcation of the slow manifold. As we have seen in a number of examples, exponentially close orbits may demonstrate very different sticking phenomena; see also Section 8.7.

8.6 Behaviour near the Slow Manifold

In this section, we consider the estimates on the regular expansion that corresponds with the slow manifold in the autonomous case. We also introduce conditions that enable us to extend the timescale of validity beyond $O(1)$.

8.6.1 Approximating the Slow Manifold

As we have seen in the discussion of Tikhonov's theorem, the solutions may leave the slow manifold (in the case of autonomous systems) or a neighbourhood of $\bar{y} = \phi(x, t)$ (for nonautonomous systems) in a relatively short time. How close do the solutions stay near $\bar{y} = \phi(x, t)$ during that time?

Example 8.8
Consider the system

$$\ddot{x} + x = 0, \ x(0) = 0, \dot{x}(0) = 1,$$
$$\varepsilon \dot{y} = -(y - x), \ y(0) = y_0.$$

Putting $\varepsilon = 0$, we find $\bar{y} = x$, and we note that this critical point - in the sense of Tikhonov's theorem - of the boundary layer equation is asymptotically stable in the linear approximation (eigenvalue -1), uniformly in the parameters x and t. The corresponding approximate two-dimensional manifold M_ε in the system is described by $y = x$.
   Introducing a regular expansion for y, we find

$$y(t) = x(t) - \varepsilon \dot{x}(t) + O(\varepsilon^2).$$

On the other hand, solving the system, we find

$$x(t) = \sin t, \ y(t) = y_0 e^{-t/\varepsilon} + x(t) - \varepsilon(\cos t - e^{-t/\varepsilon}) + O(\varepsilon^2).$$

So, putting $\bar{y} = x$ yields an $O(\varepsilon)$ (first-order) asymptotic approximation of the slow manifold, which, in this case, is valid for all time. The uniformity of the eigenvalue estimate *does not lead* to exponential closeness as would be the case for critical points of the full system.

   The $O(\varepsilon)$-estimate that we obtained above follows directly from the O'Malley-Vasil'eva theorem in Section 8.3.

Corollary (O'Malley-Vasil'eva)
Consider the system

$$\dot{x} = f(x, y, t), \ x(0) = x_0, \ x \in D \subset \mathbb{R}^n, \ t \geq 0,$$
$$\varepsilon \dot{y} = g(x, y, t), \ y(0) = y_0, \ y \in G \subset \mathbb{R}^m, \ t \geq 0.$$

We assume that the vector fields f and g are C^1 on the compact domains D and G. Suppose that $g(x, y, t) = 0$ is solved by $\bar{y} = \phi(x, t)$, and choose $y_0 = \phi(x_0, 0) + \delta(\varepsilon)$ with $\delta(\varepsilon) = O(\varepsilon)$. If

$$\text{Re Sp } g_y(x, \bar{y}, t) \leq -\mu < 0, \ x \in D, 0 \leq t \leq L,$$

we have for $x \in D, y \in G$

$$\|y(t) - \phi(x, t)\| = O(\varepsilon), \ 0 \leq t \leq L.$$

Example 8.8 shows that under these general conditions this estimate is optimal.

The size of the time-interval
If the equations are autonomous, there is no a priori restriction on the interval bound L. The restriction of the time interval arises from the conditions that x and y are in the compacta D and G. If $x(t)$ leaves D as in Example 8.5 of Section 8.2, this imposes the bound on the time interval of validity of the estimates. In this example, we also have the phenomenon of the solutions "sticking" to the x-axis, even after passing the origin. The explanation is in the exponential closeness to the slow manifold for $x < 0$ predicted by the theorem of Section 15.7.

8.6.2 Extension of the Timescale

To use regular expansions on a timescale longer than $O(1)$, we have to take care of secular terms. This is discussed extensively in textbooks; see for instance Kevorkian and Cole (1996), Holmes (1998), or Verhulst (2000). We illustrate this as follows.

Example 8.9
Consider the three-dimensional system

$$\ddot{x} + \varepsilon \dot{x} + (1 + y)x = 0, \quad x(0) = 1, \dot{x}(0) = 0,$$
$$\varepsilon \frac{dy}{dt} = -f(x, \dot{x})y, \quad y(0) = 1,$$

with $f(x, \dot{x}) \geq \mu > 0, (x, \dot{x}) \in \mathbb{R}^2$. Both Tikhonov's theorem and the O'Malley-Vasil'eva expansion technique can be applied, and we put

$$x(t) = a_0(t) + \varepsilon a_1(t) + \varepsilon^2 \cdots,$$
$$y(t) = b_0(t) + \varepsilon b_1(t) + \varepsilon^2 \cdots.$$

We find $b_0(t) = b_1(t) = \cdots = 0$; outside an initial boundary layer, the solution $y(t)$ remains exponentially close to $y = 0$ for all time. Note that the equation for x reduces to

$$\ddot{x} + \varepsilon \dot{x} + x = 0,$$

which describes damped oscillations. The expansion for x, however, leads to the set of equations

$$\ddot{a}_0 + a_0 = 0, \quad a_0(0) = 1, \dot{a}_0(0) = 0,$$
$$\ddot{a}_1 + \varepsilon \dot{a}_0 + a_1 = 0, \quad a_1(0) = 0, \dot{a}_1(0) = 0.$$

The solutions are $a_0(t) = \cos t$, $a_1(t) = \frac{1}{2}\sin t - \frac{1}{2}t\cos t$, in which $a_1(t)$ has an amplitude *increasing* with time. The expansion is valid on an $O(1)$ time interval only.

Consider again the system

$$\dot{x} = f(x, y) + \varepsilon f_1(x, y) + \varepsilon^2 \cdots, \quad x(0) = x_0, \quad x \in D \subset \mathbb{R}^n, \quad t \geq 0,$$
$$\varepsilon \dot{y} = g(x, y) + \varepsilon g_1(x, y) + \varepsilon^2 \cdots, \quad y(0) = y_0, \quad y \in G \subset \mathbb{R}^m, \quad t \geq 0.$$

The slow manifold corresponds with $y = \phi(x)$, which is a critical point of the boundary layer equation and asymptotically stable in linear approximation as long as $x \in D, y \in G$.

Suppose now that the solutions stay close to the slow manifold for at least a time interval of the order $1/\varepsilon$; we have seen such cases in earlier sections. To obtain asymptotic approximations valid on such a long timescale in the slow manifold, we have to employ two timescales: t and εt. There are various techniques, of which the method of averaging is very efficient; see Chapter 11. The procedure is illustrated by the following example.

Example 8.10
Consider the three-dimensional system

$$\ddot{x} + x = \varepsilon(1 - x^2)\dot{x} + \varepsilon f(x, \dot{x}, y)$$
$$\varepsilon \dot{y} = g(x, y) + \varepsilon \cdots.$$

We have $g(x, \phi(x)) = 0$ with $g_y(x, \phi(x)) \leq -\mu < 0$. The conditions of Tikhonov's theorem (Section 8.2) have been satisfied, and putting $x(0) = x_0, y(0) = \phi(x_0)$, we have with $x(t)$ the solution of the reduced equation $(\varepsilon = 0)$

$$y(t) - \phi(x(t)) = O(\varepsilon), \quad 0 \leq t \leq L.$$

To improve the approximation of $x(t)$, we propose the solution of the slow manifold equation

$$\ddot{X} + X = \varepsilon(1 - X^2)\dot{X} + \varepsilon f(X, \dot{X}, \phi(X))$$

with appropriate initial values. X in its turn can be approximated by averaging. We choose as an example $f(x, y) = ay\dot{x}, g(x, y) = -y + x^2$, so that $\phi(x) = x^2$; Tikhonov's theorem can be applied. We find

$$\ddot{X} + X = \varepsilon(1 - X^2)\dot{X} + \varepsilon a X^2 \dot{X}.$$

Averaging (see Chapter 11), produces that no periodic solution exists if $a \geq 1$; if $a < 1$, we have the approximation for the periodic solution $X(t)$ and so of $x(t)$ in the slow manifold in the form

$$X(t) = \frac{2}{\sqrt{1-a}} \cos(t + \psi_0) + O(\varepsilon),$$

valid on the timescale $1/\varepsilon$; ψ_0 is the initial phase. Note that if the equation for $\dot{x}$ contains y in the $O(1)$ terms, the perturbation scheme changes.

Suppose for instance that $a = -3$ and that we start outside the slow manifold at $(x(0), \dot{x}(0), y(0))$. The solution will jump to an $O(\varepsilon)$ neighbourhood of the slow manifold given by $y = x^2$, and the flow is then to first order described by

$$\ddot{X} + X = \varepsilon(1 - 4X^2)\dot{X}, y = X^2.$$

Averaging produces an approximation X_{as} of $X(t)$,

$$X_{as} = \frac{r_0}{(r_0^2 + (1 - r_0^2)e^{-\varepsilon t})^{1/2}} \cos(t + \psi_0),$$

with $r_0 = (x(0)^2 + \dot{x}(0)^2)^{1/2}$ and $\psi(0)$ determined by the initial conditions. The asymptotic approximation of the solution to $O(\varepsilon)$ contains three timescales: $t/\varepsilon, t$, and εt. For $x(t)$ and $y(t)$, they take the form

$$x(t) = X_{as} + O(\varepsilon),$$
$$y(t) = (y(0) - x(0)^2)e^{-t/\varepsilon} + X_{as}^2 + O(\varepsilon),$$

for $0 \leq t \leq c/\varepsilon$ with c a positive constant, independent of ε.

General procedure for long-time approximations

The general procedure would run as follows. Suppose we have the initial value problem

$$\dot{x} = f_0(x) + \varepsilon f_1(x, y, t) + \varepsilon^2 \cdots, \quad x(0) = x_0, \quad x \in D \subset \mathbb{R}^n, \quad t \geq 0,$$
$$\varepsilon \dot{y} = g(x, y, t) + \varepsilon \cdots, \quad y(0) = y_0, \quad y \in G \subset \mathbb{R}^m, \quad t \geq 0.$$

Suppose that $y = \phi(x, t)$ solves the equation $g(x, y, t) = 0$ and that Tikhonov's theorem applies. Assume also that we can solve the equation for x with $\varepsilon = 0$. The solutions for $\varepsilon = 0$ will be of the form $x(t) = X(t, C)$ with C a constant n-vector.

Variation of constants according to Lagrange gives the transformation

$$x(t) = X(t, z),$$

which produces the system

$$\dot{z} = \varepsilon f_1^*(z, y, t) + \varepsilon^2 \cdots,$$
$$\varepsilon \dot{y} = g^*(z, y, t) + \varepsilon \cdots.$$

If we can average the regular part of the equation for z over t, in particular the vector field $f_1^*(z, \phi(X(t, z), t), t), t)$, this produces the regular part of the expansion with validity on the timescale $1/\varepsilon$.

8.7 Periodic Solutions and Oscillations

Consider again the equations in Example 8.10,

$$\ddot{x} + x = \varepsilon(1 - x^2)\dot{x} + \varepsilon a y \dot{x},$$
$$\varepsilon \dot{y} = -y + x^2,$$

where we have found a periodic solution for $a < 1$ of the equation describing the flow in the slow manifold. Does this periodic solution exist for the full system?

In a number of cases, the existence of periodic solutions is obtained by the following theorem.

Theorem 8.3
(Anosov, 1960)
Consider the system

$$\dot{x} = f(x, y) + \varepsilon \cdots, \quad x \in D \subset \mathbb{R}^n,$$
$$\varepsilon \dot{y} = g(x, y) + \varepsilon \ldots, \quad y \in G \subset \mathbb{R}^m,$$

and again that $g(x, y) = 0$ yields $\bar{y} = \phi(x)$. Assume that: a. $g_y(x, \phi(x))$ has no eigenvalues with real part zero in D;
b. the slow manifold equation $\dot{x} = f(x, \phi(x))$ has a T_0-periodic solution P_0 with only one multiplier 1 (only one eigenvalue purely imaginary).
Then there exists a T_ε-periodic solution P_ε of the original system for which we have $P_\varepsilon \to P_0, T_\varepsilon \to T_0$ as $\varepsilon \to 0$.

A theorem under related conditions was formulated by Flatto and Levinson (1955); in their paper, condition (a) is the same and condition (b) is formulated differently.

Anosov's theorem does not apply to Example 8.10. The reason is that the slow manifold equation $\dot{x} = f(x, \phi(x))$ (here $\ddot{x} + x = 0$) is conservative, admitting a family of periodic solutions. In such a case, both resonance and dissipation may destroy the periodic solutions, as the simple example $\ddot{x} + x = -\varepsilon \dot{x}$ shows. Theorems have been formulated to allow for conservative systems ruling the regular expansion, see Alymkulov (1986).

Another way out of this is to observe that, according to Fenichel (Section 8.5), the slow manifold exists. The next step is then to determine the perturbation equations for the flow in the slow manifold. These equations may contain periodic solutions as in Example 8.10 or Example 8.11.

Example 8.11
Consider the three-dimensional system

$$\ddot{x} + x = \mu(1 - x^2)\dot{x} + \nu f(x, \dot{x}, y),$$
$$\varepsilon \dot{y} = -g(x, \dot{x})y + \varepsilon h(x, \dot{x}, y),$$

with $g(x, \dot{x}) \geq a > 0$ in $\mathbb{R}^2$.

If μ is a positive constant independent of ε and $\nu = \varepsilon$, Anosov's theorem applies, predicting the existence of a periodic solution close to the periodic solution of the Van der Pol equation.

If $\mu = \varepsilon, \nu = \varepsilon^2$, Anosov's theorem does not apply, and we have to follow the procedure outlined above. The slow manifold exists in an ε-neighbourhood of $\bar{y} = 0$. The flow in the slow manifold is described by the equation

$$\ddot{x} + x = \varepsilon(1 - x^2)\dot{x} + \varepsilon^2 f(x, \dot{x}, 0) + \varepsilon^3 \cdots .$$

As we will see in Chapter 11, this equation admits a periodic solution that is close to the periodic solution of the Van der Pol equation.

Example 8.12

There may be critical values of the parameters that necessitate the calculation of higher-order terms of the perturbations. Consider the system

$$\ddot{x} + x + yx^2 = \mu \left(1 - \frac{1}{3}\dot{x}^2\right)\dot{x},$$

$$\varepsilon \dot{y} = -y + \lambda + \varepsilon f(x, \dot{x}, y),$$

with parameters $\lambda, \mu > 0$. The flow in the slow manifold is to first order described by

$$\ddot{x} + x + \lambda x^2 = \mu \left(1 - \frac{1}{3}\dot{x}^2\right)\dot{x}.$$

In the limiting case $\lambda = 0$, we have the Rayleigh equation that admits one (stable) periodic solution. If $\lambda = O(\varepsilon)$, we have the same result (see Chapter 11).

A subtle phenomenon arises if we increase λ. Doelman and Verhulst (1994) showed that at $\lambda_h = 0.28 \cdots$ a homoclinic bifurcation takes place where the stable solution vanishes. However, there exists a number $\lambda_{SN} > \lambda_h$ such that for $\lambda_h < \lambda < \lambda_{SN}$ there exist two periodic solutions, one stable and one unstable. The surprise is that $\lambda_{SN} - \lambda_h \approx 10^{-6}$ as ε varies from 0 to 0.8. The implication is that for λ near $\lambda_h = 0.28 \cdots$ we certainly have to include higher-order terms to analyse the behaviour of the solutions.

Interesting phenomena take place when the solutions jump repeatedly from one manifold to another, nonintersecting one. This may give rise to periodic solutions or oscillations with fast and slow motions. The classical example is the Van der Pol equation in the form

$$\varepsilon \ddot{x} + x = (1 - x^2)\dot{x}.$$

This very important subject is dealt with extensively by Grasman (1987), a reason why we only mention the subject here. We end with an explicit example of jumping phenomena.

Example 8.13

Consider the system

$$\dot{x} = f(x, y),$$
$$\varepsilon \dot{y} = -xy(1 - y).$$

We assume that the equation for x contains oscillatory solutions. As an illustration we replace it by an explicit function. The slow manifolds are $y = 0$ and $y = 1$ with eigenvalues $-x$ and x, respectively.

Choosing periodic behaviour of x, we find spike-like relaxation behaviour as shown in Fig. 8.4.

Another possibility is to take for x a solution with almost-periodic or chaotic behaviour. Choosing for instance the Lorenz equations

$$\dot{x}_1 = 10(x_2 - x_1),$$
$$\dot{x}_2 = 28x_1 - x_2 - x_1 x_3,$$
$$\dot{x}_3 = x_1 x_2 - \frac{8}{3}x_3,$$

we identify x with $x_1/10$ (10 because of the $O(10)$ variations of $x_1(t)$). As expected, this leads to chaotic jumping.

8.8 Guide to the Literature

The Tikhonov (1952) theorem predicts qualitatively the behaviour of boundary layer jumps in time. Related work was done by Levinson and his students; see Flatto and Levinson (1955) or Wasow (1965). Vasil'eva (1963) and O'Malley (1968, 1971) produced actual constructions of the solutions including proofs of asymptotic validity. In the paper by Vasil'eva, the construction is carried out by calculating regular and boundary layer expansions followed by a matching process. O'Malley gave the multiple-timescale expansion of Section 8.3; this is an efficient way of obtaining approximations.

There were some imperfections in the original formulation of Tikhonov's theorem. It is interesting to note that Hoppensteadt (1967, 1969) added the uniformity requirement (Section 8.2, condition b). He also gave a modified proof using a Lyapunov function. We have added condition (c), requiring solutions to start in an interior subset, and we expect that now the formulation of Tikhonov's theorem in Section 8.2 is final.

Attention was given to the so-called critical case where a root, describing the regular expansion corresponding with the slow manifold, is not isolated. Flaherty and O'Malley (1980) called this a "singular singular perturbation problem". This actually happens in applications. For more references and examples, see O'Malley (1991), and Vasil'eva, Butuzov, and Kalachev (1995).

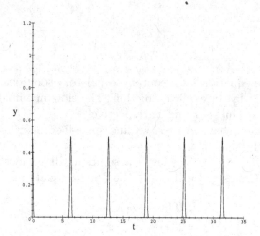

Fig. 8.4. The solution $y(t)$ of the equation $\varepsilon \dot{y} = -xy(1 - y)$ with $x(t) = \sin t$ that starts being attracted to the slow manifold $y = 0$. Although the stability between the slow manifolds $y = 0$ and $y = 1$ changes periodically, exponential sticking delays the departure from $y = 0$ and produces spike behaviour; $\varepsilon = 0.01$.

Another problem that arises in applications is that the root describing the regular expansion is stable but not stable in the linear approximation. In other words, the slow manifold is not hyperbolic. Such a case was discussed by Verhulst (1976), where it is shown that the regular expansion has to be adjusted to contain broken powers of ε and logarithmic terms.

Example 8.5 in Section 8.2 shows a case where two roots of the fast equation cross each other. Interesting here is the exchange of stabilities and what has been termed "delay of stability loss" (Neishtadt, 1991). In our example, this is simply due to sticking by exponential attraction. Delay phenomena in this context were first discovered by Shiskova (1973) and Pontrjagin; the demonstration is for a three-dimensional model equation containing focus-limit cycle transitions. Neishtadt (1991), and more references therein, shows how delay of stability loss can be tied in with temporary capture in a resonance zone of a dynamical system. Such resonance phenomena can be described by averaging methods that we shall discuss in Chapter 11. We note that regarding the problem of exchange of stabilities without delay phenomena, a detailed analysis was given by Lebovitz and Schaar (1975, 1977).

An early result on periodic solutions was obtained by Flatto and Levinson (1955). Apart from the important Anosov (1960) theorem, Hale (1963, Chapter 5), formulated a theorem combining results on slow and fast equations.

Two important subjects have only been lightly touched upon in the present chapter. First there are relaxation oscillations, for which the reader is referred to Grasman (1987) and also Mishchenko and Rosov (1980). Then there is geometric singular perturbation theory, which was prepared by Tikhonov and Anosov but really got started with Fenichel's results (Section 8.5); this theory also led to a subtle instrument, the exchange lemma; for surveys see Jones (1994), Kaper (1999), and Kaper and Jones (2001).

8.9 Exercises

Exercise 8.1 Consider a harmonic oscillator that is excited from the equilibrium position by a sudden impulse.

Model a (exponential decay):

$$\ddot{x} + x = \exp\left(\frac{-t}{\varepsilon}\right).$$

Model b (algebraic decay):

$$\ddot{x} + x = \frac{\varepsilon}{\varepsilon + t},$$

with initial conditions $x(0) = 0, \dot{x}(0) = 0$. Compute the asymptotic approximations of the solutions.

Exercise 8.2 Consider the equation

$$\frac{dx}{dt} = e^{x^3} \sin y + e^{xy^2} \sin x - y^2 - xy - 1, \quad x(0) = \alpha > 0,$$

$$\varepsilon \frac{dy}{dt} = y^2 - x^2, \quad y(o) = \beta.$$

a. Apply the theorem of Tikhonov. For which values of α and β does the theorem predict boundary layer behaviour near $t = 0$?
b. Determine in this case for $t \geq d > 0$

$$\lim_{\varepsilon \downarrow 0} x(t) \quad \text{and} \quad \lim_{\varepsilon \downarrow 0} y(t).$$

Exercise 8.3 Consider the initial value problem

$$\frac{dx}{dt} = x + 2y, \quad x(0) = 2, \quad t \geq 0,$$

$$\varepsilon \frac{dy}{dt} = x - 2y + \sin t, \quad y(0) = 3.$$

a. Are the conditions of the theorem of Tikhonov satisfied?
b. Construct a first-order formal approximation, and give a sketch of the construction of a second-order approximation.

Exercise 8.4 Consider the mathematical pendulum with large damping

$$\ddot{x} + \mu\dot{x} + \sin x = 0,$$
$$x(0) = \alpha, \quad \mu \gg 1,$$
$$\dot{x}(0) = \beta.$$

Put $\frac{1}{\mu} = \varepsilon$ and describe up to and including the second order in ε the initial boundary layer expansion that arises.

Exercise 8.5 Consider the two-body problem with variable mass given by

$$\frac{d^2\rho}{d\theta^2} + \rho = m_r + u,$$

$$\varepsilon \frac{du}{d\theta} = \frac{-u}{\rho^2}, \quad u(0) = 1 - m_r.$$

Starting in a circular orbit $\rho(0) = 1$, $\frac{d\rho}{d\theta}(0) = 0$, compute the critical masslimit up to $O(\varepsilon^2)$ for the orbits to become hyperbolic. Compare the results with numerical calculations.

Exercise 8.6 Consider the initial value problem

$$\dot{x} = f(x, y), \quad x(0) = x_0, \quad x \in D \subset \mathbb{R}^n, \quad t \geq 0,$$
$$\varepsilon\dot{y} = y^2(1 - y) + \varepsilon g(x, y), \quad y(0) = y_0, 0 < y_0 < 1, \quad t \geq 0.$$

a. Consider the two cases $y_0 = 1/2$ and $y_0 = \varepsilon$. Can one apply Tikhonov's theorem to these cases?
b. Estimate in both cases the time needed to arrive in an $O(\varepsilon)$ neighbourhood of the slow manifold $y = 1$; explain the estimates.

Exercise 8.7 In Example 8.12 we formulated the equation for the flow in the slow manifold to first order. Derive the equation in the case $\mu = \varepsilon$.

Exercise 8.8 Use Example 8.13 to construct a relaxation oscillation in which both the jumping times and the amplitude vary chaotically.

Evolution Equations with Boundary Layers

Evolution equations form a rich problem field; the purpose of this chapter is to introduce the reader to some prototype problems illustrating the analysis of expansions for parabolic and hyperbolic equations.

The analysis in Sections 9.5 and 9.2 for wave equations is well-known; see Kevorkian and Cole (1996) or De Jager and Jiang Furu (1996). We include these problems because they display fundamental aspects.

9.1 Slow Diffusion with Heat Production

In this section, we consider a case where the ideas we have met thus far can be extended directly to a spatial problem with time evolution. Slow spatial diffusion takes place in various models, for instance in material where heat conduction is not very good; see the model in this section. Other examples are the slow spreading of pollution in estuaries or seas and, in mathematical biology, the slow spreading of a population in a domain.

First, we consider a one-dimensional bar described by the spatial variable x and length of the bar L; the temperature of the bar is $T(x, t)$, and we assume slow heat diffusion. The endpoints of the bar are kept at a constant temperature. Heat is produced in the bar and is also exchanged with its surroundings, so we have that $T = T(x, t)$ is governed by the equation and conditions

$$\frac{\partial T}{\partial t} = \varepsilon \frac{\partial^2 T}{\partial x^2} - \gamma(x)[T - s(x)] + g(x)$$

with boundary conditions $T(0, t) = t_0, T(L, t) = t_1$, and initial temperature distribution $T(x, 0) = \psi(x)$. The temperature of the neighbourhood of the bar is $s(x); \gamma(x)$ is the exchange coefficient of heat and $\gamma(x) \geq d$ with d a positive constant so the bar is nowhere isolated.

In the analysis we shall follow Van Harten (1979). From Chapter 5, we know how to determine an asymptotic expansion for the stationary (time-independent) problem. Starting with the approximation of the stationary problem, we will analyse the time-dependent problem. One of the questions is: does the time-dependent expansion converge to the stationary one?

Inspired by our experience in Chapter 5, we expect boundary layers near $x = 0$ and $x = L$ with local variables

$$\xi = \frac{x}{\sqrt{\varepsilon}}, \eta = \frac{L - x}{\sqrt{\varepsilon}}.$$

The approximation of the stationary solution will be of the form

$$T(x) = V_0(x) + \varepsilon V_1(x) + \varepsilon^2 \cdots + Y_0(\xi) + \varepsilon^{1/2} Y_1(\xi) + \bar{Y}_0(\eta) + \varepsilon^{1/2} \bar{Y}_1(\eta) + \varepsilon \cdots,$$

where $V_0 + \varepsilon V_1$ is the first part of the regular expansion and the boundary layer approximations satisfy

$$\frac{d^2 Y_0}{d\xi^2} - \gamma(0) Y_0 = 0, Y_0(0) = t_0 - V_0(0), \lim_{\xi \to \infty} Y_0(\xi) = 0.$$

$\bar{Y}_0(\eta)$ will satisfy

$$\frac{d^2 \bar{Y}_0}{d\eta^2} - \gamma(L) \bar{Y}_0 = 0, \bar{Y}_0(L) = t_L - V_0(L), \lim_{\eta \to \infty} \bar{Y}_0(\eta) = 0.$$

We find

$$Y_0(\xi) = (t_0 - V_0(0)) e^{-\sqrt{\gamma(0)}\xi}, \bar{Y}_0(\eta) = (t_L - V_0(L)) e^{-\sqrt{\gamma(L)}\eta}.$$

9.1.1 The Time-Dependent Problem

Substituting a regular expansion of the form $T(x, t) = U_0(x, t) + \varepsilon U_1(x, t) \cdots$, we find in lowest order

$$\frac{\partial U_0}{\partial t} = -\gamma(x) U_0 + \gamma(x) s(x) + g(x), U_0(x, 0) = \psi(x),$$

with solution

$$U_0(x, t) = V_0(x) + (\psi(x) - V_0(x)) e^{-\gamma(x)t}.$$

This first-order regular expansion satisfies the initial condition and, as $\gamma(x) > 0$, the regular part of the time-dependent expansion tends to the regular part of the stationary solution as t tends to infinity.

This suggests proposing for the full expansion

$$T(x, t) = U_0(x, t) + \varepsilon \cdots + X_0(\xi, t) + \bar{X}_0(\eta, t) + \sqrt{\varepsilon} \cdots.$$

Substituting this expansion in the equation produces

$$\frac{\partial X_0}{\partial t} = \frac{\partial^2 X_0}{\partial \xi^2} - \gamma(0) X_0$$

with boundary condition, matching condition, and initial condition

$$X(0,t) = t_0 - U_0(0,t), \lim_{\xi \to \infty} X_0(\xi,t) = 0, X_0(\xi,0) = 0.$$

This problem is solved by putting $X_0 = Z_0 \exp(-\gamma(0)t)$, which yields

$$\frac{\partial Z_0}{\partial t} = \frac{\partial^2 Z_0}{\partial \xi^2}.$$

Applying Duhamel's principle (see, for instance, Strauss, 1992), we find

$$X_0(\xi,t) = \frac{2}{\sqrt{\pi}} e^{-\gamma(0)t} \int_{\frac{\xi}{2\sqrt{t}}}^{\infty} \phi\left(t - \frac{\xi^2}{4\tau^2}\right) e^{-\tau^2} d\tau$$

with $\phi(z) = (t_0 - U(0,z)) \exp(\gamma(0)z)$.

In the same way, we find for $\bar{X}_0(\eta,t)$ the problem

$$\frac{\partial \bar{X}_0}{\partial t} = \frac{\partial^2 \bar{X}_0}{\partial \eta^2} - \gamma(L)\bar{X}_0$$

with boundary condition, matching condition, and initial condition

$$\bar{X}(0,t) = t_L - U_0(L,t), \lim_{\eta \to \infty} \bar{X}_0(\eta,t) = 0, \bar{X}_0(\eta,0) = 0.$$

The solution is obtained as before and becomes

$$\bar{X}_0(\eta,t) = \frac{2}{\sqrt{\pi}} e^{-\gamma(L)t} \int_{\frac{\eta}{2\sqrt{t}}}^{\infty} \bar{\phi}\left(t - \frac{\eta^2}{4\tau^2}\right) e^{-\tau^2} d\tau$$

with $\bar{\phi}(z) = (t_L - U_0(L,z)) \exp(\gamma(L)z)$.

9.2 Slow Diffusion on a Semi-infinite Domain

A different problem arises on considering for $t \geq 0$ the semi-infinite domain $x \geq 0$ for the equation

$$\frac{\partial \phi}{\partial t} = \varepsilon \frac{\partial^2 \phi}{\partial x^2} - p(t)\frac{\partial \phi}{\partial x}$$

with initial condition $\phi(x,0) = f(x)$ and boundary condition $\phi(0,t) = g(t)$ (boundary input). The functions $p(t), f(x), g(t)$ are assumed to be sufficiently smooth, and $p(t)$ does not change sign. Moreover, we assume continuity at $(0,0)$ and, for physical reasons, decay of the boundary input to zero:

$$f(0) = g(0), \quad \lim_{t \to \infty} = 0.$$

We start again with a regular expansion of the form

$$\phi(x,t) = u_0(x,t) + \varepsilon u_1(x,t) + \varepsilon^2 \cdots$$

to find for the first two terms after substitution

$$\frac{\partial u_0}{\partial t} + p(t)\frac{\partial u_0}{\partial x} = 0,$$

$$\frac{\partial u_1}{\partial t} + p(t)\frac{\partial u_1}{\partial x} = \frac{\partial^2 u_0}{\partial x^2}.$$

Introducing $P(t) = \int_0^t p(s)ds$, the equation for u_0 has the characteristic $x - P(t)$, which implies that any differentiable function of $x - P(t)$ solves the equation for u_0. (The reader who is not familiar with the characteristics of first-order partial differential equations can verify this by substitution.) At this stage, the easiest way is to satisfy the initial condition by choosing

$$u_0(x,t) = f(x - P(t)).$$

A consequence is that for u_1 we have to add the initial condition $u_1(x,0) = 0$. Solving the equation

$$\frac{\partial u_1}{\partial t} + p(t)\frac{\partial u_1}{\partial x} = \frac{\partial^2}{\partial x^2}f(x - P(t))$$

with the initial condition, we find

$$u_1(x,t) = \int_0^t \frac{\partial^2}{\partial x^2}(x + P(s) - 2P(t))ds.$$

This regular expansion does not satisfy the boundary condition at $x = 0$, so we expect the presence of a spatial boundary layer there. Introducing the boundary layer variable

$$\xi = \frac{x}{\varepsilon^\nu},$$

we expect for the solution $\phi(x,t)$ an expansion of the form

$$\phi(x,t) = u_0(x,t) + \varepsilon u_1(x,t) + \varepsilon^2 \cdots + \psi(\xi,t)$$

with initial and boundary conditions

$$\psi(\xi,0) = 0, \psi(0,t) = g(t) - u_0(0,t) - \varepsilon u_1(0,t) - \varepsilon^2 \cdots$$

and matching condition

$$\lim_{\xi \to \infty} \psi(\xi,t) = 0.$$

Introducing the boundary layer variable into the equation for ϕ yields

$$\frac{\partial \psi}{\partial t} = \varepsilon^{1-2\nu}\frac{\partial^2 \phi}{\partial \xi^2} - \varepsilon^{-\nu}p(t)\frac{\partial \phi}{\partial \xi} = 0.$$

A significant degeneration arises if $1 - 2\nu = -\nu$ or $\nu = 1$. Assuming that we can expand $\psi = \psi_0 + \varepsilon\psi_1 + \varepsilon^2 \cdots$, we have

$$-\frac{\partial^2\psi_0}{\partial\xi^2} + p(t)\frac{\partial\psi_0}{\partial\xi} = 0, \quad \psi_0(0,t) = g(t) - f(-P(t)),$$

$$-\frac{\partial^2\psi_1}{\partial\xi^2} + p(t)\frac{\partial\psi_1}{\partial\xi} = \frac{\partial\psi_0}{\partial t}, \quad \psi_1(0,t) = -\int_0^t \frac{\partial^2}{\partial x^2}(P(s) - 2P(t))ds.$$

For ψ_0, we find the expression

$$\psi_0(\xi,t) = A(t)e^{p(t)\xi} + B(t)\xi + C(t)$$

with A, B, C suitably chosen functions.

From the matching condition, we find $B(t) = C(t) = 0$ and the condition

$$p(t) < 0.$$

With this condition, we have

$$\psi_0(\xi,t) = (g(t) - f(-P(t))e^{p(t)\xi},$$

which satisfies the initial and boundary conditions. It is easy to calculate ψ_1 and higher-order approximations.

9.2.1 What Happens if $p(t) > 0$?

This problem was analysed by Shih (2001); see also this paper for related references. We recall that the regular expansion starts with $u_0(x,t) = f(x - P(t))$. If $p(t) < 0$, the characteristic $x - P(t) = $ constant is not located in the quarter-plane $x \geq 0, t \geq 0$, but if $p(t) > 0$, the characteristics $x - P(t) = $ constant extend into this domain. We have $f(0) = g(0)$, but the derivatives of these functions are generally not compatible. This causes jump discontinuities along the characteristic curve $x - P(t) = 0$, which can be compensated by a boundary layer along this curve. It turns out that the appropriate boundary layer variable in this case is

$$\eta = \frac{x - P(t)}{\sqrt{\varepsilon}}.$$

For more details of the analysis, see Shih (2001).

9.3 A Chemical Reaction with Diffusion

A number of chemical reaction problems can be formulated as singularly perturbed equations. Following Vasil'eva, Butuzov, and Kalachev (1995), we consider the problem

$$\varepsilon\left(\frac{\partial u}{\partial t} - a(x,t)\frac{\partial^2 u}{\partial x^2}\right) = f(u,x,t) + \varepsilon\cdots,$$

where $0 \le x \le 1, 0 \le t \le T$, and $a(x,t) > 0$. We shall consider this as a prototype problem for the more complicated cases where u and x are vectors and we consider a system of equations.

The initial condition is

$$u(x,0) = \phi(x).$$

Natural boundary conditions in this case are the Neumann conditions

$$\frac{\partial u}{\partial x}(0,t) = \frac{\partial u}{\partial x}(1,t) = 0.$$

Putting $\varepsilon = 0$, we have $f(u,x,t) = 0$, and we note a similarity with the initial value problems of Chapter 8. We will indeed impose a similar condition as in the Tikhonov theorem (Section 8.2): assume that $f(u,x,t) = 0$ has a unique solution $u_0(x,t)$ and that

$$\frac{\partial f}{\partial u}(u_0(x,t),x,t) < 0$$

uniformly for $0 \le x \le 1, 0 \le t \le T$.

We shall determine the lowest-order terms of the appropriate expansion. A regular expansion of the form $u(x,t) = u_0(x,t) + \varepsilon u_1 \cdots$ will in general not satisfy the initial and boundary conditions. This suggests the presence of an initial layer near $t = 0$ and boundary layers near $x = 0$ and $x = 1$. We subtract the regular expansion by putting

$$u(x,t) = u_0(x,t) + \varepsilon u_1 \cdots + v(x,t,\tau,\xi,\eta),$$

in which we assume that v is of the form

$$v = P_\varepsilon(x,\tau) + Q_\varepsilon(\xi,t) + R_\varepsilon(\eta,t),$$

where $\tau = t/\varepsilon$ and ξ and η are the boundary layer variables near $x = 0$ and $x = 1$, respectively. Substitution produces for v

$$\varepsilon\left(\frac{\partial u_0}{\partial t} + \varepsilon\frac{\partial u_1}{\partial t} + \frac{\partial v}{\partial t} - a(x,t)\frac{\partial^2 u_0}{\partial x^2} - \varepsilon a(x,t)\frac{\partial^2 u_1}{\partial x^2} - a(x,t)\frac{\partial^2 v}{\partial x^2}\right) =$$

$$= f_u(u_0(x,t),x,t)(\varepsilon u_1 + v + \cdots)$$

We assume expansions for P, Q, R such as $P_\varepsilon(x,\tau) = P_0(x,\tau) + \varepsilon P_1 \cdots$. After putting $t = \varepsilon\tau$, we find for P_0

$$\frac{\partial P_0}{\partial\tau} = f(u_0(x,0) + P_0(x,\tau) + Q_0(\xi,0) + R_0(\eta,0),x,0)$$

with an initial condition for $P_0(x,0)$. It is natural to assume that $Q_0(\xi,0) = R_0(\eta,0) = 0$ so that

$$P_0(x,0) = \phi(x) - u_0(x,0).$$

We solve the equation for P_0 with x as a parameter, noting that $P_0(x,\tau) = 0$ is an equilibrium solution; we have to assume that the initial condition is located in the domain of attraction of this equilibrium solution, as we did in the Tikhonov theorem. In that case, we have immediately that the matching condition

$$\lim_{\tau \to \infty} P_0(x,\tau) = 0$$

is satisfied. As x is a parameter, we still have the freedom to put $P_0(0,\tau) = 0$. To determine the boundary layer variables, we have to rescale the equation. We abbreviate again $P + Q + R = v$ and put $\xi = x/\varepsilon^\nu$ to find near $x = 0$

$$\varepsilon \frac{\partial u_0}{\partial t} + \varepsilon \frac{\partial v}{\partial t} - \varepsilon^{1-2\nu} a(\varepsilon^\nu \xi, t) \frac{\partial^2 u_0}{\partial \xi^2} - \varepsilon^{1-2\nu} a(\varepsilon^\nu \xi, t) \frac{\partial^2 v}{\partial \xi^2} = f(u_0 + v, \varepsilon^\nu \xi, t) + \varepsilon \cdots .$$

A significant degeneration arises if $\nu = \frac{1}{2}$. We will require the boundary layer solution $R_\varepsilon(\eta, t)$ to vanish outside the boundary layer near $x = 1$, so the equation for $Q_0(\xi, t)$ becomes

$$-a(0,t) \frac{\partial^2 Q_0}{\partial \xi^2} = f(u_0(0,t) + Q_0, 0, t).$$

From the Neumann condition at $x = 0$, we have at lowest-order

$$\frac{\partial u_0}{\partial x}(0,t) + \frac{\partial P_0}{\partial x}(0,\tau) + \varepsilon^{-\frac{1}{2}} \frac{\partial Q_0}{\partial \xi}(0,t) = 0,$$

which yields

$$\frac{\partial Q_0}{\partial \xi}(0,t) = 0$$

with the matching condition

$$\lim_{\xi \to \infty} Q_0(\xi, t) = 0.$$

We conclude that $Q_0(\xi, t) = 0$ and that nontrivial solutions $Q_1(\xi, t), Q_2(\xi, t)$, etc., arise at higher order.

In the same way, we conclude that $R_0(\xi, t) = 0$ and finally that the lowest-order expansion of the solution is of the form

$$u_0(x,t) + P_0\left(x, \frac{t}{\varepsilon}\right).$$

Remark

The computation of higher-order approximations leads to linear equations for Q_1, R_1, etc. An additional difficulty is that there will be corner boundary layers at $(x,t) = (0,0)$ and $(1,0)$ involving boundary layer functions of the forms $Q^*(\xi, \tau)$ and $R^*(\eta, \tau)$ at higher order. For more details, see Vasil'eva, Butuzov, and Kalachev (1995), where a proof of asymptotic validity is also presented.

9.4 Periodic Solutions of Parabolic Equations

Following Vasil'eva, Butuzov, and Kalachev (1995), we will look for 2π-periodic solutions of the equation

$$\varepsilon \frac{\partial^2 u}{\partial x^2} = \frac{\partial u}{\partial t} + s(x,t)u + f(x,t) + \varepsilon g(x,t,u) + \varepsilon \cdots$$

with Dirichlet boundary conditions

$$u(0,t) = u(1,t) = 0$$

and periodicity condition

$$u(x,t) = u(x,t+2\pi)$$

for $0 \le x \le 1$ and $t \ge 0$. All time-dependent terms in the differential equation are supposed to be 2π-periodic. To illustrate the technique, we shall analyse an example first and discuss the more general case later.

9.4.1 An Example

Consider the equation

$$\varepsilon \frac{\partial^2 u}{\partial x^2} = \frac{\partial u}{\partial t} + a(x)u + b(x)\sin t$$

with $a(x)$ and $b(x)$ smooth functions; $a(x) > 0$ for $0 \le x \le 1$.

Based on our experience with problems in the preceding sections, we expect that the boundary layer variables

$$\xi = \frac{x}{\sqrt{\varepsilon}}, \quad \eta = \frac{1-x}{\sqrt{\varepsilon}},$$

will play a part, so we will look for a solution of the form

$$u(x,t) = u_0(x,t) + Q_0(\xi,t) + \sqrt{\varepsilon}Q_1(\xi,t) + R_0(\eta,t) + \sqrt{\varepsilon}R_1(\eta,t) + \varepsilon \cdots ,$$

where $u_0(x,t)$ is the first term of a regular expansion of the form $u_0(x,t) + \varepsilon u_1(x,t) + \varepsilon^2 \cdots$. The equation for $u_0(x,t)$ is

$$\frac{\partial u_0}{\partial t} + a(x)u_0 + b(x)\sin t = 0,$$

which has the general solution

$$c(x)e^{-a(x)t} - b(x)e^{-a(x)t}\int_0^t e^{a(x)s}\sin s\,ds.$$

After integration and applying the periodicity condition to determine the function $c(x)$, we find

$$c(x) = \frac{b(x)}{a^2(x) + 1}$$

and

$$u_0(x,t) = -\frac{a(x)b(x)}{a^2(x) + 1} \sin t + \frac{b(x)}{a^2(x) + 1} \cos t.$$

Note that $u_0(x,t)$ is 2π-periodic but u_0 will in general not satisfy the boundary conditions. Introducing the boundary layer variable ξ, we find for $Q_0(\xi, t)$ the equation

$$\frac{\partial^2 Q_0}{\partial \xi^2} = \frac{\partial Q_0}{\partial t} + a(0)Q_0,$$

where we have used that $u_0(x,t)$ satisfies the inhomogeneous equation and that $R_0(\eta, t)$ vanishes outside a boundary layer near $x = 1$. For $Q_0(\xi, t)$, we have the boundary, matching, and periodicity conditions

$$Q_0(0,t) = -u_0(0,t), \lim_{\xi \to \infty} Q_0(\xi, t) = 0, Q_0(\xi, t) = Q_0(\xi, t + 2\pi).$$

As $Q_0(\xi, t)$ is 2π-periodic in t, we propose a Fourier series for Q_0 and, the boundary condition only having two Fourier terms, we postulate

$$Q_0(\xi, t) = f_1(\xi) \sin t + f_2(\xi) \cos t.$$

Substitution in the equation for Q_0 yields

$$f_1'' = -f_2 + a(0)f_1,$$
$$f_2'' = -f_1 + a(0)f_2.$$

This is a linear system with characteristic equation $(\lambda^2 - a(0))^2 + 1 = 0$ and with eigenvalues

$$\lambda = \pm\sqrt{a(0)} + i, \pm\sqrt{a(0)} - i.$$

We assumed $a(0) > 0$; the matching condition requires us to discard the solutions corresponding with $+\sqrt{a(0)}$, and we retain the independent solutions

$$e^{-\sqrt{a(0)}\xi} \cos \xi, e^{-\sqrt{a(0)}\xi} \sin \xi.$$

We find

$$f_1(\xi) = e^{-\sqrt{a(0)}\xi} \left(\frac{a(0)b(0)}{a(0)^2 + 1} \cos \xi + \alpha \sin \xi \right),$$

$$f_2(\xi) = e^{-\sqrt{a(0)}\xi} \left(-\frac{b(0)}{a(0)^2 + 1} \cos \xi + \beta \sin \xi \right),$$

where α and β can be determined by substitution in the equations for f_1 and f_2.

Introducing the boundary layer variable η, we find for $R_0(\eta, t)$ the equation

$$\frac{\partial^2 R_0}{\partial \eta^2} = \frac{\partial R_0}{\partial t} + a(1)R_0,$$

where we have again used that $u(x, t)$ satisfies the inhomogeneous equation and that $Q_0(\xi, t)$ vanishes outside a boundary layer near $x = 0$. For $R_0(\eta, t)$, we have the boundary, matching, and periodicity conditions

$$R_0(1, t) = -u_0(1, t), \lim_{\eta \to \infty} R_0(\eta, t) = 0, R_0(\eta, t) = R_0(\eta, t + 2\pi).$$

Again we retain a finite Fourier series, and the calculation runs in the same way.

9.4.2 The General Case with Dirichlet Conditions

We return to the general case

$$\varepsilon \frac{\partial^2 u}{\partial x^2} = \frac{\partial u}{\partial t} + s(x, t)u + f(x, t) + \varepsilon g(x, t, u) + \varepsilon \cdots$$

with Dirichlet boundary conditions $u(0, t) = u(1, t) = 0$ and all time-dependent terms 2π-periodic in t. Again we expect the same boundary layer variables ξ and η and an expansion of the form

$$u(x, t) = u_0(x, t) + Q_0(\xi, t) + \sqrt{\varepsilon}Q_1(\xi, t) + R_0(\eta, t) + \sqrt{\varepsilon}R_1(\eta, t) + \varepsilon \cdots,$$

where $u_0(x, t)$ is the first term of the regular expansion. The equation for $u_0(x, t)$ is

$$\frac{\partial u_0}{\partial t} + s(x, t)u_0 + f(x, t) = 0,$$

which has to be solved with x as a parameter. After integration by variation of constants, we have a free constant - dependent on x - to apply the periodicity condition to u_0.

As before, we can derive the equation for the boundary layer solution near $x = 0$,

$$\frac{\partial^2 Q_0}{\partial \xi^2} = \frac{\partial Q_0}{\partial t} + s(0, t)Q_0,$$

with boundary, matching, and periodicity conditions

$$Q_0(0, t) = -u_0(0, t), \lim_{\xi \to \infty} Q_0(\xi, t) = 0, Q_0(\xi, t) = Q_0(\xi, t + 2\pi).$$

A Fourier expansion for Q_0 is again appropriate, and we expand

$$s(0, t) = a_0 + \sum_{k=1}^{\infty}(a_k \cos kt + b_k \sin kt)$$

and in the same way $u_0(0,t)$. We assume

$$a_0 > 0.$$

For Q_0, we substitute

$$Q_0(\xi, t) = \alpha_0(\xi) + \sum_{k=1}^{\infty}(\alpha_k(\xi)\cos kt + \beta_k(\xi)\sin kt).$$

The coefficients are obtained by substitution of the Fourier series for Q_0 and $s(0,t)$ into the differential equation for Q_0, which produces an infinite set of equations; they can be solved as they are ODE's with constant coefficients. In the same way we determine the boundary layer function $R_0(\eta, t)$. Note that higher-order approximations can be obtained by extending the regular expansion and subsequently deriving higher-order equations for $Q_k, R_k, k = 1, 2, \cdots$. These equations are linear.

9.4.3 Neumann Conditions

The problem

$$\varepsilon\frac{\partial^2 u}{\partial x^2} = \frac{\partial u}{\partial t} + s(x,t)u + f(x,t) + \varepsilon g(x,t,u) + \varepsilon^2 \cdots$$

with Neumann boundary conditions

$$\frac{\partial u}{\partial x}(0,t) = \frac{\partial u}{\partial x}(1,t) = 0$$

and all terms in the equation 2π-periodic in t is slightly easier to handle.

Determine again a regular expansion of the form $u_0(x,t) + \varepsilon \cdots$ and assume again a full expansion of the solution of the form

$$u(x,t) = u_0(x,t) + Q_0(\xi, t) + \sqrt{\varepsilon}Q_1(\xi, t) + R_0(\eta, t) + \sqrt{\varepsilon}R_1(\eta, t) + \varepsilon\cdots.$$

The regular expansion will in general not satisfy the Neumann conditions, and we require for instance at $x = 0$

$$\frac{\partial u_0}{\partial x}(0,t) + \varepsilon\frac{\partial u_1}{\partial x}(0,t) + \varepsilon^2 + \cdots + \frac{1}{\varepsilon}\frac{\partial Q_0}{\partial \xi}(0,t) + \frac{\partial Q_1}{\partial \xi}(0,t) + \varepsilon\cdots = 0.$$

Multiplying with ε and comparing coefficients, we have

$$\frac{\partial Q_0}{\partial \xi}(0,t) = 0, \frac{\partial Q_1}{\partial \xi}(0,t) = -\frac{\partial u_0}{\partial x}(0,t), \cdots.$$

The equation for Q_0 will again be

$$\frac{\partial^2 Q_0}{\partial \xi^2} = \frac{\partial Q_0}{\partial t} + s(0,t)Q_0,$$

which is satisfied by the trivial solution. A similar result holds for $R_0(\eta, t)$. Nontrivial boundary layer solutions will generally arise at higher order.

9.4.4 Strongly Nonlinear Equations

Thus far, we have considered weakly nonlinear equations. It is of interest to consider more difficult equations of the form

$$\varepsilon \frac{\partial^2 u}{\partial x^2} = \frac{\partial u}{\partial t} + f(x, t, u) + \varepsilon \cdots$$

with Dirichlet or Neumann boundary conditions.

The main obstruction for the construction of an expansion is the solvability of the lowest-order equations. For the regular expansion, we have

$$\frac{\partial u_0}{\partial t} + f(x, t, u_0) = 0,$$

where, after solving the equation, we have to apply the periodicity condition. For the boundary layer contribution, this is even nastier. Assuming again an expansion for the periodic solution $u(x, t)$ of the form $u(x, t) = u_0(x, t) + Q_0(\xi, t) + R_0(\eta, t) + \sqrt{\varepsilon} \cdots$, we have

$$\frac{\partial^2 Q_0}{\partial \xi^2} = \frac{\partial Q_0}{\partial t} + f(0, t, u_0(0, t) + Q_0) - f(0, t, u_0(0, t)),$$

which looks nearly as bad as the original problem in the case of Dirichlet conditions. At higher order, the equations are linear.

In the case of Neumann conditions, we have again that the trivial solution $Q_0(\xi, t) = 0$ satisfies the equation and the boundary condition.

9.5 A Wave Equation

As a prototype of a hyperbolic equation with initial values, we consider the equation

$$\varepsilon \left(\frac{\partial^2 \phi}{\partial x^2} - \frac{\partial^2 \phi}{\partial t^2} \right) - \left(a \frac{\partial \phi}{\partial x} + b \frac{\partial \phi}{\partial t} \right) = 0,$$

where $t \geq 0, -\infty < x < +\infty$. The initial values are

$$\phi(x, 0) = f(x), \phi_t(x, 0) = g(x), -\infty < x < +\infty.$$

The functions $f(x), g(x)$ are sufficiently smooth; a and b are constants, and in a dissipative system $b > 0$. As we shall see, the constants have to satisfy the conditions $0 < |a| < b$.

The wave operator $\partial^2 / \partial x^2 - \partial^2 / \partial t^2$ has real characteristics

$$r = t - x, s = t + x.$$

The reduced ($\varepsilon = 0$) equation

$$a\frac{\partial\phi}{\partial x} + b\frac{\partial\phi}{\partial t} = 0$$

has characteristics of the form $bx - at =$ constant; we call them subcharacteristics of the original equation. Any differentiable function of $(bx - at)$ or $(x - \frac{a}{b}t)$ satisfies the reduced equation.

We start by calculating a formal approximation. This will give us the practical experience that will lead to more general insight into hyperbolic problems. Again we start by assuming that in some part of the half-plane $t \geq 0, -\infty < x < +\infty$; a regular expansion exists of the form

$$\phi_\varepsilon(x,t) = \sum_{n=0}^{m} \varepsilon^n \phi_n(x,t) + 0(\varepsilon^{m+1}).$$

The coefficients ϕ_n satisfy

$$a\frac{\partial\phi_0}{\partial x} + b\frac{\partial\phi_0}{\partial t} = 0,$$

$$a\frac{\partial\phi_n}{\partial x} + b\frac{\partial\phi_n}{\partial t} = \frac{\partial^2\phi_{n-1}}{\partial x^2} - \frac{\partial^2\phi_{n-1}}{\partial t^2}, n = 1, 2, \cdots.$$

We find $\phi_0 = h(z), z = x - \frac{a}{b}t$, and $h(z)$ a differentiable function of its argument. The equation for ϕ_1 becomes

$$a\frac{\partial\phi_1}{\partial x} + b\frac{\partial\phi_1}{\partial t} = \frac{\partial^2 h}{\partial x^2} - \frac{\partial^2 h}{\partial t^2} = \left(1 - \frac{a^2}{b^2}\right) h''\left(x - \frac{a}{b}t\right).$$

Transforming $t, x \to t, z$, we find

$$b\frac{\partial\phi_1}{\partial t} = \left(1 - \frac{a^2}{b^2}\right) h''(z),$$

so that

$$\phi_1(z,t) = \frac{b^2 - a^2}{b^3} h''(z)t + k(z),$$

where $h(z)$ and $k(z)$ still have to be determined. It would be natural to choose $h(z) = f(z)$, but we leave this decision until later.

We cannot satisfy both initial values, so we expect boundary layer behaviour near $t = 0$. Subtracting the regular expansion

$$\psi(x,t) = \phi(x,t) - \sum_{n=0}^{m} \varepsilon^n \phi_n(x,t)$$

and substitution in the original wave equation yields

$$\varepsilon\left(\frac{\partial^2\psi}{\partial x^2} - \frac{\partial^2\psi}{\partial t^2}\right) - \left(a\frac{\partial\psi}{\partial x} + b\frac{\partial\psi}{\partial t}\right) = 0(\varepsilon^{m+1}).$$

The initial values are

$$\psi(x,0) = f(x) - \sum_0^m \varepsilon^n \phi_n(x,0),$$

$$\psi_t(x,0) = g(x) - \sum_0^m \varepsilon^n \phi_{n_t}(x,0).$$

We introduce the boundary layer variable

$$\tau = \frac{t}{\varepsilon^\nu}.$$

Transforming $x, t \to x, \tau$, the equation for ψ becomes

$$\left(\varepsilon \frac{\partial^2}{\partial x^2} - \varepsilon^{1-2\nu} \frac{\partial^2}{\partial \tau^2} - a \frac{\partial}{\partial x} - \varepsilon^{-\nu} b \frac{\partial}{\partial \tau} \right) \psi^* = 0(\varepsilon^{m+1}).$$

A significant degeneration arises when $1 - 2\nu = -\nu$ or $\nu = 1$. Near $t = 0$, we clearly have an ordinary boundary layer. Assuming the existence of a regular expansion

$$\psi_\varepsilon^* = \sum_{n=0}^m \varepsilon^n \psi_n(x,\tau) + 0(\varepsilon^{m+1}),$$

we find

$$L_0^* \psi_0 = 0,$$

$$L_0^* \psi_1 = a \frac{\partial \psi_0}{\partial x}, \quad \text{etc.,}$$

with

$$L_0^* = \frac{\partial^2}{\partial \tau^2} + b \frac{\partial}{\partial \tau}.$$

The initial conditions have to be expanded and yield

$$\psi_0(x,0) = f(x) - h(x), \quad \psi_1(x,0) = -k(x),$$

$$\frac{\partial \psi_0}{\partial \tau}(x,0) = 0, \quad \frac{\partial \psi_1}{\partial \tau}(x,0) = g(x) + \frac{a}{b} h'(x).$$

The matching conditions take the form

$$\lim_{\tau \to \infty} \psi_n(x,\tau) = 0, n = 0, 1, 2, \cdots.$$

For ψ_0, we find

$$\psi_0(x,\tau) = A(x) + B(x) e^{-b\tau}.$$

The restriction $b > 0$ is necessary to satisfy the matching conditions and, to satisfy $\partial \psi_0 / \partial \tau = 0$, we are then left with the trivial solution, $\psi_0(x,\tau) = 0$. So this determines h, as we have to take $\psi_0(x,0) = 0$:

$$h(x) = f(x).$$

For ψ_1, we find the same expression as for ψ_0. Matching produces again $A(x) = 0$, leaving

$$\psi_1(x,\tau) = C(x)e^{-b\tau}.$$

The initial values yield

$$C(x) = -k(x),$$
$$-bC(x) = g(x) + \tfrac{a}{b}f'(x),$$

which determines k:

$$k(x) = \frac{g(x)}{b} + \frac{a}{b^2}f'(x).$$

At this stage, we propose the formal expansion

$$\phi_\varepsilon(x,t) = f(x - \tfrac{a}{b}t) + \varepsilon[\tfrac{b^2-a^2}{b^3}tf''(x - \tfrac{a}{b}t) + \tfrac{1}{b}g(x - \tfrac{a}{b}t) + \tfrac{a}{b^2}f'(x - \tfrac{a}{b}t)]$$
$$-\varepsilon[\tfrac{a}{b^2}f'(x) + \tfrac{1}{b}g(x)]e^{-bt/\varepsilon} + \varepsilon^2 \cdots .$$

It is clear that, outside the boundary layer, the solution is dominated by the initial values of ϕ that propagate along the *subcharacteristic* through a given point.

This, however, opens the possibility of the following situation. The solution at a point $P(x,t)$ is determined by the propagation of initial values along the *characteristics* (i.e., no information can reach P from the initial values in $x < A$ or $x > B$). If $|a| > b$, the formal expansion consists mainly of terms carrying information from these forbidden regions. This means that in this case the formal expansion cannot be correct, resulting in the condition $0 < |a| < b$; see Fig. 9.1.

We demonstrate this somewhat more explicitly by considering the problem where $\phi(x,0) = f(x) = 0, -\infty < x < +\infty, \phi_t(x,0) = g(x)$. The solution can be written as an integral using the Riemann function R:

$$\phi(x,t) = \int_{x-t}^{x+t} R_\varepsilon(x - \tau, t)g(\tau)d\tau.$$

Riemann functions are discussed in many textbooks on partial differential equations, such as Strauss (1992). For some special values of a, b, the Riemann function can be expressed in terms of elementary functions. Suppose now that we prescribe

$$g(x) > 0, x < A, x > B,$$
$$g(x) = 0, A \le x \le B.$$

It is clear that in this case $\phi(x,t)|_P = 0$. On the other hand, the formal expansion yields

$$\phi_\varepsilon(x,t) = \frac{\varepsilon}{b}g\left(x - \frac{a}{b}t\right) - \frac{\varepsilon}{b}g(x)e^{-bt/\varepsilon} + \varepsilon^2 \cdots .$$

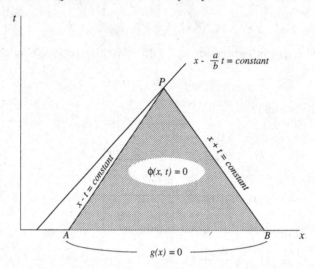

Fig. 9.1. Characteristics and subcharacteristics in the hyperbolic problem of Section 9.1.

If $|a| > b$, it is clear that $\phi_\varepsilon(x,t)|_P \neq 0$, which means that in this case the formal expansion does not produce correct results.

Remark
The condition $|a| < b$ means that the direction of the subcharacteristic is "contained" between the directions of the characteristics. In this case, one calls the subcharacteristic *time-like*.

9.6 Signalling

As an example of the part played by boundaries, we consider a so-called signalling or radiation problem. We have $x \geq 0, t \geq 0$. At $t = 0$, the medium is in a state of rest: $\phi(x,0) = \phi_t(x,0) = 0$. At the boundary, we have a source of signals or radiation:

$$\phi(0,t) = f(t), t > 0.$$

Wave propagation is described again by the equation

$$\varepsilon\left(\frac{\partial^2 \phi}{\partial x^2} - \frac{\partial^2 \phi}{\partial t^2}\right) - \left(a\frac{\partial \phi}{\partial x} + b\frac{\partial \phi}{\partial t}\right) = 0,$$

where $0 < a < b$. As ϕ is initially identically zero, the solution for $t > 0$ will be identically zero for $x > t$. (Information propagates along the characteristics.)

As before we start with a regular expansion.

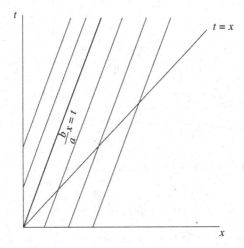

Fig. 9.2. Signalling: a source at the boundary $x = 0$.

$$\phi_\varepsilon(x,t) = \sum_{n=0}^{m} \varepsilon^n \phi_n(x,t) + 0(\varepsilon^{m+1}).$$

We find again $\phi_0 = h(z), z = x - \frac{a}{b}t$.

$$\phi_1 = \frac{b^2 - a^2}{b^3} h''(z)t + k(z)$$

with h, k sufficiently differentiable functions. We can satisfy the boundary condition by putting

$$\phi_0 = 0, \qquad t < \frac{b}{a}x$$

$$= f(t - \frac{b}{a}x), t > \frac{b}{a}x.$$

(f is defined only for positive values of its argument.)

Note that a consequence of this choice of ϕ_0 is that, unless $f(0) = 0$, the regular expansion is discontinuous along the subcharacteristic $t = \frac{b}{a}x$, so we expect a discontinuity propagating from the origin as in general $f(0) \neq 0, f'(0) \neq 0$, etc. However, the theory of hyperbolic equations tells us that such a discontinuity propagates along the characteristics, in this case $x = t$, so we expect a boundary layer along the subcharacteristic to make up for this discrepancy; see Fig. 9.2.

We transform $x, t \to \xi, t$ with

$$\xi = \frac{x - \frac{a}{b}t}{\varepsilon^\nu},$$

whereas $\phi_\varepsilon(x,t) \to \psi_\varepsilon(\xi,t)$. We have

$$\frac{\partial}{\partial x} \rightarrow \varepsilon^{-\nu} \frac{\partial}{\partial \xi},$$

$$\frac{\partial}{\partial t} \rightarrow -\varepsilon^{-\nu} \frac{a}{b} \frac{\partial}{\partial \xi} + \frac{\partial}{\partial t}.$$

The equation becomes $L_\varepsilon^* \psi = 0$ with

$$L_\varepsilon^* = \varepsilon^{1-2\nu} c \frac{\partial^2}{\partial \xi^2} + 2\varepsilon^{1-\nu} \frac{a}{b} \frac{\partial^2}{\partial \xi \partial t} - \varepsilon \frac{\partial^2}{\partial t^2} - b \frac{\partial}{\partial t},$$

where $c = 1 - a^2/b^2$. A significant degeneration arises if $\nu = \frac{1}{2}$, so that the operator becomes

$$L_0^* = c \frac{\partial^2}{\partial \xi^2} - b \frac{\partial}{\partial t}.$$

Note that $c > 0$; near the subcharacteristic $t = \frac{b}{a} x$, we have a parabolic boundary layer. For the boundary layer solution, an expansion of the form

$$\psi_\varepsilon(\xi, t) = \sum_{n=0}^{2m} \varepsilon^{n/2} \psi_n(\xi, t) + 0(\varepsilon^{m+\frac{1}{2}})$$

is taken, where ψ_0 is a solution of the diffusion equation

$$L_0^* \psi_0 = 0,$$

which has to be matched with the regular expansion. We have, moving to the right-hand side of the subcharacteristic, $(t < \frac{b}{a} x)$ that $\psi_0 \rightarrow 0$. On the left-hand side, the regular expansion becomes

$$\lim_{z \uparrow 0} f(z) = f(0^+)$$

on approaching the subcharacteristic, which should match with ψ_0 moving to the left. We find, omitting the technical details of matching, that

$$\psi_0(\xi, t) = \frac{1}{2} f(0^+) \mathrm{erfc} \left(\frac{\xi}{2\sqrt{kt/b}} \right).$$

where

$$\mathrm{erfc}(z) = \frac{2}{\sqrt{\pi}} \int_z^\infty e^{-t^2} dt; \quad \mathrm{erfc}(0) = 1.$$

The boundary layer approximation ψ_0 satisfies the equation while $\psi_0(0, t) = \frac{1}{2} f(0^+)$; moreover, moving to the right-hand side of the characteristic $\xi \rightarrow \infty$ or $z \rightarrow \infty$, we have agreement with the expression above for $\mathrm{erfc}(z)$.

Note that we did not perform the subtraction trick before carrying out the matching, but, as we have seen, ψ_0 satisfies the required conditions. Note also that, according to this analysis, the formal approximation becomes zero in the region between the characteristics $t = \frac{b}{a} x$ and $t = x$.

9.7 Guide to the Literature.

A survey of results obtained in the former Soviet Union for parabolic equations has been given by Butuzov and Vasil'eva (1983), and Vasil'eva, Butuzov, and Kalachev (1995). For expansions and proofs, see also Shih and Kellogg (1987) and Shih (2001) and further references therein. The first section on a problem with slow diffusion and heat production is actually a simplified version of a more extensive problem (Van Harten, 1979). In all of these papers, proofs of asymptotic validity are given.

We shall not consider here the case of equilibrium solutions of parabolic equations, which reduces to the study of elliptic equations; for an introduction to singular perturbations of equilibrium solutions of systems with reaction and diffusion, see Aris (1975, Vol. 1, Chapter 3.7).

A classical example of a nonlinear diffusion equation is Burgers' equation

$$u_t + uu_x = \varepsilon u_{xx}$$

on an infinite domain, which is also a typical problem for studying shock waves. An introductory treatment can be found in Holmes (1998, Chapter 2).

Applications of slow manifolds (Fenichel theory) and extensions are described by Jones (1994), Kaper (1999) and Kaper and Jones (2001). These papers contain many references and applications. One possibility is to project the solutions of the evolution equation to a finite space, which reduces the problem to the analysis given in the preceding chapter (ODE's). Another possibility is to compute travelling or solitary waves; an example is the paper by Szmolyan (1992). It is typical to start with an evolution equation such as

$$u_t + uu_x + u_{xxx} + \varepsilon(u_{xx} + u_{xxxx}) = 0,$$

(see for instance Jones, 1994) and then look for solitary waves (i.e., solutions that depend on $x - ct$ alone). This results in a system of ordinary differential equations that contains one or more slow manifolds. The reason for omitting these examples in this chapter is that in what follows one needs an extensive dynamical systems analysis with subtle reasoning also to connect the results to the original evolution equation. The results, however, are very interesting.

Another important subject that we did not discuss is combustion, which leads to interesting boundary layer problems. A basic paper is Matkowsky and Sivashinsky (1979); introductory texts are Van Harten (1982), Buckmaster and Ludford (1983), and Fife (1988). More recent results can be found in Vasil'eva, Butuzov, and Kalachev (1995), and Class, Matkowsky, and Klimenko (2003).

For hyperbolic problems, the formal construction of expansions has been discussed in Kevorkian and Cole (1996). Constructions and proofs of asymptotic validity on a timescale $O(1)$ have been given in an extensive study by Geel (1981); see also De Jager and Jiang Furu (1996). The proofs are founded on energy integral estimates and fixed-point theorems in a Banach space. An extension to larger timescales for hyperbolic problems has been given by Eckhaus and Garbey (1990), who develop a formal approximation that is shown to be valid on timescales of the order $1/\varepsilon$.

9.8 Exercises

Exercise 9.1 Consider the hyperbolic initial value problem

$$\varepsilon(u_{tt} - u_{xx}) + au_t + u_x = 0,$$
$$u(x,0) = 0,$$
$$u_t(x,0) = g(x) \quad (x \in \mathbb{R}),$$

based on Kevorkian and Cole (1996), with $a \geq 1$. We will introduce ϕ by $u(x,t) = \phi(x,t) \exp(\frac{\alpha x - \beta t}{2\varepsilon})$.

a. Compute α and β such that

$$\varepsilon(\phi_{tt} - \phi_{xx}) - \left(\frac{a^2 - 1}{4\varepsilon}\right)\phi = 0.$$

   Consider the case $a = 1$. In this special case the function $\phi(x,t)$ has to satisfy the wave equation, so write: $\phi(x,t) = \psi_l(x-t) + \psi_r(x+t)$.
b. Compute the functions ψ_l and ψ_r; use the initial conditions of $u(x,t)$.
c. Give the exact solution $u(x,t,\varepsilon)$ of the given problem for $a \geq 1$.
d. Construct an approximation $\tilde{U}(x,t,\varepsilon)$ of the form

$$\tilde{U}(x,t,\varepsilon) = U_0(x,t) + \varepsilon U_1(x,t) + \varepsilon W_1(x,\tau) + 0(\varepsilon^2)$$

   with $\tau = t/\varepsilon$.
e. Compare the approximation $\tilde{U}(x,t,\varepsilon)$ with the exact solution $u(x,t,\varepsilon)$.

Exercise 9.2 Consider the equation

$$\varepsilon u_{xx} = u_t - u_x, x \geq 0, t \geq 0,$$

with initial condition $u(x,0) = 1$ and boundary condition $u(0,t) = \exp(-t)$.

a. Produce the first two terms of a regular expansion and locate the boundary layer(s).
b. Repeat the analysis when we change the boundary condition to $u(0,t) = \sin t$.

Exercise 9.3 As an example of a parabolic problem on an unbounded domain, consider Fisher's equation (KPP)

$$u_t = \varepsilon u_{xx} + u(1-u), -\infty < x < \infty, t > 0,$$

with $u(x,0) = g(x), 0 \leq u \leq 1$.

a. Determine the stationary solutions in the case $\varepsilon = 0$. Which one is clearly unstable?

b. Introduce the regular expansion $u = u_0(x,t) + \varepsilon u_1(x,t) + \cdots$ and give the equations and conditions for u_0, u_1.

c. Solve the problem for $u_0(x,t)$.

Exercise 9.4 Consider the parabolic initial boundary value problem

$$\varepsilon(u_t - au_{xx}) + bu = f(x),$$
$$u(x,0) = 0, \ a > 0, b > 0,$$
$$u(0,t) = 0, \ (x,t) \in \mathbb{R}^2,$$
$$u(1,t) = 0, \ 0 < x < 1, 0 < t \le T.$$

We construct an approximation in four steps.

a. Compute the regular expansion $U(x,t)$:

$$U(x,t,\varepsilon) = U_0(x,t) + \varepsilon U_1(x,t) + 0(\varepsilon^2).$$

b. Compute the initial layer correction $\Pi_0(x,\tau), \tau = t/\varepsilon$, and give the equation (with initial value) for $\Pi_1(x,\tau)$ such that $V(x,t,\varepsilon)$ satisfies the initial condition at $t = 0$:

$$V(x,t,\varepsilon) = U(x,t,\varepsilon) + \Pi_0(x,\tau) + \Pi_1(x,\tau) + 0(\varepsilon^2).$$

c. Compute the boundary layer correction

$$Q_0(\xi,t), \xi = \frac{x}{\sqrt{\varepsilon}},$$

and give the equations for $Q_1(\xi,t)$ and $Q_2(\xi,t)$ such that

$$W(x,t,\varepsilon) = V(x,t,\varepsilon) + Q_0(\xi,t) + \sqrt{\varepsilon}Q_1(\xi,t) + \varepsilon Q_2(\xi,t) + 0(\varepsilon^{3/2})$$

satisfies the boundary condition at $x = 0$. In the same way, boundary layer corrections can be constructed at the boundary $x = 1$ with ($\xi^* = \frac{(1-x)}{\sqrt{\varepsilon}}$); we omit this.

Also, at the corner boundary points $(0,0)$ and $(1,0)$, corrections are needed.

d. Give the equations for the corner boundary layer corrections $P_0(\xi,\tau)$, $P_1(\xi,\tau)$, and $P_2(\xi,\tau)$ at $(0,0)$ ($\xi^* = \frac{x}{\sqrt{\varepsilon}}, \tau = t/\varepsilon$). Give the initial boundary values.

Combining (a), (b), (c) and (d) produces an approximation $\tilde{u}(x,t,\varepsilon)$ of the solution $u(x,t,\varepsilon)$ of the given problem of the form

$$\tilde{u}(x,t,\varepsilon) = U_0(x,t) + \Pi_0(x,\tau) + Q_0(\xi,t)$$
$$+ P_0(\xi,\tau) + \sqrt{\varepsilon}[Q_1(\xi,t) + P_1(\xi,\tau)]$$
$$+ \varepsilon[U_1(x,t) + \Pi_1(x,\tau) + Q_2(\xi,t) + P_2(\xi,\tau)] + 0(\varepsilon^2),$$

where $\tau = t/\varepsilon, \xi = x/\sqrt{\varepsilon}$. Note that corner boundary layer corrections at $(1,0)$ can be constructed in the same way with $\xi^* = (1-x)/\sqrt{\varepsilon}$. No correction terms are added for the boundary $x = 1$ and the corner point $(1,0)$, so the approximation only holds on

$$\{(x,t) \in \mathbb{R}^2 | 0 < x < 1 - d, \ (0 < d < 1) \ \text{and} \ 0 < t \leq T\}.$$

10

The Continuation Method

In Chapter 1 we considered an equation $L_\varepsilon y = 0$ that contains a small parameter ε. We associate with this equation an "unperturbed" problem, the equation $L_0 y = 0$. If the difference between the solutions of both equations in an appropriate norm does not tend to zero as ε tends to zero, we call the problem a singular perturbation problem.

Actually, when formulated with suitable norms, most perturbation problems are singular. This does not only apply to boundary layer problems as considered in the preceding chapters, but also to slow-time problems, as we shall see in what follows. In all of these cases, the solutions of the unperturbed problem can not be simply continued with a Taylor expansion in the small parameter to obtain an approximation of the full problem.

Simple perturbation examples can be found in the exercises of Chapter 2, where we looked at solutions of algebraic equations of the form

$$a(\varepsilon)x^2 + b(\varepsilon)x + c(\varepsilon) = 0,$$

in which a, b, and c depend smoothly on ε. Associating with this problem the "unperturbed" or "reduced" equation

$$a(0)x^2 + b(0)x + c(0) = 0,$$

we found that in the analysis the implicit function theorem plays a fundamental part. If x_0 is a solution of the unperturbed problem and if

$$2a(0)x_0 + b(0) \neq 0,$$

the implicit function theorem tells us that the solutions of the full problem depend smoothly on ε and we may expand in integral powers of ε: $x_\varepsilon = x_0 + \varepsilon \cdots$. If this condition is not satisfied, we cannot expect a Taylor series with respect to ε and there may be bifurcating solutions.

We shall discuss this application of the implicit function theorem for initial value problems for ordinary differential equations.

10.1 The Poincaré Expansion Theorem

We start with a few examples.

Example 10.1
$x(t)$ is the solution of the initial value problem

$$\dot{x} = -x + \varepsilon, \; x(0) = 1.$$

The unperturbed problem is

$$\dot{y} = -y, \; y(0) = 1.$$

We have $x(t) = \varepsilon + (1 - \varepsilon)e^{-t}, y(t) = e^{-t}$, so

$$|x(t) - y(t)| = \varepsilon - \varepsilon e^{-t} \le \varepsilon, \; t \ge \varepsilon.$$

The approximation of $x(t)$ by $y(t)$ is valid for all time.

Example 10.2
$x(t)$ is the solution of the initial value problem

$$\dot{x} = x + \varepsilon, \; x(0) = 1.$$

The unperturbed problem is

$$\dot{y} = y, \; y(0) = 1,$$

and we have

$$|x(t) - y(t)| = \varepsilon(e^t - 1),$$

so we have an approximation that is valid for $0 \le t \le 1$ (or any positive constant that does not depend on ε).

We might conjecture that in the second example the approximation breaks down because the solutions are not bounded. However, to require the solutions to be bounded is not sufficient, as the following example shows.

Example 10.3
Consider the initial value problem

$$\ddot{x} + (1 + \varepsilon)^2 x = 0, \; x(0) = 1, \dot{x}(0) = 0,$$

with solution $x(t) = \cos(1 + \varepsilon)t$. The unperturbed problem is

$$\ddot{y} + y = 0, \; y(0) = 1, \dot{y}(0) = 0$$

which is solved by $y(t) = \cos t$. So we conclude that $|x(t) - y(t)| = 2|\sin(t + \frac{1}{2}\varepsilon t)\sin(\frac{1}{2}\varepsilon t)|$, which does not vanish with ε; take for instance $t = \pi/\varepsilon$. One might expect an improvement of the timescale where the approximation is valid when expanding to higher order. However, this is generally not the case. Expand $x(t) = \cos t + \varepsilon x_1(t) + \varepsilon^2 \cdots$ and substitute this expression into the differential equation. To $O(\varepsilon)$ we obtain the problem

$$\ddot{x}_1 + 2\cos t + x_1 = 0, \quad x_1(0) = \dot{x}_1(0) = 0.$$

The solution is $x_1(t) = -t\sin t$ and is an unbounded expression, which we cannot call an improvement.

We shall now formulate the general perturbation procedure and state what we can expect in general of the accuracy of the approximation. Consider the initial value problem for $x \in \mathbb{R}^n$

$$\dot{x} = f(t, x, \varepsilon), \quad x(t_0) = a(\varepsilon).$$

We assume that $f(t, x, \varepsilon)$ can be expanded in a Taylor series with respect to x (in a neighbourhood of the initial value) and with respect to ε. We also assume that $a(\varepsilon)$ can be expanded in a Taylor series with respect to ε. The solution of the initial value problem is $x_\varepsilon(t)$.

We associate with this problem the unperturbed (or reduced) problem

$$\dot{x}_0 = f(t, x_0, 0), \quad x(t_0) = a(0),$$

with solution $x_0(t)$. The Poincaré expansion theorem tells us that

$$\|x_\varepsilon(t) - x_0(t)\| = O(\varepsilon), \quad t_0 \leq t \leq t_0 + C,$$

with C a constant independent of ε. As C is $O(1)$ with respect to ε, this is sometimes called an approximation valid on the *timescale* 1.

We can improve the result by expanding to higher order. More generally the Poincaré expansion theorem asserts that $x_\varepsilon(t)$ can be expanded in a convergent Taylor series of the form

$$x_\varepsilon(t) = x_0(t) + \varepsilon x_1(t) + \cdots + \varepsilon^n x_n(t) + \cdots .$$

The terms $x_n(t)$ are obtained by substituting the series in the differential equation and expanding the vector function f with respect to ε as

$$\dot{x}_0 + \varepsilon \dot{x}_1 + \varepsilon^2 \cdots = f(t, x_0 + x_1 + \varepsilon^2 \cdots, \varepsilon), \quad a(\varepsilon) = a_0 + \varepsilon a_1 + \varepsilon^2 \cdots ,$$

which after expansion and collecting equations at the same power of ε produces

$$\dot{x}_0 = f(t, x_0, 0), \quad x_0(t_0) = a_0,$$
$$\dot{x}_1 = \frac{\partial f}{\partial x}(t, x_0, 0)x_1 + \frac{\partial f}{\partial \varepsilon}(t, x_0.0), \quad x_1(t_0) = a_1,$$

and similar equations at higher order. Note that, apart from the unperturbed equation for x_0, all equations are linear. For the nth-order partial sum, we have the estimate

$$||x_\varepsilon(t) - (x_0(t) + \varepsilon x_1(t) + \cdots + \varepsilon^n x_n(t))|| = O(\varepsilon^{n+1})$$

for $t_0 \leq t \leq t_0 + C$. *So by this type of expansion we can improve the accuracy but not the timescale!*

Example 10.4
Consider the damped harmonic oscillator

$$\ddot{x} + 2\varepsilon\dot{x} + x = 0, \ x(0) = 1, \ \dot{x}(0) = 0.$$

Ignoring the known, exact solution, we put $x_\varepsilon(t) = x_0(t) + \varepsilon x_1(t) + \cdots$ to obtain the initial value problems

$$\ddot{x}_0 + x_0 = 0, \ \ x_0(0) = 1, \ \dot{x}_0(0) = 0,$$
$$\ddot{x}_1 + x_1 = -2\dot{x}_0, \ x_1(0) = 0, \ \dot{x}_1(0) = 0.$$

We find $x_0(t) = \cos t$ and for x_1

$$\ddot{x}_1 + x_1 = -2\cos t, \ x_1(0) = 0, \ \dot{x}_1(0) = 0,$$

with solution

$$x_1(t) = \sin t - t\cos t.$$

This looks bad. The solutions of the damped harmonic oscillator are bounded for all time and even tend to zero, while the approximation has an increasing amplitude. However, the approximation is valid only on the timescale 1.

Such terms that are increasing with time were called "secular terms" in astronomy. The discussion of how to avoid them played an important part in the classical perturbation problems of celestial mechanics. Poincaré adapted the approximation scheme to get rid of secular terms.

10.2 Periodic Solutions of Autonomous Equations

To a certain extent, periodic solutions are determined by their behaviour on a timescale 1. This characteristic enables us to obtain a fruitful application of the Poincaré expansion theorem. Usually this is called the Poincaré-Lindstedt method.

In this section, we consider autonomous equations of the form

$$\dot{x} = f(x, \varepsilon), \ x \in \mathbb{R}^n,$$

for which we assume that the conditions of the expansion theorem have been satisfied. In addition, we assume that the unperturbed equation $\dot{x} = f(x, 0)$ has one or more periodic solutions. Can we continue such a periodic solution for the equation if $\varepsilon > 0$?

The reason to make a distinction between autonomous and nonautonomous equations is that in autonomous equations the period is not a priori fixed. As a consequence, a period T_0 of a periodic solution of the unperturbed equation will in general also be perturbed. To fix the idea consider the two-dimensional equation

$$\ddot{x} + x = \varepsilon f(x, \dot{x}, \varepsilon).$$

If $\varepsilon = 0$, we have a rather degenerate case, as all solutions are periodic and even have the same period, 2π. We cannot expect that all of these periodic solutions can be continued for $\varepsilon > 0$, but maybe some periodic solutions will branch off. For such a solution, we will have a period $T(\varepsilon)$ with $T(0) = T_0 = 2\pi$. A priori we do not know for which initial conditions periodic solutions branch off (if they do!), so we assume that we find them starting at

$$x(0) = a(\varepsilon), \quad \dot{x}(0) = 0.$$

For a two-dimensional autonomous equation, putting $\dot{x}(0) = 0$ is no restriction. The expansion theorem tells us that on the timescale 1 we have

$$\lim_{\varepsilon \to 0} = a(0) \cos t.$$

It is convenient to fix the period by the transformation

$$\omega t = \theta, \quad \omega^{-2} = 1 - \varepsilon \eta(\varepsilon),$$

where $\eta(\varepsilon)$ can be expanded in a Taylor series with respect to ε and is chosen such that any periodic solution under consideration has period 2π. With the notation $x' = dx/d\theta$, the equation becomes

$$x'' + x = \varepsilon \eta(\varepsilon) x + \varepsilon (1 - \varepsilon \eta(\varepsilon)) f(x, (1 - \varepsilon \eta(\varepsilon))^{-\frac{1}{2}} x', \varepsilon)$$

with initial values $x(0) = a(\varepsilon), x'(0) = 0$.

Abbreviating the equation to

$$x'' + x = \varepsilon F(x, x', \varepsilon, \eta(\varepsilon)),$$

we are now looking for a suitable initial value $a(\varepsilon)$ and scaling of ω (or η) to obtain 2π-periodic solutions in θ. This problem is equivalent with solving the integral equation

$$x(\theta) = a(\varepsilon) \cos \theta + \varepsilon \int_0^\theta F(x(s), x'(s), \varepsilon, \eta(\varepsilon)) \sin(\theta - s) ds$$

with the periodicity condition $x(\theta + 2\pi) = x(\theta)$ for each value of θ.

Applying the periodicity condition, we find

$$\int_\theta^{\theta + 2\pi} F(x(s), x'(s), \varepsilon, \eta(\varepsilon)) \sin(\theta - s) ds = 0.$$

Expanding $\sin(\theta - s) = \sin\theta\cos s - \cos\theta\sin s$ and using that the sin- and cos-functions are independent, we find the conditions

$$I_1(a(\varepsilon), \eta(\varepsilon)) = \int_0^{2\pi} F(x(s), x'(s), \varepsilon, \eta(\varepsilon))\sin s\, ds = 0,$$

$$I_2(a(\varepsilon), \eta(\varepsilon)) = \int_0^{2\pi} F(x(s), x'(s), \varepsilon, \eta(\varepsilon))\cos s\, ds = 0.$$

For each value of ε, this is a system of two equations with two unknowns, $a(\varepsilon)$ and $\eta(\varepsilon)$. According to the implicit function theorem, this system is uniquely solvable in a neighbourhood of $\varepsilon = 0$ if the corresponding Jacobian J does not vanish for the solutions:

$$J = \left| \frac{\partial(I_1, I_2)}{\partial(a(\varepsilon), \eta(\varepsilon))} \right| \neq 0.$$

The way to handle the periodicity conditions and the corresponding Jacobian is to expand the solution $x_\varepsilon(t)$ and the parameters $a(\varepsilon)$, $\eta(\varepsilon)$ with respect to ε. At the lowest order, we find

$$F(x_0(t), x_0'(t), 0, \eta(0)) = \eta(0)a(0)\cos t + f(a(0)\cos t, -a(0)\sin t, 0)$$

and the equations

$$\int_0^{2\pi} f(a(0)\cos s, -a(0)\sin s, 0)\sin s\, ds = 0,$$

$$\pi\eta(0)a(0) + \int_0^{2\pi} f(a(0)\cos s, -a(0)\sin s, 0)\cos s\, ds = 0.$$

These equations have to be satisfied (necessary condition) to obtain a periodic solution. If the corresponding Jacobian does not vanish, a nearby periodic solution really exists. The condition derived from the lowest-order equations is

$$J_0 = \left| \frac{\partial(I_1, I_2)}{\partial(a(0), \eta(0))} \right| \neq 0.$$

We shall study this condition in a number of examples. Note that if we can satisfy the periodicity conditions but the Jacobian vanishes at lowest order, we have to calculate the Jacobian at the next order. Assuming that the Poincaré expansion theorem applies, we have the convergent series $J = J_0 + \varepsilon J_1 + \cdots + \varepsilon^n J_n + \cdots$. This calculation may decide the existence of a unique periodic solution, but it is possible that the Jacobian vanishes at all orders. This happens for instance if we have a family of periodic solutions so instead of having the existence of a unique solution, vanishing of the Jacobian can in some cases imply that there are many more periodic solutions. We shall meet examples of these phenomena.

Example 10.5
(Van der Pol equation)
A classical example is the Van der Pol equation

$$\ddot{x} + x = \varepsilon \dot{x}(1 - x^2),$$

which has a unique periodic solution for *each positive* value of ε. Of course, we consider only small values of ε. Transforming time by $\omega t = \theta$, $\omega^{-2} = 1 - \varepsilon \eta(\varepsilon)$, we find

$$x'' + x = \varepsilon \eta(\varepsilon)x + \varepsilon(1 - \varepsilon \eta(\varepsilon))^{\frac{1}{2}}x'(1 - x^2)$$

with unknown initial conditions $x(0) = a(\varepsilon), x' = 0$. The periodicity conditions at lowest order become

$$-\int_0^{2\pi} a(0) \sin s(1 - a^2(0) \cos^2 s) \sin s \, ds = 0,$$

$$\pi \eta(0) a(0) - \int_0^{2\pi} a(0) \sin s(1 - a^2(0) \cos^2 s) \cos s \, ds = 0.$$

After integration, we find

$$a(0)\left(1 - \frac{1}{4}a^2(0)\right) = 0,$$

$$\eta(0)a(0) = 0.$$

Apart from the trivial solution, we find $a(0) = 2, \eta(0) = 0$ (the solution $a(0) = -2$ produces the same approximation), so a periodic solution branches off at amplitude 2. The existence has been given, but we check this independently by computing the Jacobian at $(a(0), \eta(0)) = (2, 0)$: $J_0 = 4$ so the implicit function theorem applies.

Andersen and Geer (1982) used a formal manipulation of the expansions and obtained for the Van der Pol equation the expansion to $O(\varepsilon^{164})$.

Example 10.6
Consider the equation

$$\ddot{x} + x = \varepsilon x^3.$$

It is well-known that all the solutions of this equation are periodic in a large neighbourhood of $(0, 0)$. We follow the construction by again putting $\omega t = \theta$, $\omega^{-2} = 1 - \varepsilon \eta(\varepsilon)$ to find

$$x'' + x = \varepsilon \eta(\varepsilon)x + \varepsilon x^3$$

and at lowest order the periodicity conditions

$$\int_0^{2\pi} a^3(0) \cos^3 s \sin s \, ds = 0, \pi \eta(0)a(0) + \int_0^{2\pi} a^3(0) \cos^3 s \cos s \, ds = 0,$$

so that the first condition is always satisfied and the second condition gives

$$a(0) \left(\eta(0) + \frac{3}{4}a^2(0) \right) = 0.$$

We conclude that $a(0)$ can be chosen arbitrarily and that $\eta(0) = -\frac{3}{4}a^2(0)$. For the Jacobian, we find $J_0 = 0$, as there exists an infinite number of periodic solutions.

Remark (on the importance of existence results)
Suppose that one can apply the periodicity conditions but that the Jacobian J vanishes. Why bother about this existence question? The reason to worry about this is that higher-order terms may destroy the periodic solution. A simple example is the equation

$$\ddot{x} + x = \varepsilon x^3 - \varepsilon^n \dot{x}$$

with n a natural number ≥ 2. We can satisfy the periodicity conditions to $O(\varepsilon^{n-1})$, but the equation has no periodic solution.

10.3 Periodic Nonautonomous Equations

In this section, we consider nonautonomous, periodic equations so the period is a priori fixed. There are still many subtle problems here, as the period T_0 of a periodic solution of the unperturbed equation can be near the period of the perturbation or quite distinct. To fix the idea, we consider the two-dimensional equation

$$\ddot{x} + x = \varepsilon f(x, \dot{x}, t, \varepsilon).$$

We shall look for periodic solutions that can be continued for $\varepsilon > 0$. Let us assume that the perturbation is T-periodic with a period near 2π. To apply the periodicity condition, it is convenient to have periodicity 2π so we transform time with a factor

$$\omega^{-2} = 1 - \varepsilon\beta(\varepsilon), \ \beta(\varepsilon) = \beta_0 + \varepsilon\beta_1 + \cdots$$

with *known* constants $\beta_0, \beta_1, \cdots$.

In autonomous equations, we have the translation property that if $y(t)$ is a solution, $y(t - a)$ with a an arbitrary constant is also a solution. This is not the case in nonautonomous equations, so it is natural to introduce a phase ψ that will in general depend on ε: $\psi(\varepsilon) = \psi_0 + \varepsilon\psi_1 + \varepsilon^2 \cdots$. The time transformation becomes

$$\omega t = \theta - \psi(\varepsilon)$$

and the equation transforms with $x' = dx/d\theta$ to

$$x'' + x = \varepsilon\beta(\varepsilon)x + \varepsilon(1 - \varepsilon\beta(\varepsilon))f(x, (1 - \varepsilon\beta(\varepsilon))^{\frac{1}{2}}x', (1 - \varepsilon\beta(\varepsilon))^{\frac{1}{2}}(\theta - \psi(\varepsilon)), \varepsilon).$$

We shall look for 2π-periodic solutions starting at

$$x(0) = a(\varepsilon), \ \dot{x}(0) = 0,$$

with the expansion $a(\varepsilon) = a_0 + \varepsilon a_1 + \cdots$, which still has to be determined.

As before, the differential equation can be transformed to an integral equation, in this case of the form

$$x(\theta) = a(\varepsilon) \cos \theta + \varepsilon \int_0^\theta F(x(s), x'(s), \psi(\varepsilon), s, \varepsilon,) \sin(\theta - s) ds$$

with $F = F(x, x', \psi, \theta, \varepsilon)$ or

$$F = \beta(\varepsilon)x + (1 - \varepsilon\beta(\varepsilon))f(x, (1 - \varepsilon\beta(\varepsilon))^{\frac{1}{2}}x', (1 - \varepsilon\beta(\varepsilon))^{\frac{1}{2}}(\theta - \psi(\varepsilon)), \varepsilon)$$

and with the periodicity condition $x(\theta + 2\pi) = x(\theta)$ for each value of θ. So

$$\int_0^{2\pi} F(x(s), x'(s), \psi(\varepsilon), s, \varepsilon) \sin(\theta - s) ds = 0.$$

We find the two conditions

$$\int_0^{2\pi} F(x(s), x'(s), \psi(\varepsilon), s, \varepsilon) \sin s\, ds = 0,$$

$$\int_0^{2\pi} F(x(s), x'(s), \psi(\varepsilon), s, \varepsilon) \cos s\, ds = 0.$$

This is a system of two equations with two unknowns, $a(\varepsilon)$ and $\psi(\varepsilon)$, which we shall study in a number of examples.

Example 10.7
(forced Van der Pol equation)
Consider the case of the Van der Pol equation with a small forcing

$$\ddot{x} + x = \varepsilon\dot{x}(1 - x^2) + \varepsilon h \cos \omega t.$$

Using the transformations outlined above, we have

$$F = \beta(\varepsilon)x(\theta) + (1 - \varepsilon\beta(\varepsilon))[(1 - \varepsilon\beta(\varepsilon))^{\frac{1}{2}}x'(1 - x^2(\theta)) + h \cos(\theta - \psi(\varepsilon))],$$

which can be expanded to

$$F = \beta_0 a_0 \cos \theta - a_0 \sin \theta (1 - a_0^2 \cos^2 s) + h \cos(\theta - \psi_0) + \varepsilon \cdots.$$

From the periodicity conditions, to first order we find

$$I_1 = -a_0 \left(1 - \frac{1}{4}a_0^2\right) + h \sin \psi_0 = 0,$$

$$I_2 = \beta_0 a_0 + h \cos \psi_0 = 0.$$

For the Jacobian, we find at lowest order

$$J_0 = \left| \frac{\partial(I_1, I_2)}{\partial(a_0, \psi_0)} \right| = |h| \left| \left(1 - \frac{3}{4} a_0^2 \right) \sin \psi_0 - \beta_0 \cos \psi_0 \right|.$$

Exact 2π-periodic forcing means $\beta_0 = 0$.

Exploring this case first, we find $\psi_0 = \pi/2, 3\pi/2$, and

$$a_0 \left(1 - \frac{1}{4} a_0^2 \right) = \pm h.$$

If $|h| > h^* = 4/(3\sqrt{3})$, we have one solution; see Fig. 11.3. When $|h|$ passes the critical value h^*, there is a bifurcation producing three solutions. At the value $h = 0$, we have returned to the "ordinary" Van der Pol equation that has one periodic solution with $a_0 = 2$.

If $\beta_0 = 0$, we have $J_0 = |(1 - \frac{3}{4} a_0^2) h|$. We observe that at the bifurcation values $h = h^*, 0$, the Jacobian J_0 vanishes.

If $\beta_0 \neq 0$, we have the relation

$$\tan \psi_0 = \frac{\frac{1}{4} a_0^2 - 1}{\beta_0}$$

and a similar analysis can be made.

Example 10.8
(damped and forced Duffing equation)
A fundamental example of mechanics is an oscillator built out of a Hamiltonian system with damping and forcing added. A relatively simple but basic nonlinear case is

$$\ddot{x} + \varepsilon \mu \dot{x} + x + \varepsilon \gamma x^3 = \varepsilon h \cos \omega t$$

with damping coefficient $\mu > 0$. The equation of motion of the underlying Hamiltonian system is obtained when $\mu = h = 0$. Putting $\varepsilon \gamma = -1/6$ produces the first nonlinear term of the mathematical pendulum equation.

Using the formulas derived above, we have in this case

$$f = \mu \dot{x} - \gamma x^3 + h \cos \omega t$$

and after transformation

$$F = \beta(\varepsilon) x(\theta) + (1 - \varepsilon \beta(\varepsilon))[-(1 - \varepsilon \beta(\varepsilon))^{\frac{1}{2}} \mu x' - \gamma x^3 + h \cos(\theta - \psi(\varepsilon))],$$

which can be expanded to

$$F = \beta_0 a_0 \cos \theta + \mu a_0 \sin \theta - \gamma a_0^3 \cos^3 \theta + h \cos(\theta - \psi_0) + \varepsilon \cdots.$$

The periodicity conditions to first order produce

$$I_1 = \mu a_0 + h \sin \psi_0 = 0,$$
$$I_2 = \beta_0 a_0 - \frac{3}{4} \gamma a_0^3 + h \cos \psi_0 = 0.$$

The Jacobian to first order becomes

$$J_0 = \left| \frac{\partial(I_1, I_2)}{\partial(a_0, \psi_0)} \right| = |h| \left| \mu \sin \psi_0 + \left(\beta_0 - \frac{9}{4} \gamma a_0^2 \right) \cos \psi_0 \right|.$$

Easiest to analyse is the case without damping, $\mu = 0$. We have the periodicity conditions

$$h \sin \psi_0 = 0, \beta_0 a_0 - \frac{3}{4} \gamma a_0^3 + h \cos \psi_0 = 0,$$

so that $\psi_0 = 0, \pi$, and two possibilities,

$$a_0 \left(\beta_0 - \frac{3}{4} \gamma a_0^3 \right) = \pm h.$$

Interestingly, the product $\beta_0 \gamma$ also plays a part in the bifurcations (the existence of periodic solutions). If $\beta_0 \gamma > 0$, we have one or three solutions depending on the value of h, and if $\beta_0 \gamma < 0$, there is only one solution. This picture also emerges from the Jacobian

$$J_0 = \left| h \left(\beta_0 - \frac{9}{4} \gamma a_0^2 \right) \cos \psi_0 \right|.$$

We conclude that a small detuning of the forcing from exact 2π-periodicity is essential to obtain interesting bifurcations. On taking $\beta_0 = 0$, we find to first order the periodicity conditions

$$h \sin \psi_0 = -\mu a_0, h \cos \psi_0 = \frac{3}{4} \gamma a_0^3.$$

Again there are two possible solutions for the phase shift ψ_0 corresponding with one periodic solution each for whatever the values of h, μ, and $\gamma (\neq 0)$ are. This is also illustrated by the Jacobian, which at an exact 2π-periodic forcing, using the periodicity conditions, becomes

$$J_0 = |h| \left| \mu \sin \psi_0 - \frac{9}{4} \gamma a_0^2 \cos \psi_0 \right| = |a_0| \left| \mu^2 + \frac{27}{16} \gamma^2 a_0^4 \right|.$$

Example 10.9
(Mathieu equation)
Linear equations with periodic coefficients play an important part in physics and engineering. A typical example is the π-periodic Mathieu equation that can be written in the form

$$\ddot{x} + (\omega^2 + \varepsilon \cos 2t)x = 0.$$

Of particular interest is usually the question of for which values of ω and ε the solutions are stable (i.e., decreasing to zero) or unstable. The answer to this question is guided by Floquet theory, which tells us that for ω fixed, the (two) independent solutions of the equation can be written in the form $\exp(\lambda(\varepsilon)t)p(t)$ with $p(t)$ a π-periodic function. So the two possible expressions for the so-called characteristic exponents $\lambda(\varepsilon)$ - call them λ_1 and λ_2 - determine the stability of the solutions. Also from Floquet theory we have that $\lambda_1(\varepsilon) + \lambda_2(\varepsilon) = 0$. For extensive introductions to Floquet theory, see Magnus and Winkler (1966) and Yakubovich and Starzhinskii (1975), for summarising introductions Hale (1963) or Verhulst (2000).

The expansion theorem tells us that the exponents $\lambda_{1,2}(\varepsilon)$ can be expanded in a Taylor series with respect to ε with $\lambda_1(0) = \omega i, \lambda_2(0) = -\omega i$. This has important consequences. If ω is not close to a natural number, any perturbation of $\lambda_{1,2}(0)$ will cause it to move along the imaginary axis, so for the possibility of instability we only have to consider the cases of ω near $1, 2, 3, \cdots$.

It turns out that in an ω, ε-diagram, we have domains emerging from the ω-axis, called Floquet tongues, where the solutions are unstable; see Fig. 10.1. The boundaries of these tongues correspond with the values of ω, ε where the solutions are periodic (i.e., they are neutrally stable). We shall determine these tongues in two cases.

One should note that if we have for a linear homogeneous equation one periodic solution, we have a one-parameter family of periodic solutions and definitely no uniqueness. In this case, the boundaries of the tongues correspond with periodic solutions, so the two independent solutions are periodic and have the same period. However, this uniqueness question need not bother us, as we know a priori that in this case families of periodic solutions exist.

We assume that $\omega^2 = m^2 - \varepsilon\beta(\varepsilon)$ with $m = 1, 2, \cdots$ and $\beta(\varepsilon) = \beta_0 + \beta_1\varepsilon + \cdots$ a known Taylor series in ε. The equation becomes

$$\ddot{x} + m^2x = \varepsilon\beta(\varepsilon)x - \varepsilon\cos 2tx.$$

It will turn out that in the cases $m = 2, 3, \cdots$ we have to perform higher-order calculations. In the case of nonautonomous equations, it is then more convenient to drop the phase-amplitude representation of the solutions and use the transformation $x, \dot{x} \to y_1, y_2$:

$$x(t) = y_1(t)\cos mt + y_2(t)\sin mt,$$
$$\dot{x}(t) = -my_1(t)\sin mt + my_2(t)\cos mt.$$

The expansion of $x(t)$ will take the form $x(t) = y_1(0)\cos mt + y_2(0)\sin mt + \varepsilon\cdots$. For y_1, y_2, we find the equations

$$\dot{y}_1 = -\frac{\varepsilon}{m}(\beta(\varepsilon) - \cos 2t)(y_1\cos mt + y_2\sin mt)\sin mt,$$
$$\dot{y}_2 = \frac{\varepsilon}{m}(\beta(\varepsilon) - \cos 2t)(y_1\cos mt + y_2\sin mt)\cos mt.$$

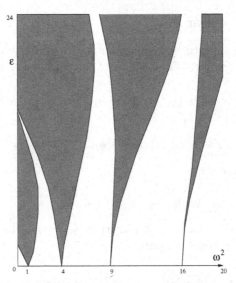

Fig. 10.1. Instability (or Floquet) tongues of the Mathieu equation $\ddot{x} + (\omega^2 + \varepsilon \cos 2t)x = 0$. The shaded domains correspond with instability.

The Mathieu equation can be transformed to an integral equation by formally integrating the equations for y_1 and y_2 and substituting them into the expression for $x(t)$.

We shall now use the periodicity conditions in various cases.

10.3.1 Frequency ω near 1

The solutions of the unperturbed (harmonic) equation are near 2π-periodic and the forcing is π-periodic; such a forcing is called subharmonic. We will look for 2π-periodic solutions, and we have the periodicity conditions

$$\int_0^{2\pi} (\beta(\varepsilon) - \cos 2s)(y_1(s)\cos s + y_2(s)\sin s)\sin s\, ds = 0,$$

$$\int_0^{2\pi} (\beta(\varepsilon) - \cos 2s)(y_1(s)\cos s + y_2(s)\sin s)\cos s\, ds = 0.$$

Expanding, we find to first order the periodicity conditions

$$y_2(0)\left(\beta_0 + \frac{1}{2}\right) = 0, \quad y_1(0)\left(\beta_0 - \frac{1}{2}\right) = 0.$$

For the boundaries of the Floquet tongue, we find $\beta_0 = \pm\frac{1}{2}$ or to first order $\omega^2 = 1 \pm \frac{1}{2}\varepsilon$.

10.3.2 Frequency ω near 2

The solutions of the unperturbed equation are π-periodic like the forcing, and so we are looking for π-periodic solutions of the Mathieu equation. We have

$$\int_0^\pi (\beta(\varepsilon) - \cos 2s)(y_1(s)\cos s + y_2(s)\sin s)\sin 2s\,ds = 0,$$

$$\int_0^\pi (\beta(\varepsilon) - \cos 2s)(y_1(s)\cos s + y_2(s)\sin s)\cos 2s\,ds = 0.$$

To first order, we find

$$y_2(0)\beta_0 = 0, \quad y_1(0)\beta_0 = 0.$$

To have nontrivial solutions, we conclude that $\beta_0 = 0$ and we have to expand to higher order. For this we compute

$$y_1(t) = y_1(0) - \frac{\varepsilon}{2}\int_0^t (-\cos 2s)(y_1(0)\cos 2s + y_2(0)\sin 2s)\cos 2s\,ds + O(\varepsilon^2)$$

$$= y_1(0) + \varepsilon\frac{3}{16}\left[y_1(0)\left(\frac{4}{3} - \cos 2t - \frac{1}{3}\cos 6t\right)\right.$$

$$\left. + y_2(0)\left(\sin 2t - \frac{1}{3}\sin 6t\right)\right] + O(\varepsilon^2).$$

Substituting this expression in the periodicity conditions produces $\beta_1 = \frac{1}{48}$ or $\beta_1 = -\frac{5}{48}$. Accordingly, the Floquet tongue is bounded by

$$\omega^2 = 4 - \frac{1}{48}\varepsilon^2 + \cdots, \omega^2 = 4 + \frac{5}{48}\varepsilon^2 + \cdots.$$

Subsequent calculations will show that the neglected terms are $O(\varepsilon^4)$.

The calculations for $m = 3, 4, \cdots$ will be even more laborious. The same holds when we want more precision (i.e., calculation of higher-order terms) for a particular value of m. In this case, computer algebra can be very helpful, especially as we know in advance here that the expansions are convergent. For a computer algebra approach, see for instance Rand (1994).

Remark

In Section 15.3, we will show that, for certain values of ω near 1, the solutions in the instability tongue are growing exponentially with $\varepsilon^{\frac{3}{2}}t$. This is not in contradiction with the Poincaré expansion theorem, as the theorem guarantees the existence of an expansion in integer powers of ε on the timescale 1. Such an expansion is clearly not valid on the timescale where the instability is developing. For periodic solutions the situation is different, as in this case timescale 1 implies "for all time" (assuming that the period does not depend on the small parameter). An extensive discussion of timescales is given in Chapter 11.

Example 10.10
(Mathieu equation with damping)
In general, dissipation effects play a part in mechanics, so it seems natural to look at the effect of damping on the Mathieu equation

$$\ddot{x} + \varepsilon\mu\dot{x} + (\omega^2 + \varepsilon\cos 2t)x = 0,$$

where μ is a positive coefficient and we consider the simplest case of ω near 1: $\omega^2 = 1 - \varepsilon\beta_0 + O(\varepsilon^2)$. Omitting the $O(\varepsilon^2)$ terms, the equation becomes

$$\ddot{x} + x = \varepsilon\beta_0 x - \varepsilon\mu\dot{x} - \varepsilon\cos(2t)x.$$

In the periodicity conditions derived in the preceding example, we have to add the term

$$\mu(y_1(0)\sin s - y_2(0)\cos s)$$

to the integrand. This results in the periodicity conditions

$$\frac{1}{2}\mu y_1(0) + \left(\beta_0 + \frac{1}{2}\right)y_2(0) = 0,$$

$$\left(\beta_0 - \frac{1}{2}\right)y_1(0) - \frac{1}{2}\mu y_2(0) = 0.$$

To have nontrivial solutions, the determinant has to be zero or

$$\beta_0 = \pm\frac{1}{2}\sqrt{1 - \mu^2},$$

and for the boundaries of the instability domain, we find

$$\omega^2 = 1 \pm \frac{1}{2}\varepsilon\sqrt{1 - \mu^2}.$$

As a consequence of damping, the instability domain of the Mathieu equation shrinks and the tongue is lifted off the ω-axis.

10.4 Autoparametric Systems and Quenching

Consider a system consisting of weakly interacting subsystems. To fix the idea, we consider a system with two degrees of freedom (four dimensions). Suppose that in one degree of freedom stable motion is possible without interaction with the other subsystem; such motion is usually called "normal mode behaviour" and in many cases this will be a (stable) periodic solution. Is this normal mode stable in the four-dimensional system? If not, the corresponding instability phenomenon is called autoparametric resonance.

This question is of particular interest in engineering problems where the normal mode may represent undesirable behaviour of flexible structures such as vibrations of overhead transmission lines, connecting cables, or chimney pipes. In the engineering context, normal modes are often called "semitrivial solutions". We may try to destabilise them by actually introducing a suitable interacting system. This may result in destabilisation or energy reduction of the undesirable normal mode; this process of permanent reduction of the amplitude of the normal mode is called "quenching", and the second oscillator, which does the destabilisation, is called the "energy absorber". For a survey and treatment of such problems, see the monograph by Tondl et al. (2000).

In the case of Hamiltonian systems, we may have autoparametric resonance and a normal mode may be destabilised; an example is the elastic pendulum. However, because of the recurrence of the phase flow, we have no quenching. For this we need energy dissipation in the second oscillator.

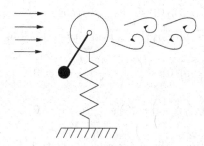

Fig. 10.2. Example of an autoparametric system with flow-induced vibrations. The system consists of a single mass on a spring to which a pendulum is attached as an energy absorber. The flow excites the mass and the spring but not the pendulum.

Example 10.11
(quenching of self-excited oscillations)
Consider the system

$$\ddot{x} - \varepsilon(1 - \dot{x}^2)\dot{x} + x = \varepsilon f(x, y),$$
$$\ddot{y} + \varepsilon\mu\dot{y} + q^2 y = \varepsilon y g(x, y),$$

where μ is the (positive) damping constant, $f(x, y)$ is an interaction term with expansion that starts with quadratic terms, and the interaction term $g(x, y)$ starts with linear terms. In Fig. 10.2, a pendulum is attached as an example of an energy absorber. The equation for x is typical for flow-induced vibrations, where $(1 - \dot{x}^2)\dot{x}$ is usually called Rayleigh self-excitation.

To fix the idea, assume that $f(x, y) = c_1 x^2 + c_2 x y + c_3 y^2, g(x, y) = d_1 x + d_2 y$. (This is different from the case where a pendulum is attached.) Putting $y = 0$ produces normal mode self-excited oscillations described by

$$\ddot{x} - \varepsilon(1 - \dot{x}^2)\dot{x} + x = \varepsilon c_1 x^2.$$

As in Example 10.5 for the Van der Pol equation, we transform time by $\omega t = \theta$, $\omega^{-2} = 1 - \varepsilon\eta(\varepsilon)$ to find

$$x'' + x = \varepsilon\eta(\varepsilon)x + \varepsilon(1 - \varepsilon\eta(\varepsilon))^{\frac{1}{2}}x'(1 - (1 - \varepsilon\eta(\varepsilon))^{-1}x'^2) + \varepsilon(1 - \varepsilon\eta(\varepsilon))c_1x^2$$

with unknown initial conditions $x(0) = a(\varepsilon), x' = 0$. The periodicity conditions at lowest order become

$$-\int_0^{2\pi} [a(0)\sin s(1 - a^2(0)\sin^2 s) + c_1 a^2(0)\cos^2 s]\sin s\, ds = 0,$$

$$\pi\eta(0)a(0) - \int_0^{2\pi} [a(0)\sin s(1 - a^2(0)\sin^2 s) + c_1 a^2(0)\cos^2 s]\cos s\, ds = 0.$$

After integration, we find as for the Van der Pol equation

$$a(0)\left(1 - \frac{1}{4}a^2(0)\right) = 0,$$

$$\eta(0)a(0) = 0,$$

so, apart from the trivial solution, we have $a(0) = 2, \eta(0) = 0$. (The solution $a(0) = -2$ produces the same approximation.) A periodic solution $\phi(t)$ branches off at amplitude 2 with first-order approximation $x_0(t) = 2\cos t$. For the periodic solution we have the estimate $\phi(t) = 2\cos t + O(\varepsilon)$.

To study the stability of this normal mode solution, we put $x(t) = \phi(t) + u$. Substitution in the equation for x and using that $\phi(t)$ is a solution, we find

$$\ddot{u} + u = \varepsilon(1 - \dot{\phi}^2)\dot{u} - \varepsilon(2\dot{\phi}\dot{u} + \dot{u}^2)(\dot{\phi} + \dot{u}) + \varepsilon(c_1(2\phi u + u^2) + c_2(\phi + u)y + c_3 y^2).$$

Also, we substitute $x(t) = \phi(t) + u$ in the equation for y. To determine the stability of $\phi(t)$, we linearise the system and replace $\phi(t)$ by its first-order approximation $x_0(t)$ to obtain

$$\ddot{u} + u = \varepsilon(1 - 12\sin^2 t)\dot{u} + \varepsilon 4c_1 u\cos t + \varepsilon 2c_2 y\cos t,$$

$$\ddot{y} + q^2 y = -\varepsilon\mu\dot{y} + \varepsilon 2d_1 y\cos t.$$

The equations are in a certain sense decoupled: first we can solve the problem for y, after which we consider the problem for u. This decoupling happens often in autoparametric systems.

As in Example 10.9, we use the transformation $y, \dot{y} \to y_1, y_2$:

$$y(t) = y_1(t)\cos qt + y_2(t)\sin qt,$$

$$\dot{y}(t) = -qy_1(t)\sin qt + qy_2(t)\cos qt.$$

The expansion will take the form $y(t) = y_1(0)\cos qt + y_2(0)\sin qt + \varepsilon \cdots$. For y_1, y_2, we find the equations

$$\dot{y}_1 = -\frac{\varepsilon}{q}[\mu q y_1(t)\sin qt - \mu q y_2(t)\cos qt + 2d_1(y_1\cos qt + y_2\sin qt)\cos t]\sin qt,$$

$$\dot{y}_2 = \frac{\varepsilon}{q}[\mu q y_1(t)\sin qt - \mu q y_2(t)\cos qt + 2d_1(y_1\cos qt + y_2\sin qt)\cos t]\cos qt.$$

Integration and application of the periodicity conditions leads at first order to nontrivial results if $q = \frac{1}{2}$. We find

$$\frac{1}{2}\mu y_1 - d_1 y_2 = 0,$$

$$d_1 y_1 - \frac{1}{2}\mu y_2 = 0.$$

We have nontrivial solutions if the determinant of the matrix of coefficients vanishes, or

$$\mu^2 = 4d_1^2.$$

If the damping coefficient μ satisfies $0 \le \mu \le 2|d_1|$, we have instability with respect to perturbations orthogonal to the normal mode (in the $y - \dot{y}$ direction); if $\mu > 2|d_1|$, we have stability.

It is an interesting question whether perturbations in the normal mode plane can destabilise the normal mode. For this we have to solve the equation for u. With the choice $q = \frac{1}{2}$ and if $c_2 \ne 0$, we find that $u = 0$ is unstable; the calculation is left to the reader.

10.5 The Radius of Convergence

When obtaining a power series expansion with respect to ε by the Poincaré-Lindstedt method, we have a convergent series for the periodic solution. So, in contrast with the results for most asymptotic expansions, it makes sense to ask the question of to what value of ε the series converges.

In the paper by Andersen and Geer (1982), where 164 terms were calculated for the expansion of the periodic solution of the van der Pol equation, the numerics surprisingly suggests convergence until $\varepsilon = O_s(1)$. These results become credible when looking at the analytic estimates by Grebenikov and Ryabov (1983). After introducing majorising equations for the expansion, Grebenikov and Ryabov give some examples. First, for the Duffing equation with forcing,

$$\ddot{x} + x - \varepsilon x^3 = \varepsilon a \sin t.$$

Grebenikov and Ryabov show that the convergence of the Poincaré-Lindstedt expansion for the periodic solution holds for

$$0 \le \varepsilon \le 1.11|a|^{-\frac{2}{3}}.$$

In the case of the Mathieu equation

$$\ddot{x} + (a + \varepsilon \cos 2t)x = 0$$

near $a = 1$, they obtain convergence for

$$0 \le \varepsilon \le 5.65.$$

For the resonances $a = 4, 9$, $O_s(1)$ estimates are also found.

Finally, we note that the radius of convergence of the power series with respect to the small parameter ε does not exclude continuation of the periodic solution beyond the radius of convergence. A simple example is the equation

$$\ddot{x} + \frac{3}{2}\varepsilon\dot{x} + x = -3\sin 2t.$$

The equation contains a unique periodic solution

$$\phi(t) = \frac{1}{1 + \varepsilon^2}\sin 2t + \frac{\varepsilon}{1 + \varepsilon^2}\cos 2t$$

that has a convergent series expansion for $0 \le \varepsilon < 1$ but exists for all values of ε.

10.6 Guide to the Literature

The techniques discussed in this chapter were already in use in the eighteenth and nineteenth centuries, but the mathematical formulation of such results for initial value problems for ordinary differential equations was given by Henri Poincaré. Usually his method of using the expansion theorem to construct periodic solutions is called the Poincaré-Lindstedt method, as Lindstedt produced a formal calculation of this type. The procedure itself is older, but Poincaré (1893, Vol. 2) was the first to present sound mathematics. His 1893 proof of the expansion theorem is based on majorising series and rather complicated, see also Roseau (1966) for an account. More recent proofs use a continuation of the problem into the complex domain in combination with contraction; an example of such a proof is given in Verhulst (2000, Chapter 9).

The analysis is in fact an example of a very general problem formulation. Consider an equation of the form

$$F(u, \varepsilon) = 0$$

with F a nonlinear operator on a linear space - a Hilbert or Banach space - and with known solution $u = u_0$ if $\varepsilon = 0$, so $F(u_0, 0) = 0$. The problem is then under what condition we can obtain for the solution a convergent series of the form

$$u = u_0 + \sum_{n \ge 1} \varepsilon^n u_n.$$

The operator F can be a function, a differential equation, or an integral equation and can take many other forms. Vainberg and Trenogin (1974) give a general discussion, with the emphasis on Lyapunov-Schmidt techniques, and many examples.

A survey of the implicit function theorem with modern extensions and applications is given by Krantz and Parks (2002). Application of the Poincaré-Lindstedt method to systems with dimension higher than two poses no fundamental problem but requires laborious formula manipulation. An example of an application to Hamiltonian systems with two degrees of freedom has been presented in two papers by Presler and Broucke (1981a, 1981b). They apply formal algebraic manipulation to obtain expansions of relatively high order. Rand (1994) gives an introduction to the use of computer algebra in nonlinear dynamics.

More general nonlinear equations, in particular integral equations, are considered by Vainberg and Trenogin (1974). The theoretical background and many applications to perturbation problems in linear continuum mechanics can be found in Sanchez Hubert and Sanchez Palencia (1989).

Apart from this well-founded work, there are applications to partial differential equations using the Poincaré-Lindstedt method formally. It is difficult to assess the meaning of these results unless one has a priori knowledge about the existence and smoothness of periodic solutions.

10.7 Exercises

Exercise 10.1 Consider the algebraic equation

$$x^3 - (3 + \varepsilon)x + 2 = 0,$$

which has three real solutions for small $\varepsilon > 0$.
a. Can one obtain the solutions in a Taylor series with respect to ε?
b. Determine a two-term expansion for the solutions.

Exercise 10.2 Kepler's equation for the gravitational two-body problem is

$$E - e \sin E = M$$

with M (depending on the period) and e (eccentricity) given. Show that the angle E, $0 \leq E \leq 2\pi$, is determined uniquely.

Exercise 10.3 Consider the equation

$$\dot{x} = 1 - x + \varepsilon x^2.$$

If $\varepsilon = 0$, $x = 1$ is a stable equilibrium and it is easy to see that if $\varepsilon > 0$, a stable equilibrium exists in a neighbourhood of $x = 1$.

a. Apply the Poincaré expansion theorem to find the first two terms of the expansion $x_0(t) + \varepsilon x_1(t) + \cdots$.

b. Compute the limit for $t \to +\infty$ of this approximation. Does this fit with the observation about the stable equilibrium?

Exercise 10.4 In Example 10.4, we showed for the damped harmonic oscillator that secular (unbounded) terms arise in the straightforward expansion. One might argue that this is caused by the presence of linear terms in the perturbation, as these are the cause of linear resonance. Consider as an example the problem

$$\ddot{x} + x = \varepsilon x^2, \quad x(0) = 1, \dot{x}(0) = 0.$$

a. Show that no secular terms arise for $x_0(t)$ and $x_1(t)$.

b. Do secular terms arise at higher order, for instance for $x_2(t)$?

Exercise 10.5 Consider again the Van der Pol equation (Example 10.5) and calculate the periodic solution to second order. The calculation to higher order becomes laborious but it is possible to implement the procedure in a computer programme. Andersen and Geer (1982) did this for the Van der Pol equation to compute the first 164 terms.

Exercise 10.6 It will be clear from Examples 10.5 and 10.6 that an interesting application arises in examples such as

$$\ddot{x} + x = \varepsilon x^3 + \varepsilon^2 \dot{x}(1 - x^2).$$

At lowest order, we find that the first-order Jacobian J_0 vanishes, but in this case a unique periodic solution exists. When going to second order in ε, the Jacobian does not vanish and this unique periodic solution is found. Check this statement and compute the approximation to this order.

11

Averaging and Timescales

In this chapter, we will consider again initial value problems for ordinary differential equations involving perturbations. In the preceding chapter we encountered the idea of "an approximation valid on a certain timescale". This is an important concept that we have to formulate more explicitly.

Suppose that we want to approximate a (vector) function $x_\varepsilon(t)$ for $t \geq 0$ by another function $y_\varepsilon(t)$. If we have for $t \geq 0$ the estimate

$$x_\varepsilon(t) - y_\varepsilon(t) = O(\varepsilon^m), \quad 0 \leq \varepsilon^n t \leq C,$$

with m, n, C constants independent of ε, we shall say that $y_\varepsilon(t)$ is an $O(\varepsilon^m)$-approximation of $x_\varepsilon(t)$ on the timescale $1/\varepsilon^n$.

Note that if necessary we can generalise somewhat by replacing ε^m by the order function $\delta_1(\varepsilon)$ and ε^n by the order function $\delta_2(\varepsilon)$.

Example 11.1
Consider the function $x_\varepsilon(t) = (1 + \varepsilon + \varepsilon^2 t) \sin t$ with $O(\varepsilon)$-approximation $\sin t$ on the timescale $1/\varepsilon$. At the same time, $\sin t$ is an $O(\varepsilon^{\frac{1}{2}})$-approximation on the timescale $1/\varepsilon^{\frac{3}{2}}$. It is also easy to see that $(1+\varepsilon) \sin t$ is an $O(\varepsilon^2)$-approximation on the timescale 1.

We shall often use this abbreviation of "an approximation on a timescale". In the preceding chapter, we approximated periodic solutions, and in such cases approximations of a quantity such as an amplitude are valid for all time. In general, the results are much more restricted; usually we will be happy to obtain an approximation on a timescale such as $1/\varepsilon$ or $1/\varepsilon^2$. As the estimates are valid for $\varepsilon \to 0$, these are already long timescales.

11.1 Basic Periodic Averaging

Suppose that we have a differential equation of the form

$$\dot{x} = f(t, x, \varepsilon)$$

with f periodic in t. We could try to approximate the solutions by simply averaging the function f over time. In general, however, this leads to incorrect results, as was realised already by Lagrange and Laplace in the eighteenth century. The correct approach is to solve the problem for $\varepsilon = 0$ and use these "unperturbed" solutions to formulate "variational" equations, also called equations "in the standard form", which can be averaged. As an example, we look again at Example 10.4, where a straightforward expansion with respect to ε produced secular terms.

Example 11.2
Consider the initial value problem

$$\ddot{x} + 2\varepsilon\dot{x} + x = 0, \ x(0) = 1, \ \dot{x}(0) = 0.$$

This equation is of the form

$$\ddot{x} + x = \varepsilon f(x, \dot{x}), \ x(0) = a, \ \dot{x}(0) = b, \tag{11.1}$$

with $f(x, \dot{x}) = -2\dot{x}$, and we can study this more general equation with the same effort.

The unperturbed equation is

$$\ddot{x} + x = 0$$

with solutions of the form

$$x(t) = r_0 \cos(t + \psi_0), \ \dot{x}(t) = -r_0 \sin(t + \psi_0),$$

in which r_0, ψ_0 are determined by the initial conditions. To study the perturbed problem, Lagrange formulated the "variation of constants" method by replacing r_0, ψ_0 by the variables $r(t), \psi(t)$, so we transform $x, \dot{x}$ to r, ψ by

$$x(t) = r(t) \cos(t + \psi(t)), \ \dot{x}(t) = -r(t) \sin(t + \psi(t)).$$

Differentiation of $x(t)$ has to produce $\dot{x}(t)$, which yields the condition

$$\dot{r} \cos(t + \psi) - \dot{\psi} r \sin(t + \psi) = 0.$$

Differentiation of $\dot{x}$ and substituting this and the expressions for $x(t), \dot{x}(t)$ into the differential equation (11.1) gives

$$-\dot{r} \sin(t + \psi) - \dot{\psi} r \cos(t + \psi) = \varepsilon f(r \cos(t + \psi), -r(t + \psi)).$$

Solving from the two algebraic equations for $\dot{r}$ and $\dot{\psi}$, we find

$$\dot{r} = -\varepsilon \sin(t + \psi) f(r \cos(t + \psi), -r(t + \psi)), \tag{11.2}$$

$$\dot{\psi} = -\varepsilon \frac{\cos(t + \psi)}{r} f(r \cos(t + \psi), -r \sin(t + \psi)). \tag{11.3}$$

In the transformation from $x, \dot{x}$ to polar coordinates r, ψ, we of course have to exclude a neighbourhood of the origin; this is also illustrated by the presence of r in one of the denominators.

The equations express that r and ψ are varying slowly with time, and the system is of the form $\dot{y} = \varepsilon g(y, t)$. This slowly varying system is called the standard form. The idea is now to consider only the nonzero average of the right-hand sides - keeping r and ψ fixed - and leave out the terms with average zero. We can think of the right-hand sides as functions periodic in t (ignoring the slow dependence of r and ψ on t), which we Fourieranalyse, and then we omit all higher order harmonics.

In our problem, we have $f(x, \dot{x}) = -2\dot{x}$ so that

$$\dot{r} = -\varepsilon 2r \sin^2(t + \psi),$$
$$\dot{\psi} = -\varepsilon 2 \cos(t + \psi) \sin(t + \psi).$$

Now we average, replacing r, ψ by their approximations r_a, ψ_a:

$$\dot{r}_a = -\varepsilon \frac{1}{2\pi} \int_0^{2\pi} 2r \sin^2(t + \psi) dt = -\varepsilon r_a,$$

$$\dot{\psi}_a = -\varepsilon \frac{1}{2\pi} \int_0^{2\pi} 2 \cos(t + \psi) \sin(t + \psi) dt = 0.$$

Solving this with $r_a(0) = r_0, \psi_a(0) = \psi_0$, the approximation takes the form

$$x_a(t) = r_0 e^{-\varepsilon t} \cos(t + \psi_0).$$

Note that secular terms are absent. We shall discuss the nature of the approximation later on.

Let us apply this procedure to the Van der Pol equation in the following example.

Example 11.3
The Van der Pol equation is

$$\ddot{x} + x = \varepsilon(1 - x^2)\dot{x},$$

and we transform again $x(t) = r(t) \cos(t + \psi(t))$, $\dot{x}(t) = -r(t) \sin(t + \psi(t))$. Using Eqs. (11.2) and (11.3) with $f = (1 - x^2)\dot{x}$, we find

$$\dot{r} = -\varepsilon \sin(t + \psi)(1 - r^2 \cos^2(t + \psi))(-r \sin(t + \psi)), \quad r(0) = r_0,$$
$$\dot{\psi} = -\varepsilon \frac{\cos(t + \psi)}{r}(1 - r^2 \cos^2(t + \psi))(-r \sin(t + \psi)), \quad \psi(0) = \psi_0.$$

The right-hand sides are 2π-periodic in t and we average over t, keeping r and ψ fixed. Replacing r, ψ by their approximations r_a, ψ_a, we have:

$$\dot{r}_a = \varepsilon \frac{r_a}{2} \left(1 - \frac{1}{4} r_a^2 \right), \quad \dot{\psi}_a = 0.$$

Clearly, the phase ψ_a is constant in this approximation, and we can solve the equation for r_a to find

$$r_a(t) = \frac{r_0 e^{\frac{1}{2}\varepsilon t}}{(1 + \frac{r_0^2}{4}(e^{\varepsilon t} - 1))^{\frac{1}{2}}}.$$

Taking the limit of this expression for $r_a(t)$ for $t \to \infty$, we find that $r_a(t) \to 2$ so this solution seems to be stable. This value of r_a corresponds with the amplitude of the periodic solution of the Van der Pol equation that we computed in Chapter 10 using the Poincaré-Lindstedt method.

There is an easier way to obtain this result. In the differential equation for r_a, the right-hand side has two zeros corresponding with stationary solutions of the amplitude: $r_a = 0$ and $r_a = 2$. The first one corresponds with the trivial solution. The second one means that choosing $r_0 = 2$, $r_a(t)$ will keep the value 2 for all time. Obviously, such a stationary solution of the amplitude corresponds with a periodic solution. We shall discuss this more extensively in Section 11.3.

Remark
The equation $\ddot{x} + x = \varepsilon f(x, \dot{x})$ is an example of an autonomous equation. (Time t does not arise explicitly in the equation.) Such equations have special features that also show up in the averaging results. Consider again the equation for ψ in Eq. (11.3):

$$\dot{\psi} = -\varepsilon \frac{\cos(t + \psi)}{r} f(r \cos(t + \psi), -r \sin(t + \psi)).$$

As ψ is kept constant during the averaging process, we might as well transform $\tau = t + \psi$ so that

$$\dot{\psi} = -\varepsilon \frac{\cos \tau}{r} f(r \cos \tau, -r \sin \tau)$$

and average over τ. The averaged equation for ψ will be of the form $\dot{\psi}_a = \varepsilon F(r_a)$ and will *not* depend on ψ_a.

The solution of the averaged equations is reduced to solving a first-order equation for r_a. In general the averaging of an autonomous equation of order n will result in a reduction to a system of order $n - 1$.

Example 11.4
(damped and forced Duffing equation)
Consider again Example 10.8 describing the forced Duffing equation

$$\ddot{x} + \varepsilon \mu \dot{x} + \varepsilon \gamma x^3 + x = \varepsilon h \cos \omega t, \tag{11.4}$$

where we assume $\omega^{-2} = 1 - \beta_0 \varepsilon$, which means that the forcing period is near 2π. It is easier when we transform $\tau = \omega t$ to obtain

$$x'' + \varepsilon \frac{\mu}{\omega} x' + \varepsilon \frac{\gamma}{\omega^2} x^3 + \frac{1}{\omega^2} x = \varepsilon \frac{h}{\omega^2} \cos \tau,$$

where a prime denotes differentiation with respect to τ. Expanding and moving the $O(\varepsilon)$ terms to the right, we get

$$x'' + x = \varepsilon \beta_0 x - \varepsilon \mu x' - \varepsilon \gamma x^3 + \varepsilon h \cos \tau + O(\varepsilon^2).$$

In deriving the system (11.2), (11.3), we did not use any explicit form of the right-hand side f, so in polar variables r, ψ we get to $O(\varepsilon)$ the slowly varying system

$$r' = -\varepsilon \sin(\tau + \psi)(\beta_0 x - \mu x' - \gamma x^3 + h \cos \tau),$$
$$\psi' = -\varepsilon \frac{\cos(\tau + \psi)}{r}(\beta_0 x - \mu x' - \gamma x^3 + h \cos \tau).$$

In the preceding equations x and x' still have to be replaced by $r(\tau)\cos(\tau + \psi(\tau))$ and $-r(\tau)\sin(\tau + \psi(\tau))$ respectively. Averaging this system over τ produces

$$r_a' = -\frac{1}{2}\varepsilon(\mu r_a + h \sin \psi_a),$$
$$\psi_a' = -\frac{1}{2}\varepsilon \left(\beta_0 - \frac{3}{4}\gamma r_a^2 + h \frac{\cos \psi_a}{r_a} \right).$$

It is interesting to note that again - as in the Van der Pol equation - stationary solutions of the right-hand side produce periodic solutions of the forced Duffing equation; the expressions are the same as those obtained by the Poincaré-Lindstedt method. However, the averaging result produces more information. We can now also study the flow near the periodic solutions and thus their stability.

11.1.1 Transformation to a Slowly Varying System

To apply averaging, we need a system in the standard form

$$\dot{x} = \varepsilon f(t, x).$$

Sometimes such a system is called the variational system of the original equations from which they were derived. The system describes the dynamics around a certain unperturbed state. The classical example is the perturbed harmonic equation

$$\ddot{x} + \omega^2 x = \varepsilon f(t, x, \dot{x}),$$

where ω is a positive constant, the frequency. As before, we can use an amplitude-phase transformation of the form

$$x(t) = r(t)\cos(\omega t + \psi(t)), \tag{11.5}$$
$$\dot{x}(t) = -r(t)\omega\sin(\omega t + \psi(t)). \tag{11.6}$$

Using variation of constants with these expressions, we find

$$\dot{r} = -\frac{\varepsilon}{\omega}\sin(\omega t + \psi)f(t, r\cos(\omega t + \psi), -r\omega\sin(\omega t + \psi)), \tag{11.7}$$
$$\dot{\psi} = -\frac{\varepsilon}{\omega r}\cos(\omega t + \psi)f(t, r\cos(\omega t + \psi), -r\omega\sin(\omega t + \psi)). \tag{11.8}$$

Of course, we can write the independent solutions of the unperturbed equation in a different form and obtain an equivalent system. This may be computationally advantageous.

The equivalent transformation

$$x(t) = y_1(t)\cos\omega t + \frac{y_2(t)}{\omega}\sin\omega t, \tag{11.9}$$
$$\dot{x}(t) = -\omega y_1(t)\sin\omega t + y_2(t)\cos\omega t, \tag{11.10}$$

leads to the standard form

$$\dot{y}_1 = -\frac{\varepsilon}{\omega}\sin\omega t f(t, x(t), \dot{x}(t)), \tag{11.11}$$
$$\dot{y}_2 = \varepsilon\cos\omega t f(t, x(t), \dot{x}(t)), \tag{11.12}$$

where the expressions for $x(t)$ and $\dot{x}(t)$ still have to be substituted. We shall consider some examples.

Example 11.5
Consider the Duffing equation with positive damping

$$\ddot{x} + \varepsilon\mu\dot{x} + x + \varepsilon\gamma x^3 = 0.$$

Using amplitude-phase variables r, ψ, we find (see the preceding subsection) for the averaged equations

$$\dot{r}_a = -\frac{1}{2}\varepsilon\mu r_a, \quad \dot{\psi}_a = \frac{3}{8}\varepsilon\gamma r_a^2.$$

Using the transformation from Eqs.(11.9), (11.10), we find, after averaging,

$$\dot{y}_{1a} = -\frac{1}{2}\varepsilon\mu y_{1a} + \frac{3}{8}\varepsilon\gamma y_{2a}(y_{1a}^2 + y_{2a}^2),$$
$$\dot{y}_{2a} = \frac{1}{2}\varepsilon\mu y_{2a} + \frac{3}{8}\varepsilon\gamma y_{1a}(y_{1a}^2 + y_{2a}^2).$$

Of course, the results are equivalent, but the expressions in amplitude-phase form are looking much simpler. We see immediately that (μ indicating positive damping) the amplitude is decreasing exponentially; the phase is changing slowly, depending on the amplitude.

Consider now a case where the second transformation by Eqs. (11.9) and (11.10) is more useful.

Example 11.6
Consider the Mathieu equation

$$\ddot{x} + (1 + 2\varepsilon \cos 2t)x = 0.$$

In phase-amplitude form, from the transformation Eqs. (11.5) and (11.6), we find

$$\dot{r} = 2\varepsilon \sin(t + \psi) \cos 2t\, r \cos(t + \psi),$$
$$\dot{\psi} = 2\varepsilon \cos^2(t + \psi) \cos 2t.$$

Averaging produces

$$\dot{r}_a = \frac{1}{2}\varepsilon r_a \sin 2\psi, \quad \dot{\psi}_a = \frac{1}{2}\varepsilon \cos 2\psi_a.$$

This system is not convenient to analyse, and it is even nonlinear.
    Trying the second transformation from Eqs. (11.9) and (11.10), we have

$$\dot{y}_1 = 2\varepsilon \sin t \cos 2t(y_1 \cos t + y_2 \sin t),$$
$$\dot{y}_2 = 2\varepsilon \cos t \cos 2t(y_1 \cos t + y_2 \sin t).$$

Averaging produces

$$\dot{y}_{1a} = -\frac{1}{2}\varepsilon y_{2a}, \quad \dot{y}_{2a} = -\frac{1}{2}\varepsilon y_{1a}.$$

These are linear equations with constant coefficients and are easier to analyse; the eigenvalues are $\pm\frac{1}{2}\varepsilon$ so that the approximations of y_1, y_2 are a linear combination of $\exp(\pm\frac{1}{2}\varepsilon t)$. If for instance $x(0) = x_0, \dot{x}(0) = 0$, we have

$$y_{1a}(t) = \frac{1}{2}x_0 e^{-\frac{1}{2}\varepsilon t} + \frac{1}{2}x_0 e^{\frac{1}{2}\varepsilon t}, \quad y_{2a}(t) = \frac{1}{2}x_0 e^{-\frac{1}{2}\varepsilon t} - \frac{1}{2}x_0 e^{\frac{1}{2}\varepsilon t},$$

and for $x_a(t)$:

$$x_a(t) = \frac{1}{2}x_0 e^{-\frac{1}{2}\varepsilon t}(\cos t + \sin t) + \frac{1}{2}x_0 e^{\frac{1}{2}\varepsilon t}(\cos t - \sin t).$$

Note that these solutions of the Mathieu equation show the instability of the trivial solution.

    This process of variation of constants to produce a slowly varying standard form can be applied to a system of equations provided we know the solutions for $\varepsilon = 0$ more or less explicitly. For instance, for the n-dimensional initial value problem

$$\dot{x} = A(t)x + \varepsilon f(t, x), \; x(0) = x_0,$$

we can produce the standard form by variation of constants if we know the fundamental matrix of the system $\dot{y} = A(t)y$.

More generally, considering a system of the form $\dot{x} = f(t, x, \varepsilon)$, all depends on the knowledge we have of the solutions of the unperturbed system $\dot{x} = f(t, x, 0)$. For instance, in the case of the perturbed mathematical pendulum equation

$$\ddot{x} + \sin x = \varepsilon \cdots,$$

the unperturbed solutions are elliptic functions. The analysis becomes rather technical but can still be carried out.

11.1.2 The Asymptotic Character of Averaging

It is important to know the nature of the averaging approximations.

Theorem 11.1
Consider the n-dimensional system

$$\dot{x} = \varepsilon f(t, x) + \varepsilon^2 g(t, x, \varepsilon), \; x(0) = x_0, \; t \geq 0.$$

Suppose that $f(t, x)$ is T-periodic in t with $T > 0$ a constant independent of ε. Perform the averaging process

$$f^0(y) = \frac{1}{T} \int_0^T f(t, y) dt,$$

where y is considered as a parameter that is kept constant during integration. Consider the associated initial value problem

$$\dot{y} = \varepsilon f^0(y), \; y(0) = x_0.$$

Then we have $y(t) = x(t) + O(\varepsilon)$ on the timescale $1/\varepsilon$ under fairly general conditions:

- the vector functions f and g are continuously differentiable in a bounded n-dimensional domain D, x_0 an interior point, on the timescale $1/\varepsilon$;
- $y(t)$ remains interior to the domain D on the timescale $1/\varepsilon$ to avoid boundary effects.

The smoothness conditions can be weakened somewhat; see the notes on the literature.

It is interesting to know that in some cases the timescale of validity can be extended, for instance to the timescale $1/\varepsilon^2$ or sometimes even for all time. Boundedness of the solutions or the presence of a classical attractor can be used. However, these are exceptions that require stronger conditions on the differential equation.

To show that *in general* the time estimate is optimal we consider the following example.

Example 11.7

Consider the equation

$$\dot{x} = 2\varepsilon x^2 \sin^2 t, \ x(0) = 1.$$

The right-hand side is 2π-periodic in t, and averaging over t produces the associated initial value problem

$$\dot{y} = \varepsilon y^2, \ y(0) = 1,$$

with solution

$$y(t) = \frac{1}{1 - \varepsilon t}.$$

The function $y(t)$ becomes unbounded when t approaches $1/\varepsilon$. Note that if we choose for instance $t = 1/(2\varepsilon)$, $y(t)$ takes the value 2. For the domain D, we can take any compact segment containing $x(0) = 1$ as an interior point but $y(t)$ will leave D on the timescale $1/\varepsilon$. We have

$$x(t) - y(t) = O(\varepsilon)$$

on the timescale $1/\varepsilon$. In this example, we can also compute the solution $x(t)$ in the form of elementary functions:

$$x(t) = \frac{1}{1 - \varepsilon t + \frac{1}{2}\varepsilon \sin 2t}.$$

Such a comparison with the exact solution is only possible in special cases, and it is remarkable that we can still prove the asymptotic validity theorem in general.

One might think that in this example the fact that the solution becomes unbounded on a timescale $1/\varepsilon$ is essential for the restriction on the time interval. This is not the case, as the following very simple example shows.

Example 11.8

Consider the harmonic equation with a slightly perturbed frequency

$$\ddot{x} + (1 + 2\varepsilon)x = 0.$$

Amplitude-phase variables produce the standard form

$$\dot{r} = 2\varepsilon \sin(t + \psi)r\cos(t + \psi),$$
$$\dot{\psi} = 2\varepsilon \cos^2(t + \psi).$$

Averaging produces

$$\dot{r}_a = 0, \ \dot{\psi}_a = -\varepsilon.$$

Putting for instance $r(0) = 1$, $\psi(0) = 0$, we find for the solution and the approximation

$$x(t) = \cos(\sqrt{1 + 2\varepsilon t}), \ x_a(t) = \cos(t + \varepsilon t).$$

As $x(t) = \cos(\sqrt{1 + 2\varepsilon t}) = \cos(t + \varepsilon t + O(\varepsilon^2 t))$, the validity of the approximation cannot be extended to $O(1/\varepsilon^2)$.

In our statement about the asymptotic character of the averaging process, we assumed that the period T does not depend on ε. What is important here is that T does not become unbounded as ε tends to zero. The following simple example shows this.

Example 11.9
Consider the equation

$$\dot{x} = \varepsilon \cos(\varepsilon t), \ x(0) = x_0.$$

Crude averaging produces $\dot{x}_a = 0$ and so $x_a(t) = x_0$. Integration yields the solution $x(t) = \sin(\varepsilon t) + x_0$ and so $x_a(t)$ is not an asymptotic approximation on the timescale $1/\varepsilon$.

This kind of assumption becomes important when integrating orbits near a separatrix, a homoclinic or heteroclinic orbit.

11.1.3 Quasiperiodic Averaging

In a number of applications, more than one period plays a part. If there is a common period, we can use this to average, but this is not always the case.

Example 11.10
Consider for instance

$$\ddot{x} + (1 + 2\varepsilon \cos 2t)x = \varepsilon f(t),$$

which is a Mathieu equation with small external forcing. The Mathieu equation contains a π-periodic term, and the unperturbed equation is 2π-periodic. If $f(t)$ is T-periodic with T/π rational, we can find a common period. But suppose for instance that $T = \pi\sqrt{3}$. In this case, there is no common period and we might, after putting the problem in the standard form, try to average over both periods. Using amplitude-phase variables $x, \dot{x} \to r, \psi$ by the transformation Eqs. (11.5) and (11.6), we find

$$\dot{r} = 2\varepsilon \sin(t + \psi) \cos 2t \, r \cos(t + \psi) - \varepsilon \sin(t + \psi)f(t),$$
$$\dot{\psi} = 2\varepsilon \cos^2(t + \psi) \cos 2t + \frac{\varepsilon}{r} \cos(t + \psi)f(t).$$

Taking for instance $f(t) = \cos(\frac{2}{\sqrt{3}}t)$, we can write for $\sin(t + \psi)f(t)$

$$\sin(t + \psi) \cos\left(\frac{2}{\sqrt{3}}t\right) = \frac{1}{2} \sin\left(t + \frac{2}{\sqrt{3}}t + \psi\right) + \frac{1}{2} \sin\left(t - \frac{2}{\sqrt{3}}t + \psi\right),$$

which consists of the sum of two functions with different periods. A similar splitting applies to the term $\cos(t + \psi)f(t)$ in the equation for ψ. We can now average the two terms over their respective periods, which produces an asymptotic approximation when we solve the averaged equation.

Theorem 11.2
To formulate the result more generally, consider the n-dimensional initial value problem in the standard form

$$\dot{x} = \varepsilon \sum_{n=1}^{N} f_n(t, x), \; x(0) = x_0,$$

where the vector functions $f_n(t, x)$ have N independent periods $T_n, n = 1, \cdots, N$. Averaging over the respective periods, we obtain the associated system

$$\dot{y} = \varepsilon \sum_{n=1}^{N} \frac{1}{T_n} \int_0^{T_n} f_n(t, y) dt, \; y(0) = x_0.$$

As before, we have $y(t) = x(t) + O(\varepsilon)$ on the timescale $1/\varepsilon$.

There are two important restrictions here. First N has to be finite, and second it should be possible to write the right-hand side of the equation as a sum of vector functions where each of them is periodic with one period.

In a number of interesting problems, it is not possible to satisfy these conditions. We shall discuss such cases in the next section and in the subsequent chapter.

11.2 Nonperiodic Averaging

Suppose now that we have obtained a slowly varying system $\dot{x} = \varepsilon f(t, x)$, where the right-hand side is not periodic and also not a finite sum of periodic vector fields.

Example 11.11
Consider the equation

$$\dot{x} = \varepsilon(-1 + \frac{1}{1+t})x, \; x(0) = 1,$$

with solution
$$x(t) = e^{-\varepsilon t + \varepsilon \ln(1+t)}.$$

Suppose that we could not solve the equation. The right-hand side is not periodic, but we might try a more general form of averaging by taking the limit

$$\lim_{T\to\infty} \frac{1}{T} \int_0^T \left(-1 + \frac{1}{1+t}\right) x\, dt = -x$$

and considering the associated (averaged) system

$$\dot{y} = -\varepsilon y, \; y(0) = 1,$$

with solution $y(t) = \exp(-\varepsilon t)$. As we know the exact solution, we can explicitly compare

$$|x(t) - y(t)| = e^{-\varepsilon t}(e^{\varepsilon \ln(1+t)} - 1).$$

On the timescale $1/\varepsilon$, we have $|x(t) - y(t)| = O(\varepsilon|\ln\varepsilon|)$, so in this case more general averaging produces an asymptotic approximation on a long timescale but with a somewhat bigger error than ε. Note that if a vector field is periodic in t, this general averaging limit exists and equals the (ordinary) average; see Section 15.8. In the exercises, we repeat this calculation for different right-hand sides.

There is a theory behind this more general averaging, which we summarise as follows.

Theorem 11.3
Consider the n-dimensional system

$$\dot{x} = \varepsilon f(t, x) + \varepsilon^2 g(t, x, \varepsilon), \; x(0) = x_0, \; t \geq 0.$$

Suppose that $f(t, x)$ can be averaged over t in the sense that the limit

$$f^0(y) = \lim_{T\to\infty} \frac{1}{T} \int_0^T f(t, y)\, dt$$

exists, where y is again considered a parameter that is kept constant during integration. Consider the associated initial value problem

$$\dot{y} = \varepsilon f^0(y), \; y(0) = x_0.$$

Then we have $y(t) = x(t) + O(\delta(\varepsilon))$ on the timescale $1/\varepsilon$ under fairly general conditions:

- the vector functions f and g are continuously differentiable in a bounded n-dimensional domain D with x_0 an interior point on the timescale $1/\varepsilon$;
- $y(t)$ remains interior to the domain D on the timescale $1/\varepsilon$ to avoid boundary effects.

For the error $\delta(\varepsilon)$, we have the explicit estimate

$$\delta(\varepsilon) = \sup_{x\in D} \sup_{0\leq \varepsilon t\leq C} \varepsilon \left\| \int_0^t (f(s, x) - f^0(x))\, ds \right\|$$

with C a constant independent of ε.

In Verhulst (2000, Chapter 11), there is the following, more realistic example.

Example 11.12
Consider an oscillator with friction that, through wear and tear, increases slowly with time. The equation is

$$\ddot{x} + \varepsilon f(t)\dot{x} + x = 0.$$

Assume that $f(0) = 1$ and $f(t)$ increases monotonically with time. Examples are

$$f_1(t) = 2 - e^{-t}, \ f_2(t) = 2 - \frac{1}{1+t}.$$

In amplitude-phase variables r, ψ, we have with Eqs. (11.5) and (11.6)

$$\dot{r} = -\varepsilon r \sin^2(t + \psi)f(t),$$
$$\dot{\psi} = -\varepsilon \sin(t + \psi)\cos(t + \psi)f(t).$$

We can perform averaging in the more general sense, and both for $f_1(t)$ and $f_2(t)$ we find the approximating system

$$\dot{r}_a = -\varepsilon r_a, \ \dot{\psi}_a = 0.$$

In both cases, we have the approximation

$$x_a(t) = r(0)e^{-\varepsilon t}\cos(t + \psi(0)).$$

The error, however, is different. For the function $f_1(t)$, the integral $\int_0^t (f(s, x) - f^0(x))ds$ is uniformly bounded, so the error is $O(\varepsilon)$. Choosing $f_2(t)$, the integral is logarithmically growing and we have $\delta(\varepsilon) = O(\varepsilon|\ln \varepsilon|)$.

Such a difference is not unexpected. Note that the approximation corresponds with a constant friction coefficient 2, which is the limit for $t \to \infty$ of $f_1(t)$ and $f_2(t)$. The function $f_1(t)$ reaches this limit much faster than $f_2(t)$.

The more general version of averaging is highly relevant for the treatment of so-called almost-periodic functions. We will not give an explicit definition, but we note that for almost-periodic functions there exists an analogue of Fourier theory so that a vector field $f(t, x)$ that is almost-periodic in t, can be written as a uniformly convergent series

$$f(t, x) = \sum_{n=0}^{\infty} (A_n(x)\cos \lambda_n t + B_n(x)\sin \lambda_n t).$$

If $f(t, x)$ is periodic, we would have $\lambda_n = n$, but in this more general case the λ_n can be any sequence of real numbers. The collection of these numbers is called the generalised Fourier spectrum. For an almost-periodic vector function, the following average exists:

$$f^0(x) = \lim_{T \to \infty} \frac{1}{T} \int_0^T f(t, x) dt.$$

A strong contrast with periodic Fourier theory is that if the average $f^0(x)$ vanishes, the primitive of $f(t, x)$ need not be bounded for all time. This weakens the error estimate, as we shall see in an explicit example.

Example 11.13
Think of an equation that contains a function of the form

$$f(t) = \sum_{n=0}^{\infty} \frac{1}{(2n+1)^2} \sin\left(\frac{t}{2n+1}\right).$$

A simple example would be $\dot{x} = \varepsilon f(t)x$.

We shall make a number of statements that are proved in Section 15.8. First, we note that the average of $f(t)$ vanishes. To estimate the error induced by using this average, we find, interchanging summation and integration,

$$\int_0^t f(s)ds = \sum_0^{\infty} \frac{1}{(2n+1)^2} \int_0^t \sin\left(\frac{s}{2n+1}\right) ds$$

$$= \sum_0^{\infty} \frac{1}{2n+1}\left(1 - \cos\left(\frac{t}{2n+1}\right)\right) = \sum_0^{\infty} \frac{2}{2n+1} \sin^2 \frac{t}{2(2n+1)}.$$

The series converges for each fixed value of t but is not bounded for all time. In Section 15.8 we estimate

$$\sup_{0 \le \varepsilon t \le C} \int_0^t f(s)ds = O(|\ln \varepsilon|),$$

and so the error when using the averaged equation is $O(\varepsilon|\ln \varepsilon|)$.

The error estimate in this example is typical for almost-periodic functions with a Fourier spectrum that accumulates near zero. To find the actual error estimates is in general not so easy, as we have to estimate growth rates of diverging series.

11.3 Periodic Solutions

When averaging the Van der Pol equation in Example 11.3, we obtained for the amplitude r the approximate equation

$$\dot{r}_a = \varepsilon \frac{r_a}{2}\left(1 - \frac{1}{4}r_a^2\right).$$

We noted that choosing $r_a = 0$ corresponds with the trivial solution and choosing $r_a = 2$ yields the periodic solution.

If we find a stationary solution r_0 of an (approximate) amplitude equation $\dot{r}_a = f(r_a)$, we know that $r(t) = r_0 + O(\varepsilon)$ on the timescale $1/\varepsilon$. If we have additional information about the existence of corresponding periodic solutions, we can use the stationary solutions (critical points of f) to find these solutions. This shows the importance of connecting quantitative information with qualitative results. We consider a simple, artificial example.

Example 11.14

The one-dimensional equation

$$\dot{x} = \varepsilon(1 + \sin t - x)$$

can be averaged to produce

$$\dot{y} = \varepsilon(1 - y).$$

The exact solution is

$$x(t) = 1 - \frac{\varepsilon}{1 + \varepsilon^2} \cos t + \frac{\varepsilon^2}{1 + \varepsilon^2} \sin t + ce^{-\varepsilon t},$$

where c is determined by the initial condition. The last term vanishes for $t \to \infty$ and what remains is a 2π-periodic solution in a neighbourhood of $x = 1$. At the same time, $y = 1$ is a stationary solution of the averaged equation.

For slowly varying systems in the standard form, these results are part of a very general theorem:

Theorem 11.4

(Bogoliubov and Mitropolsky)
Consider the n-dimensional system $\dot{x} = \varepsilon f(t, x)$ with $f(t, x)$ T-periodic and the averaged system $\dot{y} = \varepsilon f^0(y)$; y_0 is a stationary solution of the averaged equation $(f^0(y_0) = 0)$. If

- $f(t, x)$ is a smooth vector field;
- for the Jacobian in y_0 we have

$$\left| \frac{\partial f^0(y)}{\partial y} \Big|_{y=y_0} \right| \neq 0,$$

then a T-periodic solution of the equation $\dot{x} = \varepsilon f(t, x)$ exists in an ε-neighbourhood of $x = y_0$.

We can even establish the stability of the periodic solution, as it matches exactly the stability of the stationary solution of the averaged equation. This reduces the stability problem of the periodic solution to determining the eigenvalues of a matrix.

Remark

We have noted before that averaging an autonomous equation or system eliminates one angle or, more generally, reduces the system with one dimension.
The implication is that a stationary solution obtained from an autonomous
equation always has one eigenvalue zero and so $J = 0$. This is for instance
the case for the Van der Pol equation. In accordance with this observation,
we have an important modification of the existence requirement for periodic
solutions in the case of n-dimensional autonomous equations: the Jacobian in
y_0 must have rank $n - 1$.

Example 11.15

Consider again the forced Duffing quation (11.4)

$$\ddot{x} + \varepsilon\mu\dot{x} + \varepsilon\gamma x^3 + x = \varepsilon h \cos \omega t,$$

with $\mu \geq 0, h \neq 0, \omega^{-2} = 1 - \beta_0\varepsilon$. This is a fundamental example representing
nonlinear oscillation with damping and forcing. Transforming $\tau = \omega t$ we found
in amplitude-phase variables the averaged system

$$r_a' = -\frac{1}{2}\varepsilon(\mu r_a + h \sin \psi_a),$$

$$\psi_a' = -\frac{1}{2}\varepsilon\left(\beta_0 - \frac{3}{4}\gamma r_a^2 + h\frac{\cos \psi_a}{r_a}\right).$$

Stationary solutions $r_a > 0$ satisfy the transcendental equations

$$\mu r_a + h \sin \psi_a = 0, \ \beta_0 - \frac{3}{4}\gamma r_a^2 + h\frac{\cos \psi_a}{r_a} = 0.$$

For the derivative of the right-hand side, leaving out the scalefactor $-\frac{1}{2}\varepsilon$, we
find

$$\begin{pmatrix} \mu & h \cos \psi_a \\ -\frac{3}{2}\gamma r_a - h\frac{\cos \psi_a}{r_a^2} & -h\frac{\sin \psi_a}{r_a} \end{pmatrix}.$$

Leaving out the scalefactor $\frac{1}{4}\varepsilon^2$, the Jacobian J becomes

$$J = \left(-\mu \sin \psi_a + \frac{3}{2}\gamma r_a^2 \cos \psi_a + h\frac{\cos^2 \psi_a}{r_a}\right)\frac{h}{r_a}.$$

For periodic solutions to exist, we have the requirement $J \neq 0$ when the
stationary solutions are substituted.

To be more explicit, we will look at the case of exact resonance $\beta_0 = 0$.
For the stationary solutions, we have

$$h \sin \psi_a = -\mu r_a, \ h \cos \psi_a = \frac{3}{4}\gamma r_a^3.$$

Using that $\cos^2 \psi_a + \sin^2 \psi_a = 1$, we find

$$\frac{9}{16}\gamma^2 r_a^6 + \mu^2 r_a^2 - h^2 = 0,$$

which always admits at least one positive solution for r_a. For the Jacobian, we have, eliminating ψ_a,

$$J = r_a \left(\mu + \frac{27}{16}\gamma^2 r_a^4 \right).$$

If μ and γ do not vanish simultaneously, we have $J > 0$, so a stationary solution $r_a > 0$ corresponds with a periodic solution of the original forced Duffing equation.

11.4 The Multiple-Timescales Method

It is time to discuss a very popular method and its relation with averaging. When we considered initial value problems in the context of averaging for an equation such as

$$\ddot{x} + x = \varepsilon f(x, \dot{x}),$$

we first introduced the standard form $\dot{y} = \varepsilon g(t, y)$ (with y two-dimensional) by an amplitude-phase transformation $x = r\cos(t + \psi)$, etc., or something related. After averaging, we have an equation of the form $\dot{y}_a = \varepsilon g^0(y_a)$. It is clear that we can absorb ε in time by putting $\tau = \varepsilon t$ so the approximation y_a will be a function of τ. Using again the transformation for the standard form, we then obtain an approximation

$$x(t) = x_a(t, \tau) + o(1) = x_a(t, \varepsilon t) + o(1).$$

Of course, similar reasoning holds for more general equations. If we have the n-dimensional equation $\dot{x} = f(t, x, \varepsilon)$, we suppose that we can solve the unperturbed problem $\dot{x} = f(t, x, 0)$, which produces a variation of constants transformation to a slowly varying system of the form $\dot{y} = \varepsilon g(t, y)$. Again the approximation of y will be a function of τ and the approximation of $x(t)$ will be a function of t and τ.

This experience with approximations of slowly varying systems suggests an elegant idea: introduce directly an approximation of the form $x(t) = x_a(t, \tau)$ into the original equation. As this means that we have two variables, the equations to be solved are partial differential equations. As it turns out, these are easy to solve. We explore this for the Van der Pol equation.

Example 11.16
A multiple-timescale analysis of the Van der Pol equation

$$\ddot{x} + x = \varepsilon \dot{x}(1 - x^2)$$

means that we will look for an approximation of the form

$$x(t) = x_0(t, \tau) + \varepsilon x_1(t, \tau) + \varepsilon^2 \cdots .$$

To replace $\dot{x}$ and $\ddot{x}$, we differentiate

$$\dot{x} = \frac{\partial x}{\partial t} + \varepsilon \frac{\partial x}{\partial \tau}, \ \ddot{x} = \frac{\partial^2 x}{\partial t^2} + 2\varepsilon \frac{\partial^2 x}{\partial t \partial \tau} + \varepsilon^2 \frac{\partial^2 x}{\partial \tau^2}.$$

We substitute this into the equation and collect terms of equal size in ε to find at lowest order

$$\frac{\partial^2 x_0}{\partial t^2} + x_0 = 0.$$

The variable τ is absent in the equation at this order; integration yields

$$x_0(t, \tau) = r(\tau) \cos(t + \psi(\tau)).$$

To determine r and ψ, we have to look at the next order. The equation at $O(\varepsilon)$ is

$$\frac{\partial^2 x_1}{\partial t^2} + 2 \frac{\partial^2 x_0}{\partial t \partial \tau} + x_1 = \frac{\partial x_0}{\partial t}(1 - x_0^2)$$

or

$$\frac{\partial^2 x_1}{\partial t^2} + x_1 = 2 \frac{dr}{d\tau} \sin(t + \psi) + 2r \cos(t + \psi) \frac{d\psi}{d\tau} - r \sin(t + \psi)(1 - r^2 \cos^2(t + \psi)).$$

Again, at this order, we have no derivatives with respect to τ, so we may treat this as an ordinary differential equation with "constants" depending on τ. The solution will be given by

$$x_1(t) = R(\tau) \cos(t + \phi(\tau)) + \int_0^t \left(2 \frac{dr}{d\tau} \sin(s + \psi) + 2r \cos(s + \psi) \frac{d\psi}{d\tau} \right.$$
$$\left. -r \sin(s + \psi)(1 - r^2 \cos^2(s + \psi)) \right) \sin(t - s) ds.$$

At this point, we may wonder whether the method leads to a full determination of terms, as we have introduced new unknown functions of τ without determining the first ones. The answer to this is that our approximations are expected to be bounded on the timescale $1/\varepsilon$. This inspires us to apply the *secularity conditions* of the Poincaré-Lindstedt method discussed in Chapter 10. Terms resonating with $\sin t$ and $\cos t$ have to vanish from the integral. From the coefficient of $\sin(t + \psi)$, we find that necessarily

$$2 \frac{dr}{d\tau} - r + \frac{1}{4} r^3 = 0.$$

From the coefficient of $\cos(t + \psi)$,

$$2r \frac{d\psi}{d\tau} = 0.$$

We conclude that $\psi(\tau)$ is constant, for instance 0. We can solve the equation for r,

$$r(\tau) = \frac{r_0 e^{\frac{1}{2}\tau}}{(1 + \frac{r_0^2}{4}(e^\tau - 1))^{\frac{1}{2}}},$$

which is the same result we obtained by averaging in Example 11.3. It is also clear that on choosing $r_0 = 2$ we obtain the same approximation for the periodic solution as before.

To show that *in general* first-order averaging corresponds with a first-order multiple-timescale calculation is quite easy, We assume again that we can solve an unperturbed ($\varepsilon = 0$) problem to obtain a slowly varying system of the form

$$\dot{x} = \varepsilon f(t, x)$$

with f T-periodic. A multiple-timescale expansion will be of the form

$$x(t) = x_0(t, \tau) + \varepsilon x_1(t, \tau) + \varepsilon^2 \cdots,$$

where we can take $x_0(0,0) = x(0)$, $x_1(0,0) = 0$, etc. Differentiation transforms to

$$\frac{d}{dt} = \frac{\partial}{\partial t} + \varepsilon \frac{\partial}{\partial \tau}.$$

Substituting the expansion into the equation and transforming yields

$$\frac{\partial x_0}{\partial t} + \varepsilon \frac{\partial x_0}{\partial \tau} + \varepsilon \frac{\partial x_1}{\partial t} + \cdots = \varepsilon f(t, x_0 + \varepsilon x_1 + \cdots).$$

We Taylor expand

$$f(t, x_0 + \varepsilon x_1 + \cdots) = f(t, x_0) + \varepsilon Df(t, x_0)x_1 + \varepsilon^2 \cdots$$

with Df an $n \times n$ matrix, the first derivative of the n-dimensional vector field f with respect to the n-dimensional variable x.

Collecting terms of order 1 and order ε, we find

$$\frac{\partial x_0}{\partial t} = 0,$$

$$\frac{\partial x_1}{\partial t} = -\frac{\partial x_0}{\partial \tau} + f(t, x_0).$$

From the first equation, we find

$$x_0(t, \tau) = A(\tau), \ A(0) = x(0),$$

in which the function $A(\tau)$ is still unknown. It is also easy to integrate the second equation; we find

$$x_1(t, \tau) = \int_0^t \left(-\frac{dA}{d\tau} + f(s, A(\tau)) \right) ds + B(\tau).$$

$B(\tau)$ is again an unknown function; to fit the initial condition, we have $B(0) = 0$. We apply again the secularity condition by requiring that the integral be bounded. This is the case if the Fourier expansion of the integrand has no constant term or its average is zero,

$$\frac{1}{T} \int_0^T \left(-\frac{dA}{d\tau} + f(s, A(\tau)) \right) ds = 0,$$

which is the case if we have

$$\frac{dA}{d\tau} = \frac{1}{T} \int_0^T f(s, A(\tau)) ds,$$

With this requirement and $A(0) = x(0)$, we have determined $A(\tau)$. The approximation obtained in this way is exactly the same as the one obtained by averaging. At the same time, we observe that multiple-timescales analysis produces, at least to first order, an $O(\varepsilon)$ approximation on the timescale $1/\varepsilon$ if $f(t, x)$ is periodic in t.

Conclusion
The use of timescales t and $\tau = \varepsilon t$ helps us to develop an elegant and intuitively clear perturbation method. It is understandable that the method is rather popular. Natural questions are whether we can develop higher-order approximations in the same way and whether this requires more than two timescales. The answer is that this is possible, but a fundamental problem is that such expansions are based on an a priori assumption regarding the form of the timescales. This is a very serious objection, as in open research problems we do not know which timescales are prominent. We will return to this question in the subsequent chapter, where we will see that at the next order unexpected timescales may appear.

11.5 Guide to the Literature

The method of averaging goes back to the eighteenth century, in particular to Lagrange and Laplace; see Lagrange (1788), which contains a clear description of the method. It took a long time for the first proof of asymptotic validity to appear: in 1928 in a paper by Fatou. An influential book has been the monograph by Bogoliubov and Mitropolsky (1961); in this book one also finds general averaging and the first formulation and proof of the existence and approximation of periodic solutions by averaging (Section 11.3). A number of new results and proofs can be found in the monograph by Sanders and Verhulst (1985); see this book for more historical details and see also the introductions in Murdock (1999) and Verhulst (2000).

The idea of "elimination of secular terms" arose in the early work of Lagrange and Laplace. It was taken up again by Poincaré (1893) in his treatment of periodic solutions; for this idea Poincaré refers to Lagrange. Nowadays, it plays an essential part in all asymptotic methods for long-time dynamics.

The method of multiple-timescales was invented by Kryloff and Bogoliubov (1935). Another paper in this school was by Kuzmak (1959), but they did not pursue the idea as "they thought multiple timing was not a good method" (Mitropolsky, 1981). The method of multiple-timescales was independently discovered by Kevorkian (1961), Cochran (1962), and Mahony (1962) and promoted by Nayfeh (1973). Using the method, Kevorkian ingeniously solved a number of difficult problems. We shall discuss the method again in the next chapter.

11.6 Exercises

Exercise 11.1 An oscillator with nonlinear damping is described by

$$\ddot{x} + 2\varepsilon\dot{x} + \varepsilon a\dot{x}^3 + x = 0.$$

If $a > 0$, we call the nonlinear damping progressive. Compute an asymptotic approximation of the solutions valid on $1/\varepsilon$.

Exercise 11.2 The Duffing equation is an example of the more general (potential) equation

$$\ddot{x} + x + \varepsilon f(x) = 0$$

with $f(x)$ a function that has a convergent Taylor series near $x = 0$. Compute an asymptotic approximation of the solutions valid on $1/\varepsilon$.

Exercise 11.3 An interesting extension of Example 11.10 is the nonlinear Mathieu equation

$$\ddot{x} + (1 + 2\varepsilon\cos 2t)x + \varepsilon a x^3 = 0$$

with a a constant. Study this equation by computing asymptotic approximations.

Exercise 11.4 Consider again the equation in Example 11.10:

$$\ddot{x} + (1 + 2\varepsilon\cos 2t)x = \varepsilon f(t)$$

with $f(t) = \cos(\frac{2}{\sqrt{3}}t)$. Choose initial values $x(0) = 1, \dot{x}(0) = 0$, and compute an asymptotic approximation of the solution.

Exercise 11.5 Consider the initial value problem

$$\dot{x} = \varepsilon x \sin t^2, \ x(0) = 1.$$

Compute an approximation valid on the timescale $1/\varepsilon$. Give an error estimate.

Exercise 11.6 Consider the initial value problem

$$\dot{x} = \varepsilon x \sin\left(\frac{1}{1+t}\right), \quad x(0) = 1.$$

Compute an approximation valid on the timescale $1/\varepsilon$. Give an error estimate.

Exercise 11.7 The Rayleigh equation is

$$\ddot{x} + x = \varepsilon \dot{x}(1 - \dot{x}^2).$$

Compute an asymptotic approximation of the periodic solution.

Exercise 11.8 Using averaging, find 2π-periodic solutions of the system

$$\dot{x}_1 = \varepsilon x_1(2x_2 \cos^2 t + 2x_3 \sin^2 t - 2),$$
$$\dot{x}_2 = \varepsilon x_2(2x_3 \cos^2 t + 2x_1 \sin^2 t - 2),$$
$$\dot{x}_3 = \varepsilon x_3(2x_1 \cos^2 t + 2x_2 \sin^2 t - 2).$$

Exercise 11.9 Consider the equation

$$\ddot{x} + x = \varepsilon \dot{x}(1 - 3x^2 - 5\dot{x}^2).$$

Prove that the equation has a periodic solution, and compute an asymptotic approximation.

Exercise 11.10 Consider the forced Duffing equation from Example 11.15. Determine the stability of the periodic solution(s) obtained in the case $\beta_0 = 0$.

Exercise 11.11 Consider for the forced Duffing equation from Example 11.15 the special case of no friction ($\mu = 0$) and the cases of a softening spring ($\gamma < 0$) and a hardening spring ($\gamma > 0$). Do periodic solutions exist? If they do, determine their stability.

12

Advanced Averaging

In this chapter, we will look at a number of useful and important extensions of averaging. Also, at the end there will again be a discussion of timescales.

12.1 Averaging over an Angle

In oscillations, we can identify amplitudes and angles and also other quantities such as energy, angular momentum, etc. An angle can often be used as a time-like quantity that can be used for averaging. A simple example is the Duffing equation in Example 11.5.

Example 12.1
Consider the Duffing equation with positive damping and basic (positive) frequency ω,

$$\ddot{x} + \varepsilon \mu \dot{x} + \omega^2 x + \varepsilon \gamma x^3 = 0.$$

Using amplitude-phase variables r, ψ, we found slowly varying equations that could be averaged. Instead, we put $\dot{x} = \omega y$ and propose an amplitude-angle transformation of the form

$$x(t) = r(t) \sin \phi(t), \quad y(t) = r(t) \cos \phi(t). \tag{12.1}$$

When transforming $x, \dot{x}(y) \rightarrow r, \phi$, we use the differential equation and the relation between $\dot{x}$ and y to find

$$\dot{r} = \varepsilon \frac{\cos \phi}{\omega}(-\mu \omega r \cos \phi - \gamma r^3 \sin^3 \phi),$$

$$\dot{\phi} = \omega - \varepsilon \frac{\sin \phi}{\omega r}(-\mu \omega r \cos \phi - \gamma r^3 \sin^3 \phi).$$

These equations are 2π-periodic in ϕ, and the equation for r looks like the standard form for averaging except that time is not explicitly present. We average the first equation over ϕ to find

$$\dot{r}_a = -\frac{1}{2}\varepsilon\mu r_a$$

with solution $r_a(t) = r(0)\exp(-\frac{1}{2}\varepsilon t)$.

What does $r_a(t)$ represent? We have averaged over ϕ but are interested in the behaviour with time. Below we will formulate a theorem that states that $r_a(t)$ is an $O(\varepsilon)$ approximation of $r(t)$ on the timescale $1/\varepsilon$ if ω is bounded away from zero ($\omega = O_s(1)$).

The last condition is quite natural. In averaging over ϕ, we treated ϕ as a time-like variable. This would not hold in an asymptotic sense if for instance ω is $O(\varepsilon)$. In such a case, a different approach is needed.

We will now look at a classical example, a linear oscillator with slowly varying (prescribed) frequency.

Example 12.2
Consider the equation

$$\ddot{x} + \omega^2(\varepsilon t)x = 0.$$

We put $\dot{x} = \omega(\varepsilon t)y, \tau = \varepsilon t$, and use transformation (12.1) to find

$$\dot{r} = -\varepsilon\frac{1}{\omega(\tau)}\frac{d\omega}{d\tau}r\cos^2\phi,$$

$$\dot{\phi} = \omega(\tau) + \varepsilon\frac{1}{\omega(\tau)}\frac{d\omega}{d\tau}\sin\phi\cos\phi,$$

$$\dot{\tau} = \varepsilon.$$

If we assume that $0 < a < \omega(\tau) < b$ (with a, b constants independent of ε), we have a three-dimensional system periodic in ϕ with two equations slowly varying. The function $\omega(\tau)$ is smooth and its derivative is bounded.

Averaging the slowly varying equations over ϕ, we obtain

$$\dot{r}_a = -\varepsilon\frac{1}{2\omega(\tau)}\frac{d\omega}{d\tau}r_a,$$

$$\dot{\tau} = \varepsilon.$$

We write τ instead of τ_a, as the equation for τ does not change. From this system, we get

$$\frac{dr_a}{d\tau} = -\frac{1}{2\omega(\tau)}\frac{d\omega}{d\tau}r_a,$$

which can be integrated to find

$$r_a(\tau) = \frac{r(0)\sqrt{\omega(0)}}{\sqrt{\omega(\tau)}}.$$

Note that the quantity $r_a(\varepsilon t)\sqrt{\omega(\varepsilon t)}$ is conserved in time. The same techniques can be applied to equations of the form

$$\ddot{x} + \varepsilon\mu\dot{x} + \omega^2(\varepsilon t)x + \varepsilon f(x) = 0.$$

The application is straightforward and we leave this as an exercise.

We now summarise the result behind these calculations:

Theorem 12.1
Consider the system

$$\dot{x} = \varepsilon X(\phi, x) + \varepsilon^2 \cdots,$$
$$\dot{\phi} = \omega(x) + \varepsilon \cdots,$$

where the dots stand for higher-order terms. We assume that $x \in D \subset \mathbb{R}^n$, ϕ is one-dimensional, and $0 < \phi < 2\pi$. Averaging over the angle ϕ produces

$$X^0(y) = \int_0^{2\pi} X(\phi, y)d\phi.$$

Assuming that

- the right-hand sides of the equations for x and ϕ are smooth;
- the solution of

$$\dot{y} = \varepsilon X^0(y), \ y(0) = x(0)$$

  is contained in an interior subset of D;
- $\omega(x)$ is bounded away from zero by a constant independent of ε,

then $x(t) - y(t) = O(\varepsilon)$ on the timescale $1/\varepsilon$.

What happens if the frequency in Example 12.2 is not bounded away from zero? We consider a well-known example that can be integrated exactly.

Example 12.3
A spring with a stiffness that wears out in time is described by

$$\ddot{x} + e^{-\varepsilon t}x = 0, \ x(0) = 0, \dot{x}(0) = 1.$$

In the analysis of Example 12.2, we considered such an equation with the condition that the variable frequency $\omega(\varepsilon t)$ is bounded away from zero. On the *timescale* $1/\varepsilon$, this is still the case, so we conclude from Example 12.2 that the amplitude $r(t)$ on this timescale is approximated by

$$r_a(\varepsilon t) = \frac{r(0)\sqrt{\omega(0)}}{\sqrt{\omega(\varepsilon t)}}.$$

With $r(0) = 1, \omega(\varepsilon t) = e^{-\frac{1}{2}\varepsilon t}$, we find

$$r(t) = e^{\frac{1}{4}\varepsilon t} + O(\varepsilon)$$

on the timescale $1/\varepsilon$. Of course, this result may not carry through on longer timescales.

The exact solution can be obtained by transforming $s = 2\varepsilon^{-1}e^{-\varepsilon t/2}$, which leads to the Bessel equation of index zero and the solution

$$x(t) = \frac{\pi}{\varepsilon}Y_0\left(\frac{2}{\varepsilon}\right)J_0\left(\frac{2}{\varepsilon}e^{-\frac{1}{2}\varepsilon t}\right) - \frac{\pi}{\varepsilon}J_0\left(\frac{2}{\varepsilon}\right)Y_0\left(\frac{2}{\varepsilon}e^{-\frac{1}{2}\varepsilon t}\right),$$

where J_0, Y_0 are the well-known Bessel functions of index zero. On the timescale $1/\varepsilon$, all of the arguments are large and we can use the known behaviour of Bessel functions for arguments tending to infinity. We find that the solution is approximated by

$$x(t) = e^{\frac{\varepsilon t}{4}}\sin\left(\frac{2}{\varepsilon}(1 - e^{-\varepsilon t/2})\right) + O(\varepsilon).$$

Naturally, this agrees with the results obtained by averaging. If time runs beyond $1/\varepsilon$, the arguments of the Bessel functions J_0 and Y_0 tend to zero. Using the corresponding known expansions, we find that the solution behaves as $c_1 t + c_2$ with

$$c_1 = \sqrt{\frac{\varepsilon}{\pi}}\cos\left(\frac{2}{\varepsilon} - \frac{\pi}{4}\right), \ c_2 = \frac{\pi}{\varepsilon}\left(\sin\left(\frac{2}{\varepsilon} - \frac{\pi}{4}\right) - \frac{2}{\pi}\cos\left(\frac{2}{\varepsilon} - \frac{\pi}{4}\right)\left(\ln\frac{1}{\varepsilon} + \gamma\right)\right),$$

where γ is Euler's constant. Interestingly, on timescales larger than $1/\varepsilon$, there are no oscillations anymore, while the velocity c_2 becomes large with ε.

In the examples until now, we have introduced a different way of averaging but not a real improvement on elementary averaging as discussed in the preceding chapter. An improvement can arise when the equation for ϕ is rather intractable and we are satisfied with an approximation of the quantity x.

As we have seen, interesting problems arise when $\omega(x)$ is not bounded away from zero. In the next section, this will become a relevant issue in problems with more angles. To prepare for this, we consider an example from Arnold (1965).

Example 12.4
Consider the two scalar equations

$$\dot{x} = \varepsilon(1 - 2\cos\phi),$$
$$\dot{\phi} = x.$$

If $x(0) > 0$, independent of ε, $\omega(x) = x$ will remain bounded away from zero and averaging produces

$$\dot{y} = \varepsilon, \ y(0) = x(0),$$

so that $y(t) = x(0) + \varepsilon t$ and $x(t) - y(t) = O(\varepsilon)$ on the timescale $1/\varepsilon$. What happens if $x(0) \leq 0$? The system is conservative (compute the divergence or differentiate the equation for ϕ to eliminate x), and the original system has two stationary solutions (critical points of the right-hand side vector field): $(x, \phi) = (0, \frac{\pi}{3})$ and $(x, \phi) = (0, \frac{5\pi}{3})$. So for these solutions the expression $y(t) = x(0) + \varepsilon t$ is clearly not an approximation. However, this expression represents an approximation of $x(t)$ if $x(0) < 0$ until $x(t)$ enters a neighbourhood of $x = 0$. This is usually called the *resonance zone* and the set corresponding with $x = 0$ the *resonance manifold* for reasons that will become clear in the next section. A picture of the phase-plane of the solutions helps us to understand the dynamics (see Fig. 12.1). Starting away from $x = 0$, $x(t)$ increases as a linear function with small modulations. In the resonance zone, the solutions may be captured for all time or pass through the resonance zone.

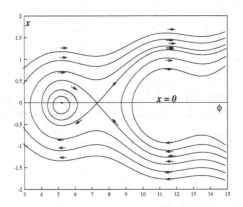

Fig. 12.1. Passage through resonance and capture into resonance in Example 12.4; the resonance zone is located near the resonance manifold $x = 0$.

This resonance zone is really a boundary layer in the sense we discussed in Chapter 4. We can see that by rescaling near $x = 0$: $\xi = x/\delta(\varepsilon)$. Transforming $x, \phi \to \xi, \phi$, the equations become

$$\dot{\xi} = \frac{\varepsilon}{\delta(\varepsilon)}(1 - 2\cos\phi),$$

$$\dot{\phi} = \delta(\varepsilon)\xi.$$

A significant degeneration arises when the terms on the right-hand side are of the same order or

$$\frac{\varepsilon}{\delta(\varepsilon)} = \delta(\varepsilon).$$

We conclude that $\delta(\varepsilon) = \sqrt{\varepsilon}$, which is the size of the resonance zone.

This example of a conservative equation is more typical than it seems. Consider for instance the two much more general scalar equations

$$\dot{x} = \varepsilon f(\phi, x),$$
$$\dot{\phi} = g(x),$$

with f and g smooth functions and $f(\phi, x)$ 2π-periodic in ϕ. Averaging over ϕ is possible outside the zeros of g. Suppose $g(a) = 0$ and rescale

$$\xi = \frac{x - a}{\delta(\varepsilon)}.$$

The equations become

$$\delta(\varepsilon)\dot{\xi} = \varepsilon f(\phi, a + \delta(\varepsilon)\xi),$$
$$\dot{\phi} = g(a + \delta(\varepsilon)\xi).$$

Expanding, we find

$$\delta(\varepsilon)\dot{\xi} = \varepsilon f(\phi, a) + O(\delta(\varepsilon)\varepsilon),$$
$$\dot{\phi} = \delta(\varepsilon)g'(a)\xi + O(\delta^2(\varepsilon)).$$

Again, a significant degeneration arises if we choose $\delta(\varepsilon) = \sqrt{\varepsilon}$, which is the size of the resonance zone. The equations in the resonance zone are to first order

$$\dot{\xi} = \sqrt{\varepsilon}f(\phi, a),$$
$$\dot{\phi} = \sqrt{\varepsilon}g'(a)\xi,$$

which is again a conservative system.

12.2 Averaging over more Angles

When more angles are present, many subtle problems and interesting phenomena arise. We start with a few simple examples.

Example 12.5
Consider first the system

$$\dot{x} = \varepsilon x(\cos\phi_1 + \cos\phi_2),$$
$$\dot{\phi}_1 = 1,$$
$$\dot{\phi}_2 = 2.$$

We average the equation for x over ϕ_1 and ϕ_2, which produces the averaged equation $\dot{x}_a = 0$ so that $x_a(t) = x(0)$. The solution $x(t)$ can easily be obtained and reads

$$x(t) = x(0)e^{\varepsilon(\sin(t+\phi_1(0)) + \frac{1}{2}\sin(2t+\phi_2(0)))},$$

so $x_a(t) = x(0)$ is an approximation and this correct answer seems quite natural, as the right-hand sides of the angle equations do not vanish.

We modify the equations slightly in the next example.

Example 12.6
Consider the system

$$\dot{x} = \varepsilon x(\cos \phi_1 + \cos \phi_2 + \cos(2\phi_1 - \phi_2)),$$
$$\dot{\phi}_1 = 1,$$
$$\dot{\phi}_2 = 2.$$

We average the equation for x over ϕ_1 and ϕ_2, which produces again the averaged equation $\dot{x}_a = 0$ so that $x_a(t) = x(0)$. Also, the solution $x(t)$ can easily be obtained and reads

$$x(t) = x(0)e^{\varepsilon(\sin(t+\phi_1(0))+\frac{1}{2}\sin(2t+\phi_2(0))+t\cos(2\phi_1(0)-\phi_2(0)))},$$

so $x_a(t) = x(0)$ is in general not an approximation valid on the timescale $1/\varepsilon$. What went wrong?

The right-hand side of the equation for x actually contains three angle combinations, ϕ_1, ϕ_2, and $2\phi_1 - \phi_2$. Averaging should take place over three angles instead of two. Adding formally the (dependent) equation for this third angle, we find that the right-hand side vanishes. The experience from the preceding section tells us that we might expect trouble because of this resonance.

Note that this example is rather extreme in its simplicity. The angles vary with a constant rate so that once we have resonance we have for all values of x a resonance manifold, it fills up the whole x-space. In general, the angles do not vary with a constant rate and behave more as in the following modification.

Example 12.7
Consider the system

$$\dot{x} = \varepsilon x(\cos \phi_1 + \cos \phi_2 + \cos(2\phi_1 - \phi_2)),$$
$$\dot{\phi}_1 = x,$$
$$\dot{\phi}_2 = 2.$$

We have still the same three angle combinations in the equation for x as in the preceding example. Resonance can be expected if $\dot{\phi}_1 = 0$, $\dot{\phi}_2 = 0$, or $2\dot{\phi}_1 - \dot{\phi}_2 = 2(x - 1) = 0$. This leads to resonance if $x = 0$ and if $x = 1$ with corresponding resonance manifolds. Outside the resonance zones around these manifolds, we can average over the three angles to find the approximation $x_a(t) = x(0)$. How do we visualise the flow?

Outside the resonance manifolds $x = 0$ and $x = 1$, the variable x is nearly constant. To see what happens for instance in the resonance zone near $x = 1$, we put $\psi = 2\phi_1 - \phi_2$ and rescale,

$$\xi = \frac{x - 1}{\delta(\varepsilon)}.$$

Introducing this into the equations and expanding, we obtain

$$\delta(\varepsilon)\dot{\xi} = \varepsilon(\cos\phi_1 + \cos\phi_2 + \cos\psi) + O(\varepsilon\delta(\varepsilon)),$$
$$\dot{\psi} = 2\delta(\varepsilon)\xi,$$
$$\dot{\phi}_1 = 1 + O(\delta(\varepsilon)),$$
$$\dot{\phi}_2 = 2.$$

The three angles are dependent and can be replaced by two angles, for instance ϕ_1 and ψ, but this makes little difference in the outcome. A significant degeneration arises on choosing $\delta(\varepsilon) = \sqrt{\varepsilon}$, which is the size of the resonance zone. We find locally

$$\dot{\xi} = \sqrt{\varepsilon}(\cos\phi_1 + \cos\phi_2 + \cos\psi) + O(\varepsilon),$$
$$\dot{\psi} = 2\sqrt{\varepsilon}\xi,$$
$$\dot{\phi}_1 = 1 + O(\sqrt{\varepsilon}),$$
$$\dot{\phi}_2 = 2.$$

The system we obtained has two slowly varying variables, ξ and ψ, and two angles with $O(1)$ variation, so we can average over ϕ_1 and ϕ_2 to obtain equations for the first-order approximations ξ_a and ψ_a:

$$\dot{\xi}_a = \sqrt{\varepsilon}\cos\psi_a,$$
$$\dot{\psi}_a = 2\sqrt{\varepsilon}\xi_a.$$

Differentiation yields that ψ_a satisfies the (conservative) pendulum equation

$$\ddot{\psi}_a - 2\varepsilon\cos\psi_a = 0,$$

which describes oscillatory motion in the resonance zone. It is remarkable that we have found a conservative equation at first order, which is of course very sensitive to perturbations. The original system of equations is not conservative, so this suggests that we have to compute a second-order approximation to obtain a structurally stable result.

Before presenting a more general formulation, we consider the phenomenon of *resonance locking*.

Example 12.8
Consider the four-dimensional system

$$\dot{x}_1 = \varepsilon,$$
$$\dot{x}_2 = \varepsilon\cos(\phi_1 - \phi_2),$$
$$\dot{\phi}_1 = 2x_1,$$
$$\dot{\phi}_2 = x_1 + x_2.$$

There is one angle combination, $\phi_1 - \phi_2$, and we expect resonance if $\dot{\phi}_1 - \dot{\phi}_2 = x_1 - x_2$ vanishes. Outside the resonance zone around the line $x_1 = x_2$ in the x_1, x_2-plane, we can average over the angles to obtain

$$\dot{x}_{1a} = \varepsilon, \ \dot{x}_{2a} = 0,$$

so that $x_{1a} = \varepsilon t + x_1(0), x_{2a} = x_2(0)$. Inside the resonance zone, we can analyse the original equations by putting

$$x = x_1 - x_2, \ \psi = \phi_1 - \phi_2,$$

which leads to the reduced system

$$\dot{x} = \varepsilon(1 - \cos \psi),$$
$$\dot{\psi} = x.$$

Differentiation of the equation for ψ produces

$$\ddot{\psi} + \varepsilon \cos \psi = \varepsilon,$$

which is a forced pendulum equation.

Note that we have resonance locking as the solutions $x = 0$, $\cos \psi = 1$ are equilibrium solutions of the reduced system.

This is typical for many near-integrable Hamiltonian systems where we have in general an infinite number of resonance zones in which resonance locking can take place.

12.2.1 General Formulation of Resonance

We will now give a general formulation for the case of two or more angles. Consider the system

$$\dot{x} = \varepsilon X(\phi, x),$$
$$\dot{\phi} = \Omega(x),$$

with $x \in \mathbb{R}^n, \phi \in T^m$; T^m is the m-dimensional torus described by m angles. Suppose that the vector function X is periodic in the m angles ϕ and that we have the multiple (complex) Fourier expansion

$$X(\phi, x) = \sum_{k_1, \cdots, k_m = -\infty}^{+\infty} c_{k_1, \cdots, k_m}(x) e^{i(k_1 \phi_1 + k_2 \phi_2 + \cdots k_m \phi_m)}$$

with $(k_1, \cdots, k_m) \in \mathbb{Z}^m$. The resonance manifolds in $\mathbb{R}^n$ (x-space) are determined by the relations

$$k_1 \Omega_1(x) + \cdots k_m \Omega_m(x) = 0,$$

assuming that the Fourier coefficient c_k with $k = (k_1, \cdots, k_m)$ does not vanish.

In applications, there are usually order of magnitude variations in the Fourier coefficients so that we can neglect some. If the resonance manifolds do not fill up the whole x-space, we can average *outside* the resonance manifolds to obtain the equation for the approximation $x_a(t)$,

$$\dot{x}_a = \varepsilon c_{0,\cdots,0}(x_a).$$

One can prove that outside the resonance manifolds, assuming that $x(0) = x_a(0)$, we have the estimate

$$x(t) - x_a(t) = O(\varepsilon) \text{ on the timescale } 1/\varepsilon.$$

12.2.2 Nonautonomous Equations

In practice, it happens quite often that time t enters explicitly into the equations. Consider the system

$$\dot{x} = \varepsilon X(\phi, t, x),$$
$$\dot{\phi} = \Omega(x),$$

with the vector function X periodic in t. The scaling to the same period of angles and time is important to make the variations comparable. Suppose that ϕ is m-dimensional; put

$$\phi_{m+1} = t, \ \dot{\phi}_{m+1} = 1,$$

and consider averaging over $m + 1$ angles.

This procedure is correct, but of course the dependence on t may produce many additional resonances.

Example 12.9
Consider the system

$$\dot{x} = \varepsilon X(\phi_1, t, x),$$
$$\dot{\phi}_1 = x,$$

with $X(\phi_1, t, x) = 2x \sin t \sin \phi_1$. Putting

$$\phi_2 = t, \ \dot{\phi}_2 = 1, X(\phi_1, t, x) = x(\cos(\phi_1 - t) - \cos(\phi_1 + t)),$$

we obtain the system with two angles,

$$\dot{x} = \varepsilon x(\cos(\phi_1 - \phi_2) - \cos(\phi_1 + \phi_2)),$$
$$\dot{\phi}_1 = x,$$
$$\dot{\phi}_2 = 1.$$

The resonance manifolds correspond with the zeros of the right-hand sides of $\dot{\phi}_1 - \dot{\phi}_2$ and $\dot{\phi}_1 + \dot{\phi}_2$, so we find $x = 1$ and $x = -1$. Outside these resonance zones, we can average over the angles to find $\dot{x}_a = 0$. In the resonance zones, the flow is again described by pendulum equations.

Note that the Fourier expansion of X contains only two terms. If there were an infinite number of terms, we would have resonance relations like

$$k_1 x + k_2 = 0,$$

which would produce resonance manifolds for an infinite number of rational values of x.

12.2.3 Passage through Resonance

We have seen an example of locking into resonance. Interesting phenomena arise when we have passage through resonance; in Example 12.11, we shall see an application.

To start with, we discuss an interesting example of *forced* passage through resonance that was constructed by Arnold (1965).

Example 12.10
A seemingly small variation of an earlier example is the system

$$\dot{x}_1 = \varepsilon,$$
$$\dot{x}_2 = \varepsilon \cos(\phi_1 - \phi_2),$$
$$\dot{\phi}_1 = x_1 + x_2,$$
$$\dot{\phi}_2 = x_2,$$

with initial values $x_1(0) = -a, x_2(0) = 1, \phi_1(0) = \phi_2(0) = 0$, with a a constant independent of ε. We have one angle combination that can lead to resonance (i.e., if $\dot{\phi}_1 - \dot{\phi}_2 = x_1 = 0$). Integration produces

$$x_1(t) = -a + \varepsilon t, \ \phi_1(t) - \phi_2(t) = -at + \frac{1}{2}\varepsilon t^2,$$

and so we have

$$. \ x_2(t) = 1 + \varepsilon \int_0^t \cos\left(-as + \frac{1}{2}\varepsilon s^2\right) ds.$$

If $x_1(0) > 0$ (a negative), the solution does not pass through the resonance zone around $x_1 = 0$. Partial integration produces that $x_2(t) = 1 + O(\varepsilon)$ for all time. If $x_1(0) < 0$ (a positive), we have forced crossing of the resonance zone. In the case $a = 0$, we start in the resonance zone and we can use the well-known integral

$$\int_0^\infty \cos s^2 ds = \frac{1}{2}\sqrt{\frac{\pi}{2}},$$

so if $a = 0$ we find the long-term effect of this crossing by taking the limit $t \to \infty$:

$$\lim_{t\to\infty} x_2(t) = 1 + \frac{1}{2}\sqrt{\pi\varepsilon}.$$

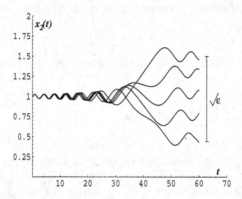

Fig. 12.2. Dispersion of orbits by passage through resonance; the five orbits started at $x_1(0) = -a$ with $a = 2, 2 \pm \varepsilon, 2 \pm \frac{1}{2}\varepsilon, \varepsilon = 0.1$.

More generally, we have to calculate or estimate

$$\lim_{t\to\infty} x_2(t) = 1 + \varepsilon \int_0^\infty \cos\left(-as + \frac{1}{2}s^2\right) ds.$$

Transforming

$$s = \sqrt{\frac{2}{\varepsilon}}u + \frac{a}{\varepsilon},$$

we find

$$\varepsilon \int_0^\infty \cos\left(-as + \frac{1}{2}s^2\right) ds = \sqrt{2\varepsilon} \int_{-\frac{a}{\sqrt{2\varepsilon}}}^\infty \cos\left(-\frac{a^2}{2\varepsilon} + u^2\right) du,$$

which can be split into

$$\sqrt{2\varepsilon}\left(\cos\left(\frac{a^2}{2\varepsilon}\right) \int_{-\frac{a}{\sqrt{2\varepsilon}}}^\infty \cos u^2\, du + \sin\left(\frac{a^2}{2\varepsilon}\right) \int_{-\frac{a}{\sqrt{2\varepsilon}}}^\infty \sin u^2\, du\right).$$

The two integrals equal $\sqrt{(\pi/2)} + o(1)$, so that we have an estimate for the long-term effect of passing through resonance,

$$\lim_{t\to\infty} x_2(t) = 1 + \sqrt{\pi\varepsilon}\left(\cos\left(\frac{a^2}{2\varepsilon}\right) + \sin\left(\frac{a^2}{2\varepsilon}\right) + o(1)\right).$$

This shows that the effect of passing through resonance is $O(\sqrt{\varepsilon})$ and remarkably that the solution displays sensitive dependence on the initial condition. Small changes of a produce relatively large changes of the solution. This "dispersion" of orbits is illustrated in Fig. 12.2.

Finally, we shall briefly discuss an application in mechanics that displays both passage through and (undesirable) capture into resonance.

Example 12.11
Consider a spring that can move in the vertical x direction on which a rotating wheel is mounted; the rotation angle is ϕ. The wheel has a small mass fixed on the edge that makes it slightly eccentric, a flywheel. The vertical displacement x and the rotation ϕ are determined by the equations

$$\ddot{x} + x = \varepsilon(-x^3 - \dot{x} + \dot{\phi}^2 \cos \phi) + O(\varepsilon^2),$$

$$\ddot{\phi} = \varepsilon\left(\frac{1}{4}(2 - \dot{\phi}) + (1 - x)\sin \phi\right) + O(\varepsilon^2).$$

See Evan-Ewanowski (1976) for the equations; we have added an appropriate scaling, assuming that the friction, the nonlinear restoring force, the eccentric mass, and several other forces are small.

To obtain a standard form suitable for averaging, we transform

$$x = r \sin \phi_2, \quad \dot{x} = r \cos \phi_2, \quad \phi = \phi_1, \quad \dot{\phi}_1 = \Omega,$$

with $r > 0, \Omega > 0$. This introduces two angles and two slowly varying quantities:

$$\dot{r} = \varepsilon \cos \phi_2(-r^3 \sin^3 \phi_2 - r \cos \phi_2 + \Omega^2 \cos \phi_1),$$

$$\dot{\Omega} = \varepsilon\left(\frac{1}{4}(2 - \Omega) + \sin \phi_1 - r \sin \phi_1 \sin \phi_2\right),$$

$$\dot{\phi}_1 = \Omega,$$

$$\dot{\phi}_2 = 1 + \varepsilon\left(r^2 \sin^4 \phi_2 + \frac{1}{2}\sin 2\phi_2 - \frac{\Omega^2}{r}\cos \phi_1 \sin \phi_2\right).$$

The $O(\varepsilon^2)$ terms have been omitted. Resonance zones exist if

$$m\Omega + n = 0, \quad m, n \in \mathbb{Z}.$$

In the equation for r and Ω to $O(\varepsilon)$, the angles are $\phi_1, \phi_2, \phi_1 + \phi_2, \phi_1 - \phi_2$. As ϕ_1 and ϕ_2 are monotonically increasing, the only resonance zone that can arise is when $\phi_1 - \phi_2 = 0$, which determines the resonance manifold $\Omega = 1$. Outside the resonance zone, a neighbourhood of $\Omega = 1$, we average over the angles to find the approximations r_a and Ω_a given by

$$\dot{r}_a = -\frac{1}{2}\varepsilon r_a,$$

$$\dot{\Omega}_a = \frac{1}{4}\varepsilon(2 - \Omega_a).$$

This is already an interesting result. Outside the resonance zone, $r(t) = r_a(t) + O(\varepsilon)$ will decrease exponentially with time; on the other hand, $\Omega(t)$ will tend to the value 2. If we start with $\Omega(0) < 1$, $\Omega(t)$ will after some time enter the resonance zone around $\Omega = 1$. How does this affect the dynamics? Will the system pass in some way through resonance or will it stay in the resonance zone, resulting in vertical oscillations that are undesirable for a mounted flywheel.

The way to answer these questions is to analyse what is going on in the resonance zone and find out whether there are attractors present. Following the analysis of localising into the resonance zone as before, we introduce the resonant combination angle $\psi = \phi_1' - \phi_2$ and the local variable

$$\omega = \frac{\Omega - 1}{\sqrt{\varepsilon}}.$$

Transforming the equations for r, Ω and the angles, the leading terms are $O(\sqrt{\varepsilon})$; we find

$$\dot{r} = \varepsilon \cdots,$$
$$\dot{\omega} = \sqrt{\varepsilon}\left(\frac{1}{4} + \sin\phi_1 - \frac{1}{2}r\cos\psi + \frac{1}{2}r\cos(2\phi_1 - \psi)\right) + \varepsilon \cdots,$$
$$\dot{\psi} = \sqrt{\varepsilon}\omega + \varepsilon \cdots,$$
$$\dot{\phi}_1 = 1 + \sqrt{\varepsilon}\omega.$$

We can average over the remaining angle ϕ_1; as the equation for r starts with $O(\varepsilon)$ terms, we have in the resonance zone that $r(t) = r(0) + O(\sqrt{\varepsilon})$. The equations for the approximations of ω and ψ are

$$\dot{\omega}_a = \sqrt{\varepsilon}\left(\frac{1}{4} - \frac{1}{2}r\cos\psi\right),$$
$$\dot{\psi} = \sqrt{\varepsilon}\omega.$$

By differentiation of the equation for ψ_a, we can write this as the pendulum equation

$$\ddot{\psi}_a + \frac{1}{2}\varepsilon r(0)\cos\psi_a = \frac{1}{4}\varepsilon.$$

The timescale of the dynamics is clearly $\sqrt{\varepsilon}t$; there are two equilibria, one a centre point and the other a saddle. They correspond with periodic solutions of the original system. The saddle is definitely unstable, and for the centre point we have to perform higher-order averaging, to $O(\varepsilon)$, to determine the stability. This analysis was carried out by Van den Broek (1988); see also Van den Broek and Verhulst (1987). The result is that by adding $O(\varepsilon)$ terms, the centre point in the resonance zone becomes an attracting focus so that the corresponding periodic solution is stable.

The implication is that for certain initial values the oscillator-flywheel might pass into resonance and stay there. Van den Broek (1988) identified three sets of initial values leading to capture into resonance.

1. **Remark**

   An extension of the theory of averaging over angles is possible for systems of the form

   $$\dot{x} = \varepsilon X(\phi, x),$$
   $$\dot{\phi} = \Omega(\phi, x).$$

   This generalisation complicates the calculations; see Section 5.4 in Sanders and Verhulst (1985).

2. **Remark**

   In the examples studied here, we obtained pendulum equations describing the flow in the resonance zones of the respective cases. This was observed by many authors in examples. In Section 11.7 of Verhulst (2000), it is shown that a first-order (in ε) computation in a resonance zone always leads to a conservative equation - often a pendulum equation or system of pendulum equations - describing the flow. This is remarkable, as the original system need not be conservative at all and the first-order result will probably change qualitatively under perturbation. The result stresses again the importance of second-order calculations in these cases.

12.3 Invariant Manifolds

An important problem is to determine invariant manifolds such as tori or cylinders in nonlinear equations. Consider a system such as

$$\dot{x} = f(x) + \varepsilon R(t, x, \varepsilon).$$

Suppose for instance that we have found an isolated torus T_a by first-order averaging. Does this manifold persist, slightly deformed as a torus T, when considering the original equation? Note that the original equation can be seen as a perturbation of the averaged equation, and the question can then be rephrased as the question of persistence of the torus T_a under perturbation. If the torus in the averaged equation is *normally hyperbolic*, the answer is affirmative. Normally hyperbolic means, loosely speaking, that the strength of the flow along the manifold is weaker than the rate of attraction to the manifold. We have used such results in Chapter 9 for Tikhonov-Fenichel problems. In many applications, however, the approximate manifold that one obtains is hyperbolic but not normally hyperbolic. In the Hamiltonian case, the tori arise in families and they will not even be hyperbolic.

We will look at different scenarios for the emergence of tori in some examples. A torus is generated by various independent rotational motions - at least two - and we shall find different timescales characterising these rotations.

12.3.1 Tori in the Dissipative Case

First, we look at cases where the branching off of tori is similar to the emergence of periodic solutions in the examples we have seen before. The theory of such questions was considered extensively by Bogoliubov and Mitropolsky (1961) and uses basically continuation of quasiperiodic motion under perturbations; for a summary and other references, see also Bogoliubov and Mitropolsky (1963). Another survey and new results can be found in Hale (1969); see the references therein.

There are many interesting open problems in this field, as the bifurcation theory of invariant manifolds is clearly even richer than for equilibria or periodic solutions. We present a few illustrative examples.

Example 12.12
Consider the system

$$\ddot{x} + x = \varepsilon\left(2x + 2\dot{x} - \frac{8}{3}\dot{x}^3 + y^2x^2 + \dot{y}^2x^2\right) + \varepsilon^2 R_1(x, y),$$

$$\ddot{y} + \omega^2 y = \varepsilon(\dot{y} - \dot{y}^3 + x^2y^2 + \dot{x}^2y^2) + \varepsilon^2 R_2(x, y),$$

where R_1 and R_2 are smooth functions. Introducing amplitude-phase coordinates by $x = r_1\cos(t + \psi_1), \dot{x} = -r_1\sin(t + \psi_1), y = r_2\cos(\omega t + \psi_2), \dot{y} = -\omega r_2\sin(\omega t + \psi_1)$, and after first-order averaging, we find, omitting the subscripts a, the system

$$\dot{r}_1 = \varepsilon r_1(1 - r_1^2), \dot{\psi}_1 = -\varepsilon,$$

$$\dot{r}_2 = \varepsilon\frac{r_2}{2}\left(1 - \frac{3}{4}r_2^2\right), \dot{\psi}_2 = 0.$$

The averaged equations contain a torus T_a in phase-space described by

$$x_a(t) = \cos(t - \varepsilon t + \psi_1(0)), \dot{x}_a(t) = -\sin(t - \varepsilon t + \psi_1(0)),$$

$$y_a(t) = \frac{2}{3}\sqrt{3}\cos(\omega t + \psi_2(0)), \dot{y}_a(t) = -\frac{2\omega}{3}\sqrt{3}\sin(\omega t + \psi_2(0)).$$

From linearisation of the averaged equations, it is clear that the torus is attracting: it is hyperbolic but not normally hyperbolic, as the motion along the torus has $O(1)$ speed and the attraction rate is $O(\varepsilon)$. If the ratio of $1 - \varepsilon$ and ω is rational, the torus T_a is filled up with periodic solutions. If the ratio is irrational, we have a quasiperiodic (two-frequency) flow over the torus. Remarkably enough, the theorems in the literature cited above tell us that in the original equations a torus T exists in an $O(\varepsilon)$ neighbourhood of T_a with the same stability properties. The torus is two-dimensional and the timescales of rotation are in both directions $O(1)$.

The next example was formulated as an exercise by Hale (1969). It is a rich problem and we cannot discuss all of its aspects.

Example 12.13
Consider the system

$$\ddot{x} + x = \varepsilon(1 - x^2 - ay^2)\dot{x},$$
$$\ddot{y} + \omega^2 y = \varepsilon(1 - y^2 - \alpha x^2)\dot{y},$$

with ε-independent positive constants a, α, ω. Using the same amplitude-phase transformation as in the preceding example, we find the slowly varying system

$$\dot{r}_1 = \varepsilon r_1 \sin(t + \psi_1)(1 - r_1^2 \cos^2(t + \psi_1) - ar_2^2 \cos^2(\omega t + \psi_2))\sin(t + \psi_1),$$
$$\dot{\psi}_1 = \varepsilon \cos(t + \psi_1)(1 - r_1^2 \cos^2(t + \psi_1) - ar_2^2 \cos^2(\omega t + \psi_2))\sin(t + \psi_1),$$
$$\dot{r}_2 = \varepsilon r_2 \sin(\omega t + \psi_2)(1 - r_2^2 \cos^2(\omega t + \psi_2) - ar_1^2 \cos^2(t + \psi_1))\sin(\omega t + \psi_2),$$
$$\dot{\psi}_2 = \varepsilon \cos(\omega t + \psi_2)(1 - r_2^2 \cos^2(\omega t + \psi_2) - ar_1^2 \cos^2(t + \psi_1))\sin(\omega t + \psi_2).$$

First-order averaging yields different results in two cases, $\omega \neq 1$ and $\omega = 1$. However, in all cases we have the following periodic solutions that are also present as solutions of the original equations.

Normal modes
Putting $r_1 = 0, r_2 = 2$ produces a normal mode periodic solution P_2 in the $y, \dot{y}$ coordinate plane. In the same way, we obtain a normal mode periodic solution P_1 in the $x, \dot{x}$ coordinate plane by putting $r_2 = 0, r_1 = 2$. These normal modes in the coordinate planes also exist in the original system. Their stability is studied by linearisation of the averaged equations.

$\omega \neq 1$
The averaged equations are (we omit again the subscript a)

$$\dot{r}_1 = \frac{\varepsilon}{2} r_1 \left(1 - \frac{1}{4}r_1^2 - \frac{1}{2}ar_2^2\right),$$
$$\dot{\psi}_1 = 0,$$
$$\dot{r}_2 = \frac{\varepsilon}{2} r_2 \left(1 - \frac{1}{4}r_2^2 - \frac{1}{2}ar_1^2\right),$$
$$\dot{\psi}_2 = 0.$$

Linearisation around the normal modes produces matrices with many zeros. In the case $r_1 = 0, r_2 = 2$ (P_2), we find for the derivative in the $y, \dot{y}$-plane $-\varepsilon$, which means attraction in this plane; in the $x, \dot{x}$-plane, we find for the derivative $\frac{1}{2}\varepsilon(1 - 2a)$, which means attraction if $a > \frac{1}{2}$ and repulsion if $a < \frac{1}{2}$. In the case of repulsion, we have instability of the $r_1 = 0, r_2 = 2$ normal mode. In the same way, we find instability of the $r_2 = 0, r_1 = 2$ normal mode P_1 in the $x, \dot{x}$-plane if $\alpha < \frac{1}{2}$.

We will now study the flow outside the coordinate planes. Stationary solutions outside the normal modes can be found if simultaneously

$$1 - \frac{1}{4}r_1^2 - \frac{1}{2}ar_2^2 = 0, \ 1 - \frac{1}{4}r_2^2 - \frac{1}{2}ar_1^2 = 0.$$

These relations correspond with quadrics in the r_1, r_2-plane. They intersect,

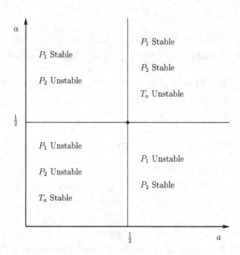

Fig. 12.3. Diagram of invariant manifolds, periodic solutions, and tori in Example 12.13; the point $a = \frac{1}{2}, \alpha = \frac{1}{2}$ is exceptional.

producing one solution, in the cases $a < \frac{1}{2}, \alpha < \frac{1}{2}$ and $a > \frac{1}{2}, \alpha > \frac{1}{2}$. In these two cases, the stationary solutions of the averaged equations correspond with a torus T_a. By linearisation of the averaged equations, we can establish the instability of the torus if $a > \frac{1}{2}, \alpha > \frac{1}{2}$. Stability cannot be deduced from the averaged system in these coordinates, as the matrix is singular. However, it is interesting to write in this case the original system in amplitude-angle coordinates and perform averaging over two angles.

Putting $x = r_1 \sin\phi_1, \dot{x} = r_1 \cos\phi_1, y = r_2 \sin(\omega\phi_2), \dot{y} = r_2\omega \cos(\omega\phi_2)$, we find when averaging over ϕ_1, ϕ_2

$$\dot{r}_1 = \frac{\varepsilon}{2}r_1\left(1 - \frac{1}{4}r_1^2 - \frac{1}{2}ar_2^2\right), \dot{\phi}_1 = 1 + O(\varepsilon),$$

$$\dot{r}_2 = \frac{\varepsilon}{2}r_2\left(1 - \frac{1}{4}r_2^2 - \frac{1}{2}ar_1^2\right), \dot{\phi}_2 = 1 + O(\varepsilon).$$

For the system in amplitude-angle coordinates, we can apply the theory cited above with the conclusion that the torus T_a, when it exists, corresponds with a torus T of the original equations. The torus is stable if $a < \frac{1}{2}, \alpha < \frac{1}{2}$ and unstable if $a > \frac{1}{2}, \alpha > \frac{1}{2}$; see Fig. 12.3. One can see immediately that a critical case is $a = \alpha = \frac{1}{2}$. In this case, the averaged equations contain an invariant sphere $r_1^2 + r_2^2 = 4$ in four-dimensional phase-space. For the question of persistence of this invariant sphere (slightly deformed) in the original equations, first-order averaging is sufficient. The sphere is a first-order approximation of a centre manifold. For the flow *on* the sphere, we need higher-order approximations. Outside this special point, we can cross the lines $a = \frac{1}{2}, \alpha = \frac{1}{2}$ to pass from a phase-space with two periodic solutions to a phase-space of two periodic solutions and a torus. When crossing these lines, one of the normal modes changes stability and produces the torus by a so-called branching bifurcation. The results are summarised in the diagram of Fig. 12.3. In this example, the torus again is two-dimensional and the timescales of rotation are in both directions $O(1)$.

$\omega = 1$

We consider now the special resonance case when the basic frequencies are equal. The averaged equations are

$$\dot{r}_1 = \frac{\varepsilon}{2}r_1\left(1 - \frac{1}{4}r_1^2 - \frac{1}{2}ar_2^2 + \frac{1}{4}ar_2^2\cos 2(\psi_1 - \psi_2)\right),$$

$$\dot{\psi}_1 = -\frac{\varepsilon}{8}ar_2^2\sin 2(\psi_1 - \psi_2),$$

$$\dot{r}_2 = \frac{\varepsilon}{2}r_2\left(1 - \frac{1}{4}r_2^2 - \frac{1}{2}\alpha r_1^2 + \frac{1}{4}\alpha r_1^2\cos 2(\psi_1 - \psi_2)\right),$$

$$\dot{\psi}_2 = \frac{\varepsilon}{8}\alpha r_1^2\sin 2(\psi_1 - \psi_2).$$

Using the combination angle $\chi = 2(\psi_1 - \psi_2)$, the equations become

$$\dot{r}_1 = \frac{\varepsilon}{2}r_1\left(1 - \frac{1}{4}r_1^2 - \frac{1}{2}ar_2^2 + \frac{1}{4}ar_2^2\cos\chi\right),$$

$$\dot{r}_2 = \frac{\varepsilon}{2}r_2\left(1 - \frac{1}{4}r_2^2 - \frac{1}{2}\alpha r_1^2 + \frac{1}{4}\alpha r_1^2\cos\chi\right),$$

$$\dot{\chi} = -\frac{\varepsilon}{4}(\alpha r_1^2 + ar_2^2)\sin\chi.$$

Apart from the normal modes P_1 and P_2, we find phase-locked periodic solutions in a general position by putting $\sin\chi = 0$. Introducing this condition into the equations for the amplitudes, we find that for these solutions to exist we have

$$1 - \frac{1}{4}r_1^2 - \frac{1}{2}ar_2^2 + \pm\frac{1}{4}ar_2^2 = 0,$$

$$1 - \frac{1}{4}r_2^2 - \frac{1}{2}\alpha r_1^2 + \pm\frac{1}{4}\alpha r_1^2 = 0.$$

The analysis of these equations and the corresponding stability is a lot of work but straightforward and is left to the reader.

As stated in the introduction to this section, when we start off with a normally hyperbolic torus, small perturbations will only deform the torus. A simple example follows.

Example 12.14

$$\ddot{x} + x = \mu(1 - x^2)\dot{x} + \varepsilon f(x, y),$$
$$\ddot{y} + \omega^2 y = \mu(1 - y^2)\dot{y} + \varepsilon g(x, y),$$

with ε-independent positive constant ω, μ a fixed large, positive number, and smooth perturbations f, g. Omitting the perturbations f, g, we have two normally hyperbolic relaxation oscillations. If ω is irrational, the combined oscillations attract to a torus filled with quasiperiodic motion. Adding the perturbations f, g cannot destroy this torus but only deforms it. Also, in this example the torus is two-dimensional but the timescales of rotation are in both directions determined by the timescales of relaxation oscillation (see Grasman, 1987) and so $O(1/\mu)$.

12.3.2 The Neimark-Sacker Bifurcation

Another important scenario for creating a torus arises from the Neimark-Sacker bifurcation. Suppose that we have obtained an averaged equation $\dot{x} = \varepsilon f(x, a)$ with dimension 3 or higher by variation of constants and subsequent averaging; a is a parameter or a set of parameters. We have discussed before that if this equation contains a hyperbolic critical point, the original equation contains a periodic solution. The first-order approximation of this periodic solution is characterised by the timescales t and εt.

Suppose now that by varying the parameter a a pair of eigenvalues of the critical point becomes purely imaginary. For this value of a, the averaged equation undergoes a Hopf bifurcation, producing a periodic solution of the averaged equation; the typical timescale of this periodic solution is εt, so the period will be $O(1/\varepsilon)$. As it branches off an existing periodic solution in the original equation, it will produce a torus, and the bifurcation has a different name: the Neimark-Sacker bifurcation. The result will be a two-dimensional torus that contains two-frequency oscillations, one on a timescale of order 1 and the other with timescale $O(1/\varepsilon)$. A typical example runs as follows.

Example 12.15
A special case of a system studied by Bakri et al. (2004) is

$$\ddot{x} + \varepsilon\kappa\dot{x} + (1 + \varepsilon\cos 2t)x + \varepsilon xy = 0,$$
$$\ddot{y} + \varepsilon\dot{y} + 4(1 + \varepsilon)y - \varepsilon x^2 = 0.$$

This is a system with parametric excitation and nonlinear coupling; κ is a positive damping coefficient that is independent of ε. Away from the coordinate planes, we may use amplitude-phase variables with $x = r_1\cos(t + \psi_1), \dot{x} = -r_1\sin(t + \psi_1), y = r_2\cos(2t + \psi_2), \dot{y} = -2r_2\sin(2t + \psi_1)$; after first-order averaging, we find, omitting the subscripts a, the system

$$\dot{r}_1 = \varepsilon r_1 \left(\frac{r_2}{4} \sin(2\psi_1 - \psi_2) + \frac{1}{4} \sin 2\psi_1 - \frac{1}{2}\kappa \right),$$

$$\dot{\psi}_1 = \varepsilon \left(\frac{r_2}{4} \cos(2\psi_1 - \psi_2) + \frac{1}{4} \cos 2\psi_1 \right),$$

$$\dot{r}_2 = \varepsilon \frac{r_2}{2} \left(\frac{r_1^2}{4r_2} \sin(2\psi_1 - \psi_2) - 1 \right),$$

$$\dot{\psi}_2 = \frac{\varepsilon}{2} \left(-\frac{r_1^2}{4r_2} \cos(2\psi_1 - \psi_2) + 2 \right).$$

Putting the right-hand sides equal to zero produces a nontrivial critical point corresponding with a periodic solution of the system for the amplitudes and phases and so a quasiperiodic solution of the original coupled system in x and y. We find for this critical point the relations

$$r_1^2 = 4\sqrt{5}r_2, \cos(2\psi_1 - \psi_2) = \frac{2}{\sqrt{5}}, \sin(2\psi_1 - \psi_2) = \frac{1}{\sqrt{5}}, r_1 = \sqrt{2\kappa + \sqrt{5 - 16\kappa^2}}.$$

This periodic solution exists if the damping coefficient is not too large: $0 \leq \kappa < \frac{\sqrt{5}}{4}$. Linearisation of the averaged equations at the critical point while using these relations produces the matrix

$$A = \begin{pmatrix} 0 & 0 & \frac{r_1}{4\sqrt{5}} & -\frac{r_1^3}{40} \\ 0 & -\kappa & \frac{1}{2\sqrt{5}} & \frac{r_1^2}{80} \\ \frac{r_1}{4\sqrt{5}} & \frac{r_1^2}{2\sqrt{5}} & -\frac{1}{2} & -\frac{r_1^2}{4\sqrt{5}} \\ -\frac{2}{r_1} & 1 & \frac{4\sqrt{5}}{r_1^2} & -\frac{1}{2} \end{pmatrix}.$$

Another condition for the existence of the periodic solution is that the critical point is hyperbolic (i.e., the eigenvalues of the matrix A have no real part zero). It is possible to express the eigenvalues explicitly in terms of κ by using a software package such as MATHEMATICA. However, the expressions are cumbersome. Hyperbolicity is the case if we start with values of κ just below $\frac{\sqrt{5}}{4} = 0.559$. Diminishing κ, we find when $\kappa = 0.546$ that the real part of two eigenvalues vanishes. This value corresponds with a Hopf bifurcation that produces a nonconstant periodic solution of the averaged equations. This in turn corresponds with a torus in the orginal equations (in x and y) by a Neimark-Sacker bifurcation. As stated before, the result will be a two-dimensional torus that contains two-frequency oscillations, one frequency on a timescale of order 1 and the other with timescale $O(1/\varepsilon)$.

12.3.3 Invariant Tori in the Hamiltonian Case

Integrability of a Hamiltonian system means, loosely speaking, that the system has at least as many independent first integrals as the number of degrees of freedom. As each degree of freedom corresponds with one position and one momentum, n degrees of freedom will mean a $2n$-dimensional system of differential equations of motion. With n independent integrals, the special structure of Hamiltonian systems will then invoke a complete foliation of phase-space into invariant manifolds.

Most Hamiltonian systems are nonintegrable, but in practice most are near an integrable system. A fundamental question is then how many of these invariant manifolds survive the nonintegrable perturbation. This question was solved around 1960 in the celebrated KAM theorem, which tells us that near a stable equilibrium point under rather general conditions, an infinite subset of invariant tori will survive. There is now extensive literature on KAM theory; for introductions, see Arnold (1978) or Verhulst (2000). This is an extensive subject and we restrict ourselves to an example here.

Example 12.16
A classical mechanical example is the elastic pendulum. Consider a spring that can both oscillate in the vertical z direction and swing like a pendulum with angular deflection ϕ, where the corresponding momenta are p_z and p_ϕ. The model is discussed in many places, see for instance Van der Burgh (1975) or Tuwankotta and Verhulst (2000). It is shown there that near the vertical rest position, the Hamiltonian can be expanded as $H = H_0 + H_2 + \varepsilon H_3 + \varepsilon^2 H_4 + O(\varepsilon^3)$ with H_0 a constant; ε measures the deflection from the rest position. We have

$$H_2 = \frac{1}{2}\omega_z \left(z^2 + p_z^2\right) + \frac{1}{2}\omega_\phi \left(\phi^2 + p_\phi^2\right),$$

$$H_3 = \frac{\omega_\phi}{\sqrt{\sigma\omega_z}} \left(\frac{1}{2}z\phi^2 - zp_\phi^2\right),$$

$$H_4 = \frac{1}{\sigma}\left(\frac{3}{2}\frac{\omega_\phi}{\omega_z}z^2 p_\phi^2 - \frac{1}{24}\phi^4\right),$$

with positive frequencies ω_z, ω_ϕ; σ is a positive constant depending on the mass and the length of the pendulum. As expected from the physical setup, the relatively few terms in the Hamiltonian are symmetric in the second degree of freedom and also in p_z. Due to the physical restrictions, we will have for the frequency ratio $\omega_z/\omega_\phi > 1$.

Fixing $\omega_z/\omega_\phi \neq 2$, we find that by averaging to first order all terms vanish. For these frequency ratios, interesting phenomena take place either on a longer timescale or on a smaller scale with respect to ε. The important lower-order resonance is the 2 : 1 resonance, which has been intensively studied. This resonance is the one with resonant terms of the lowest degree.
The case $\omega_z = 2, \omega_\phi = 1$.
The equations of motion derived from the Hamiltonian can, after rescaling of parameters, be written as

$$\ddot{z} + 4z = \varepsilon \left(\dot{\phi}^2 - \frac{1}{2}\phi^2 \right) + O(\varepsilon^2),$$

$$\ddot{\phi} + \phi = \varepsilon(z\phi - 2\dot{z}\dot{\phi}) + O(\varepsilon^2).$$

Introduce the transformation $z = r_1 \cos(2t + \psi_1), \dot{z} = -2r_1 \sin(2t + \psi_1), \phi = r_2 \cos(t + \psi_2), \dot{\phi} = -r_2 \sin(t + \psi_2)$.

An objection against these amplitude-phase transformations is that they are not canonical (i.e., they do not preserve the Hamiltonian structure of the equations of motion). However, as we know, they yield asymptotically correct results; also, the averaging process turns out to conserve the energy. We omit the standard form and give directly the first-order averaging result, leaving out the approximation index a:

$$\dot{r}_1 = -\varepsilon \frac{3}{16} r_2^2 \sin \chi,$$

$$\dot{\phi}_1 = \varepsilon \frac{3r_2^2}{16r_1} \cos \chi,$$

$$\dot{r}_2 = \varepsilon \frac{3}{4} r_1 r_2 \sin \chi,$$

$$\dot{\phi}_2 = \varepsilon \frac{3r_1}{4} \cos \chi,$$

with $\chi = 2\psi_2 - \psi_1$. Actually, because of the presence of the combination angle χ, the system can be reduced to three equations. It is easy to find two integrals of this system. First is the approximate energy integral

$$4r_1^2 + r_2^2 = 2E,$$

with E indicating the constant, initial energy. The second integral is cubic and reads

$$z\phi^2 \cos \chi = I,$$

with I a constant determined by the initial conditions.

Note that the approximate energy integral represents a family of ellipsoids in four-dimensional phase-space around stable equilibrium. For each value of the energy, the second integral induces a foliation of the energy manifold into invariant tori. In this approximation, the foliation is a continuum, but for the original system the KAM theorem guarantees the existence of an infinite number of invariant tori with gaps in between. The gaps cannot be "seen" in a first-order approximation, and remarkably enough, they cannot be found at any algebraic order of approximation. The gaps turn out to be exponentially small (i.e., of size $\varepsilon^a \exp(-b/\varepsilon^c)$ with suitable constants a, b, c). This analysis is typical for time-independent Hamiltonian systems with two degrees of freedom and to some extent for more degrees of freedom. A description and references can be found in Lochak et al. (2003).

In this case of an infinite set of invariant tori, we have basically two timescales. Embedded on the energy manifold are periodic solutions with period $O(1)$. Around the stable periodic solutions, the tori are nested, with the periodic solutions as guiding centers. In the direction of the periodic solution, the timescale is $O(1)$, and in the other direction(s) it is $O(1/\varepsilon)$.

The case of two degrees of freedom is easiest to visualise. The energy manifold is three-dimensional, and a section perpendicular to a stable periodic solution is two-dimensional. In this section, called a Poincaré section, the periodic solution shows up in a few points where it subsequently hits the section. Around these fixed points, we find closed curves corresponding with the tori. Orbits moving on such a torus hit a particular closed curve recurrently with circulation time on the closed curve $O(1/\varepsilon)$.

12.4 Adiabatic Invariants

In Example 12.2, we considered a linear oscillator with slowly varying (prescribed) frequency:

$$\ddot{x} + \omega^2(\varepsilon t)x = 0.$$

Putting $\tau = \varepsilon t$ and assuming that $0 < a < \omega(\tau) < b$ (with a, b constants independent of ε), we found

$$r_a(\tau) = \frac{r(0)\sqrt{\omega(0)}}{\sqrt{\omega(\tau)}},$$

so that the quantity $r_a(\varepsilon t)\sqrt{\omega(\varepsilon t)}$ is conserved in time with accuracy $O(\varepsilon)$ on the timescale $1/\varepsilon$. Such a quantity we call an adiabatic invariant for the equation. More generally, consider the n-dimensional equation $\dot{x} = f(x, \varepsilon t, \varepsilon)$ and suppose that we have found a function $I(x, \varepsilon t)$ with the property

$$I(x, \varepsilon t) = I(x(0), 0) + o(1)$$

on a timescale that tends to infinity as ε tends to zero. In this case, we will call $I(x, \varepsilon t)$ an *adiabatic invariant* of the equation.

Note that this concept is a generalisation of the concept of a "first integral" of a system. The equation $\dot{x} = f(x, \varepsilon t, \varepsilon)$ may not have a first integral but, to a certain approximation, it behaves as if it has the adiabatic invariant as a first integral.

In a number of cases in the literature, the concept of an adiabatic invariant is also used for equations of the type $\dot{x} = f(x, \varepsilon)$. However, in such a case we prefer the term "asymptotic integral".

In some cases, the simple trick using $\tau = \varepsilon t$ as a dependent variable produces results. Consider the following nearly trivial example.

Example 12.17

Replace the equation

$$\ddot{x} + x = \varepsilon a(\varepsilon t)x^3,$$

with $a(\varepsilon t)$ a smooth function, by the system

$$\ddot{x} + x = \varepsilon a(\tau)x^3, \ \dot{\tau} = \varepsilon.$$

First-order averaging in amplitude-angle coordinates r, ϕ as in Example 12.1 for the Duffing equation can easily be carried out. Supposing that $r(0) = r_0$, we have the approximation $r(t) = r_0 + O(\varepsilon)$ valid on the timescale $1/\varepsilon$. The implication is that the quadratic integral $x^2(t) + \dot{x}^2(t) = r_0^2$ is conserved to $O(\varepsilon)$ on the timescale $1/\varepsilon$. This is a simple but nontrivial adiabatic invariant for the system.

A much more difficult example arises when the perturbation term depends on εt but is not small. Consider for instance the equation

$$\ddot{x} + x = a(\varepsilon t)x^2.$$

Huveneers and Verhulst (1997) studied this equation in the case where $a(0) = 1$ and the smooth function $a(\varepsilon t)$ vanishes as $t \to \infty$. In this case, there exist bounded and unbounded solutions and the analysis leans heavily on averaging over elliptic functions that are the solutions of the equation

$$\ddot{x} + x = x^2.$$

The analysis is too technical to include here.

The example of the Duffing equation is typical for the treatment of Hamiltonian systems with one degree of freedom and slowly varying coefficients and can be found in many texts on classical mechanics; see for instance Arnold (1978). With additional conditions, it is sometimes possible to extend the timescale of validity of the adiabatic invariant beyond $1/\varepsilon$.

Such examples can also be extended to more degrees of freedom, but the analysis becomes much more subtle. The extension of the timescale beyond $1/\varepsilon$ is then not easy to reach. Fascinating problems arise when the slowly varying coefficient leads the system through a bifurcation value. Unfortunately, this topic is beyond the scope of this book, but we give some references in our guide to the literature at the end of this chapter.

12.5 Second-Order Periodic Averaging

To calculate a second-order or even higher-order approximation, generally takes a much larger effort than in first order. One reason for this extra effort could be to obtain an approximation with higher precision but a more fundamental motivation derives from the fact that in quite a number of cases essential qualitative features are not described by first order. Let us consider a few simple examples. For the Van der Pol equation (Example 11.3)

$$\ddot{x} + x = \varepsilon\dot{x}(1 - x^2),$$

we have obtained a first-order approximation of the unique periodic solution. Suppose we add damping of a slightly larger magnitude. A model equation would be

$$\ddot{x} + \varepsilon\mu\dot{x} + x = \varepsilon^2\dot{x}(1 - x^2),$$

where the damping constant μ is positive. At first order, the solutions are damped, but do we recover a periodic solution at second order? In this case the answer is "no," as we know from qualitative information about this particular type of equation; see for instance Verhulst (2000). In most research problems, we do not have much a priori knowledge. A slightly less trivial modification arises when at first order we have just a phase shift, changing the "basic" period. A model equation could be

$$\ddot{x} + x - \varepsilon ax^3 = \varepsilon^2\dot{x}(1 - x^2)$$

with a a suitable constant. We shall analyse this equation later on.

12.5.1 Procedure for Second-Order Calculation

Consider the n-dimensional equation in the standard form

$$\dot{x} = \varepsilon f(t, x) + \varepsilon^2 g(t, x) + \varepsilon^3 R(t, x, \varepsilon),$$

in which the vector fields f and g are T-periodic in t with averages f^0 and g^0 (T independent of ε). The vector fields f and g have to be sufficiently smooth; in particular, f has to be expanded, as we shall see. The higher-order term $R(t, x, \varepsilon)$ is smooth and bounded as ε tends to zero. If we are looking for T-periodic solutions, R in addition has to be T-periodic.

The interpretation of results of second-order averaging is more subtle than in first order, so we have to show some of the technical details of the construction. We denote with $\nabla f(t, x)$ the derivative with respect to x only; this is an $n \times n$ matrix. For instance, if $n = 2$, we have $x = (x_1, x_2)$, $f = (f_1, f_2)$, and

$$\nabla f(t, x) = \begin{pmatrix} \frac{\partial f_1(t, x_1, x_2)}{\partial x_1} & \frac{\partial f_1(t, x_1, x_2)}{\partial x_2} \\ \frac{\partial f_2(t, x_1, x_2)}{\partial x_1} & \frac{\partial f_2(t, x_1, x_2)}{\partial x_2} \end{pmatrix}.$$

The reason we need this nabla operator (∇) is that to obtain a second-order approximation we have to Taylor expand the vector field. We need another vector field:

$$u^1(t, x) = \int_0^t (f(s, x) - f^0(x))ds - a(x),$$

where of course $f(s,x) - f^0(x)$ has average zero, but this does not hold necessarily for the integral (think of the function $\sin t \cos t$); $a(x)$ is chosen such that the average of u^1, $u^{10}(x)$ vanishes. We now introduce the *near-identity transformation*

$$x(t) = w(t) + \varepsilon u^1(t, w(t)). \qquad (12.2)$$

This is also called the *averaging* or *normalising transformation*. Substituting Eq. (12.2) into the equation for x, we get

$$\dot{w}(t) + \varepsilon \frac{\partial u^1}{\partial t}(t, w(t)) + \varepsilon \nabla u^1(t, w(t))\dot{w}(t)$$

$$= \varepsilon f(t, w(t) + \varepsilon u^1(t, w(t))) + \varepsilon^2 g(t, w(t) + \varepsilon u^1(t, w(t))) + \varepsilon^3 \cdots.$$

Using the definition of u^1, the left-hand side of this equation becomes

$$(I + \varepsilon \nabla u^1(t, w))\dot{w} + \varepsilon f(t, w) - \varepsilon f^0(w),$$

where I is the identity $n \times n$ matrix. Inverting the matrix $(I + \varepsilon \nabla u^1(t, w))$ and expanding f and g, we obtain

$$\dot{w} = \varepsilon f^0(w) + \varepsilon^2 \nabla f(t, w)u^1(t, w) + \varepsilon^2 g(t, w) + \varepsilon^3 \cdots.$$

We put

$$f_1(t, x) = \nabla f(t, x)u^1(t, x),$$

the product of a matrix and a vector. Introducing now also the average f_1^0 and the equation

$$\dot{v} = \varepsilon f^0(v) + \varepsilon^2 f_1^0(v) + \varepsilon^2 g^0(v), \;\; v(0) = x(0),$$

we can prove that

$$x(t) = v(t) + \varepsilon u^1(t, v(t)) + O(\varepsilon^2)$$

on the timescale $1/\varepsilon$.

Note, that we did not simply expand with the first-order approximation as a first term; $v(t)$ contains already terms $O(\varepsilon)$ and $O(\varepsilon^2)$. What does this mean for the timescales? One would expect εt and $\varepsilon^2 t$, but this conclusion is too crude, as we shall see later on.

This second-order calculations has another interesting aspect. As u^1 is uniformly bounded, $v(t)$ is an $O(\varepsilon)$-approximation of $x(t)$. In general, it is different from the first-order approximation that we obtained before. This illustrates the nonuniqueness of asymptotic approximations. In some cases, however, the interpretation will be that $v(t)$ is an $O(\varepsilon)$-approximation on a different timescale. We shall return to this important point in a later section.

As mentioned above, we are now able to compute more accurate approximations but, more excitingly, we are able to discover new qualitative phenomena. A simple example is given below.

Example 12.18

Consider the equation

$$\ddot{x} + x - \varepsilon a x^3 = \varepsilon^2 \dot{x}(1 - x^2)$$

with a a suitable constant independent of ε. Introducing amplitude-phase variables $x(t) = r(t)\cos(t + \psi(t))$, $\dot{x}(t) = -r(t)\sin(t + \psi(t))$, we find

$$f(t, r, \psi) = \begin{pmatrix} -ar^3 \sin(t + \psi)\cos^3(t + \psi) \\ -ar^2 \cos^4(t + \psi) \end{pmatrix}, \quad f^0(r, \psi) = \begin{pmatrix} 0 \\ -\frac{3}{8}ar^2 \end{pmatrix}.$$

Using initial values $r(0) = r_0, \psi(0) = 0$, we have a first-order approximation $x_a(t) = r_0 \cos(t - \varepsilon\frac{3}{8}ar_0^2 t)$, $\dot{x}_a(t) = -r_0 \sin(t - \varepsilon\frac{3}{8}ar_0^2 t)$ corresponding with a set of periodic solutions with an $O(\varepsilon)$ shifted period that also depends on the initial r_0. Does a periodic solution branch off from one of the first-order approximations when considering second-order approximations? We have to compute and average $f_1 = \nabla f u^1$. Abbreviating $t + \psi = \alpha$, we find

$$\nabla f(t, r, \psi) = \begin{pmatrix} -3ar^2 \sin\alpha \cos^3\alpha & -ar^3 \cos^4\alpha + 3ar^3 \sin^2\alpha \cos^2\alpha \\ -2ar\cos^4\alpha & 4ar^2 \cos^3\alpha \sin\alpha \end{pmatrix},$$

and, using $\cos^4\alpha = \frac{3}{8} + \frac{1}{2}\cos 2\alpha + \frac{1}{8}\cos 4\alpha$,

$$u^1(t, r, \psi) = \begin{pmatrix} \frac{1}{8}ar^3 \cos 2\alpha + \frac{1}{32}ar^3 \cos 4\alpha \\ -\frac{1}{4}ar^2 \sin 2\alpha - \frac{1}{32}ar^2 \sin 4\alpha \end{pmatrix}.$$

Multiplying, we find f_1, and after averaging

$$f_1^0(r, \psi) = \begin{pmatrix} 0 \\ -\frac{51}{256}a^2 r^4 \end{pmatrix}.$$

Now we can write down the equation for v as formulated in the procedure for second-order calculation. Note that $g^0(v)$ was calculated in Example 11.3 and we obtain

$$\dot{v}_1 = \varepsilon^2 \frac{v_1}{2}\left(1 - \frac{1}{4}v_1^2\right)$$

$$\dot{v}_2 = -\varepsilon\frac{3}{8}av_1^2 - \varepsilon^2 \frac{51}{256}a^2 v_1^4.$$

If $v_1(0) = r_0 = 2$, we have a stationary solution $v_1(t) = 2$, and with $v_2(0) = \psi(0) = 0$

$$v_2(t) = -\varepsilon\frac{3}{2}at - \varepsilon^2\frac{51}{16}a^2 t.$$

An $O(\varepsilon^2)$-approximation of the periodic solution is obtained by inserting this into u^1 so that

$$r(t) = 2 + \varepsilon a \cos 2(t + v_2(t)) + \varepsilon\frac{a}{4}\cos 4(t + v_2(t)) + O(\varepsilon^2),$$

$$\psi(t) = v_2(t) - \varepsilon a \sin 2(t + v_2(t)) - \varepsilon\frac{a}{8}\sin 4(t + v_2(t)) + O(\varepsilon^2),$$

valid on the timescale $1/\varepsilon$. At this order of approximation, the timescales for the periodic solution are $t, \varepsilon t$, and $\varepsilon^2 t$. Can you find the timescales for the other solutions?

12.5.2 An Unexpected Timescale at Second-Order

As another example, we consider the following Mathieu equation.

Example 12.19
Consider

$$\ddot{x} + (1 + \varepsilon a + \varepsilon^2 b + \varepsilon \cos 2t)x = 0$$

with free parameters a, b, which is a slight variation of the Mathieu equation we studied in Examples 10.9 and 11.6; see also the discussion in Section 15.3. Transforming by Eqs. 11.9

$$x(t) = y_1(t)\cos t + y_2(t)\sin t, \quad \dot{x}(t) = -y_1(t)\sin t + y_2(t)\cos t,$$

we obtain the slowly varying system

$$\dot{y}_1 = \sin t(\varepsilon a + \varepsilon \cos 2t + \varepsilon^2 b)(y_1(t)\cos t + y_2(t)\sin t),$$
$$\dot{y}_2 = -\cos t(\varepsilon a + \varepsilon \cos 2t + \varepsilon^2 b)(y_1(t)\cos t + y_2(t)\sin t).$$

We use again the terminology of the procedure for second-order calculation. We have to first order

$$f^0(y_1, y_2) = \begin{pmatrix} \frac{1}{2}(a - \frac{1}{2})y_2 \\ -\frac{1}{2}(a + \frac{1}{2})y_1 \end{pmatrix}$$

so that the first-order approximation is described by

$$\dot{y}_{1a} = \varepsilon \frac{1}{2}\left(a - \frac{1}{2}\right)y_{2a},$$

$$\dot{y}_{2a} = -\varepsilon \frac{1}{2}\left(a + \frac{1}{2}\right)y_{1a}.$$

This is a system of linear equations with constant coefficients; the solutions are of the form $c\exp(\lambda t)$ with λ an eigenvalue of the matrix of coefficients. We find

$$\lambda_{1,2} = \pm\frac{1}{2}\sqrt{\frac{1}{4} - a^2}.$$

It is clear that we have stability of the trivial solution if $a^2 > \frac{1}{4}$ and instability if $a^2 < \frac{1}{4}$; $a = \pm\frac{1}{2}$ determines the boundary of the instability domain, which is called a Floquet tongue. On this boundary, the solutions are periodic to first approximation. What happens when we look more closely at the Floquet tongue? We find for ∇f and u^1

$$\nabla f(t, y_1, y_2) = \begin{pmatrix} \sin t \cos t(a + \cos 2t) & \sin^2 t(a + \cos 2t) \\ -\cos^2 t(a + \cos 2t) & -\sin t \cos t(a + \cos 2t) \end{pmatrix},$$

$$u^1(t, y_1, y_2) = \begin{pmatrix} -y_1(\frac{a}{4}\cos 2t + \frac{1}{16}\cos 4t) + y_2(-\frac{a}{4}\sin 2t + \frac{1}{4}\sin 2t - \frac{1}{16}\sin 4t) \\ -y_1(\frac{a}{4}\sin 2t + \frac{1}{4}\sin 2t + \frac{1}{16}\sin 4t) + y_2(\frac{a}{4}\cos 2t + \frac{1}{16}\cos 4t) \end{pmatrix}.$$

After some calculations, we find the average

$$f_1^0(y_1, y_2) = \begin{pmatrix} (\frac{a}{8} - \frac{a^2}{8} - \frac{1}{64})y_2 \\ (\frac{a}{8} + \frac{a^2}{8} + \frac{1}{64})y_1 \end{pmatrix}.$$

When adding g^0, we can write down the equation for v as formulated in the procedure for second-order calculation:

$$\dot{v}_1 = \varepsilon \frac{1}{2}\left(a - \frac{1}{2}\right)v_2 + \varepsilon^2\left(\frac{a}{8} - \frac{a^2}{8} - \frac{1}{64} + \frac{1}{2}b\right)v_2,$$

$$\dot{v}_2 = -\varepsilon\frac{1}{2}\left(a + \frac{1}{2}\right)v_1 + \varepsilon^2\left(\frac{a}{8} + \frac{a^2}{8} + \frac{1}{64} - \frac{1}{2}b\right)v_1.$$

Choosing for instance $a = \frac{1}{2}$, we have one of the boundaries of the Floquet tongue consisting of periodic solutions. The equations for v become

$$\dot{v}_1 = \varepsilon^2\left(\frac{1}{64} + \frac{1}{2}b\right)v_2,$$

$$\dot{v}_2 = -\varepsilon\frac{1}{2}v_1 + \varepsilon^2\left(\frac{7}{64} - \frac{1}{2}b\right)v_1.$$

For the eigenvalues of the matrix of coefficients to second order, we find

$$\lambda_{1,2} = \pm\sqrt{-\frac{1}{4}\left(\frac{1}{32} + b\right)\varepsilon^3 + \left(\frac{1}{64} + \frac{1}{2}b\right)\left(\frac{7}{64} - \frac{1}{2}b\right)\varepsilon^4}.$$

We conclude that if $\frac{1}{32} + b > 0$, we have stability, and if $\frac{1}{32} + b < 0$, we have instability. The value $b = -\frac{1}{32}$ gives us the second-order approximation of the Floquet tongue. This result has an interesting consequence for the timescales of the solutions in the case $a = \frac{1}{2}$; they are $t, \varepsilon t, \varepsilon^{\frac{3}{2}}t, \varepsilon^2 t$. The timescale $\varepsilon^{\frac{3}{2}}t$ is quite unexpected.

Remark
Second-order general averaging, without periodicity assumptions, runs along the same lines as demonstrated in the periodic case. For the theory and examples, see Sanders and Verhulst (1985).

12.6 Approximations Valid on Longer Timescales

In a number of problems, we have a priori knowledge that the solutions of equations we are studying exist on a longer timescale than $1/\varepsilon$ or even exist for all time. Is it not possible in these cases to obtain approximations valid on such a longer timescale? For instance, when calculating an $O(\varepsilon^2)$-approximation on the timescale $1/\varepsilon$ as in the preceding section, can we not as a trade-off consider this as an $O(\varepsilon)$-approximation on the timescale $1/\varepsilon^2$? It turns out that in general the answer is "no," as can easily be seen from examples. However, we shall consider an important case where this idea carries through.

12.6.1 Approximations Valid on $O(1/\varepsilon^2)$

We formulate the following result.

Theorem 12.2
Consider again the n-dimensional equation in the standard form

$$\dot{x} = \varepsilon f(t, x) + \varepsilon^2 g(t, x) + \varepsilon^3 R(t, x, \varepsilon)$$

in which the vector fields f and g are T-periodic in t with averages f^0 and g^0 (T independent of ε); f, g, and R are sufficiently smooth. Suppose that

$$f^0(x) = 0.$$

We have from the first-order approximation $x(t) = x(0) + O(\varepsilon)$ on the timescale $1/\varepsilon$. Following the construction of the second-order approximation of the preceding section, we have with $f^0(x) = 0$ the equation

$$\dot{v} = \varepsilon^2 g^0(v), \ v(0) = x(0).$$

It is easy to prove that

$$x(t) = v(t) + O(\varepsilon)$$

on the timescale $1/\varepsilon^2$ (Van der Burgh, 1975).

A simple example is given as follows.

Example 12.20

$$\ddot{x} + x - \varepsilon a x^2 = 0, \ r(0) = r_0, \psi(0) = 0,$$

with a an ε-independent parameter. In amplitude-phase variables r, ψ, we find with transformation (11.5)

$$\dot{r} = -a r^2 \sin(t + \psi) \cos^2(t + \psi),$$
$$\dot{\psi} = -a r \cos^3(t + \psi),$$

so with $f(t, r, \psi)$ for the right-hand side we have $f^0(r, \psi) = 0$. For the second-order approximation, we abbreviate $t + \psi = \alpha$ and calculate

$$\nabla f(t, r, \psi) = \begin{pmatrix} -2ar \sin \alpha \cos^2 \alpha & -ar^2 \cos^3 \alpha + 2ar^2 \sin^2 \alpha \cos \alpha \\ -a \cos^3 \alpha & 3ar \cos^2 \alpha \sin \alpha \end{pmatrix},$$

$$u^1(t, r, \psi) = \begin{pmatrix} \frac{1}{3}ar^2\cos^3\alpha \\ -ar\sin\alpha + \frac{1}{3}ar\sin^3\alpha \end{pmatrix}.$$

With $f_1 = \nabla f u^1$, we find after averaging

$$f_1^0(r, \psi) = \begin{pmatrix} 0 \\ -\frac{5}{12}a^2r^2 \end{pmatrix}.$$

We conclude that $\dot{v}_1 = 0, \dot{v}_2 = -\varepsilon^2\frac{5}{12}a^2v_1^2$, and we have $x(t) = r_0\cos(t - \varepsilon^2\frac{5}{12}a^2r_0^2t) + O(\varepsilon)$, valid on the timescale $1/\varepsilon^2$.

This is already a nontrivial result, but a more interesting example arises when considering the Mathieu equation for other resonance values than those studied in Example 12.19.

Example 12.21
Consider

$$\ddot{x} + (n^2 + \varepsilon a + \varepsilon^2 b + \varepsilon\cos mt)x = 0,$$

again with free parameters a, b. Transforming by Eq. (11.10)

$$x(t) = y_1(t)\cos nt + \frac{1}{n}y_2(t)\sin nt, \quad \dot{x}(t) = -ny_1(t)\sin nt + y_2(t)\cos nt,$$

we obtain the slowly varying system

$$\dot{y}_1 = \frac{\varepsilon}{n}\sin nt(a + \cos mt)\left(y_1(t)\cos nt + \frac{1}{n}y_2(t)\sin nt\right) + O(\varepsilon^2),$$

$$\dot{y}_2 = -\frac{\varepsilon}{n}\cos nt(a + \cos mt)\left(y_1(t)\cos nt + \frac{1}{n}y_2(t)\sin nt\right) + O(\varepsilon^2).$$

It is easy to see (by averaging) that f^0 is nontrivial if $2n - m = 0$, which is the case of Example 12.19. This is the most prominent resonance of the Mathieu equation. For other rational ratios of m and n, we find other resonances with different sizes of the resonance tongues. As an example, we explore the case $m = n = 2$. If $2n - m \neq 0$, the averaged equations are dominated by the parameter a so it makes sense to choose $a = 0$, as the Floquet tongue will be narrower than in the case $2n - m = 0$. The equation becomes

$$\ddot{x} + (4 + \varepsilon^2 b + \varepsilon\cos 2t)x = 0.$$

We omit the expressions for ∇f and u^1 and produce f_1 directly:

$$f_1(t, y_1, y_2) = \begin{pmatrix} \frac{1}{192}\sin 4t(-6y_1 + 2y_1\cos 4t + y_2\sin 4t) \\ \frac{-1}{48}\cos^2 2t(-6y_1 + 2y_1\cos 4t + y_2\sin 4t) \end{pmatrix}.$$

So we find

$$f_1^0(y_1, y_2) = \begin{pmatrix} \frac{1}{384} y_2 \\ -\frac{1}{96} y_1 \end{pmatrix}$$

and for the second-order equations

$$\dot{v}_1 = \varepsilon^2 \frac{y_2}{8} \left(\frac{1}{48} + b \right), \ \dot{v}_2 = \varepsilon^2 \frac{y_1}{2} \left(\frac{5}{48} - b \right).$$

The solutions $v_1(t)$ and $v_2(t)$ are $O(\varepsilon)$-approximations of $y_1(t)$ and $y_2(t)$ valid on the timescale $1/\varepsilon^2$. From the eigenvalues, we conclude instability if

$$-\frac{1}{48} < b < \frac{5}{48}.$$

The Floquet tongue in the case $m = n = 2$ is determined by the boundary values $4 - \frac{1}{48}\varepsilon^2$ and $4 + \frac{5}{48}\varepsilon^2$.

When assuming $m \neq 2n$ and $m \neq n$, we can study higher order Floquet tongues. See also the discussion in Example 10.9, where the tongues are determined by the continuation (Poincaré-Lindstedt) method, and in particular Fig. 10.1.

12.6.2 Timescales near Attracting Solutions

Another natural idea to obtain extension of the timescale of validity is attraction. Suppose the solutions are attracted exponentially fast to a special solution, equilibrium or periodic, for which the process will also result in a kind of compression of the neighbouring solutions. In this case, we have the following result.

Theorem 12.3
Consider the n-dimensional equation in the standard form

$$\dot{x} = \varepsilon f(t, x) + \varepsilon^2 R(t, x, \varepsilon)$$

in which the vector field f is T-periodic in t with average f^0 (T independent of ε); f and R are sufficiently smooth. Suppose that $f^0(x)$ contains a critical point (equilibrium of the averaged equation) $x = a$, so $f^0(a) = 0$. We assume that all the eigenvalues in $x = a$ have a negative real part. The solution $x(t)$ starting in $x(0)$, which is located in an interior subset of the domain of attraction of $x = a$, is approximated by the solution $x_a(t)$ of the averaged equation starting in $x(0)$ as

$$x(t) - x_a(t) = O(\varepsilon), \ 0 \leq t < \infty.$$

Example 12.22
A simple example is the one-dimensional equation

$$\dot{x} = -\varepsilon 2 \sin^2 tx + \varepsilon^2 R(t, x)$$

with averaged equation

$$\dot{x}_a = -x_a.$$

We have $x(t) = x(0) \exp(-t) + O(\varepsilon)$ for all time.

Example 12.23
More important is the forced Duffing equation in Example 11.4

$$\ddot{x} + \varepsilon\mu\dot{x} + \varepsilon\gamma x^3 + x = \varepsilon h \cos\omega t,$$

with $\mu > 0$; we choose the case of exact resonance $\omega = 1$. In this case, the system averaged to first order is

$$\dot{r}_a = -\frac{1}{2}\varepsilon(\mu r_a + h \sin\psi_a),$$
$$\dot{\psi}_a = -\frac{1}{2}\varepsilon\left(-\frac{3}{4}\gamma r_a^2 + h\frac{\cos\psi_a}{r_a}\right).$$

The critical points are determined by the equations

$$h\sin\psi_a = -\mu r_a, \quad h\cos\psi_a = \frac{3}{4}\gamma r_a^3.$$

Using these equations, it is not difficult to find the eigenvalues

$$\lambda_{1,2} = -\frac{1}{2}\mu \pm i\frac{3\sqrt{3}}{8}|\gamma|r_a^2,$$

so if we find critical points, they are asymptotically stable and the solutions attracting to the corresponding periodic solution are approximated by the solutions of the averaged equation for all time.

A problem arises when the original equation is autonomous. In this case, a periodic solution has at least one eigenvalue zero, a feature that is inherited by the averaged equation.

Example 12.24
Consider again the Van der Pol equation

$$\ddot{x} + x = \varepsilon(1 - x^2)\dot{x}$$

with averaged equation (amplitude-phase)

$$\dot{r}_a = \varepsilon\frac{r_a}{2}\left(1 - \frac{1}{4}r_a^2\right), \quad \dot{\psi}_a = 0.$$

As we remarked in the discussion of Example 11.3, we can reduce such an autonomous equation by introducing $\tau = t + \psi$ and average over τ. We find

$$\frac{dr_a}{d\tau} = \varepsilon \frac{r_a}{2}\left(1 - \frac{1}{4}r_a^2\right),$$

and we can apply the theory as follows. Solutions $r(\tau)$ starting outside a neighbourhood of $r = 0$ are approximated for all times τ or t by the solutions of the averaged equation. Such a result is not valid for the phase ψ.

This example can easily be generalised for second-order autonomous equations.

12.7 Identifying Timescales

A fundamental question that comes up very often is whether we have the right expansion coefficients with respect to ε and, in the context of evolution problems, whether we have the right timescales. In the framework of boundary layer problems, we have developed a technique in Chapter 4 to identify local or boundary layer variables. This technique works fine in many cases but not always. When studying evolution problems, the situation is worse; we shall first review some clarifying examples.

12.7.1 Expected and Unexpected Timescales

In Chapters 10 and 11, we have seen that in equations of the form $\dot{x} = f(t, x, \varepsilon)$ where the right-hand side depends smoothly on ε, the solutions can sometimes be expanded in powers of ε while we have timescales such as $t, \varepsilon t$, and $\varepsilon^2 t$. In fact, as we have shown before, when first-order averaging is possible, we have definitely the timescales t and εt. However, in general and certainly at second order, the situation is not always as simple as that.

Example 12.25
Consider for $t \geq 0$ the linear initial value problem

$$\ddot{x} + \frac{1}{1+t}\dot{x} = \varepsilon\frac{2}{(1+t)^2}, \quad x(0) = \dot{x}(0) = 0,$$

with solution

$$x(t) = \varepsilon \ln^2(1+t),$$

so $x(t)$ is characterised by the timescale $\varepsilon \ln^2(1+t)$, and $\dot{x}(t)$ by the timescales t and $\varepsilon \ln(1+t)$.

Example 12.26
Consider the initial value problem for the Bernoulli equation

$$\dot{x} = x^\alpha - cx, \quad x(0) = 1.$$

For the constants, we have $0 < \alpha < 1, c > 0$. The solution of the initial value problem is

$$x(t) = \left(\frac{1}{c} + \left(1 - \frac{1}{c}\right) e^{-c(1-\alpha)t}\right)^{\frac{1}{1-\alpha}}.$$

The solutions starting with $x(0) > 0$ tend to stable equilibrium $c^{-\frac{1}{1-\alpha}}$. We consider two cases.

1. Choose $c = \varepsilon$, where α does not depend on ε:

$$\dot{x} = x^\alpha - \varepsilon x, \ x(0) = 1,$$

with solution

$$x(t) = \left(\frac{1}{\varepsilon} + \left(1 - \frac{1}{\varepsilon}\right) e^{-\varepsilon(1-\alpha)t}\right)^{\frac{1}{1-\alpha}}.$$

One of the expansion coefficients is $\varepsilon^{-\frac{1}{1-\alpha}}$, and the relevant timescale is εt.

2. Choose $c = \varepsilon, 1 - \alpha = \varepsilon$, producing

$$\dot{x} = x^{1-\varepsilon} - \varepsilon x, \ x(0) = 1,$$

with solution

$$x(t) = \left(\frac{1}{\varepsilon} + \left(1 - \frac{1}{\varepsilon}\right) e^{-\varepsilon^2 t}\right)^{\frac{1}{\varepsilon}}.$$

The relevant timescale is $\varepsilon^2 t$, and $1/\varepsilon$ plays a part in the expansion.

The Mathieu equation in Example 12.19 is important to show that unexpected timescales occur in practical problems. We summarise as follows.

Example 12.27
Consider

$$\ddot{x} + (1 + \varepsilon a + \varepsilon^2 b + \varepsilon \cos 2t)x = 0.$$

On choosing $a = \pm\frac{1}{2}$, the solutions are located near or on the Floquet tongue. A second-order approximation has shown that the timescales in this case are $t, \varepsilon t, \varepsilon^{\frac{3}{2}} t, \varepsilon^2 t$. They naturally emerge from the averaging process.

Other important examples of unexpected timescales can be found in the theory of Hamiltonian systems. These examples are much more complicated.

12.7.2 Normal Forms, Averaging and Multiple Timescales

The Mathieu equation that we discussed above is a linear equation where the calculation to second order yields an unexpected timescale. In Section 15.3, we show for matrices A with constant entries that, when arising in equations of the form $\dot{x} = Ax$, in particular the bifurcation values produce such unexpected timescales. For nonlinear equations, we have no general theory, only examples and local results.

Where does this leave us in constructing approximations for initial value problems? When given a perturbation problem, we always start with transformations and other operations that we think suitable for the problem. The main directive is that there should be *no a priori assumptions on the timescales*. The general framework for this is the theory of normal forms, of which averaging is one of the parts (see for instance Sanders and Verhulst, 1985). In this framework, suitable transformations are introduced, and from the analysis of the resulting equations the timescales follow naturally. Also, the corresponding proofs of validity yield confirmation and natural restrictions on the use of these timescales.

It should be clear by now that the method of multiple timescales for initial value problems of ordinary differential equations is not suitable for research problems except in the case of simple problems that are accessible also to first-order averaging. For other problems, this method presupposes the presence of certain timescales, a knowledge we simply do not have.

Multiple timing, as it is often called, is of course suitable and an elegant method in the cases of solved problems where we know a priori the structure of the approximations. It should also be mentioned that the formal calculation schemes of multiple timing are easier to extend to problems for partial differential equations; we will return to this in Chapter 14. However, proofs of validity of such extensions are often lacking.

12.8 Guide to the Literature

There are thousands of papers associated with this chapter. We will mention here basic literature with good literature sections for further study. Several times we touched upon the relation between averaging and normalisation. Averaging can be seen as a special normal-form method with the advantage of explicitly formulated normal forms. On the other hand, normal-form methods are approximation tools, using in some form localisation around a special point or another solution. More about this relation can be found in Arnold (1982) and Sanders and Verhulst (1985). The same references can be used for the theory of averaging over angles. A monograph by Lochak and Meunier (1988) is devoted to this topic.

Invariant manifolds represent a subject that is still very much in development. The idea to obtain tori by continuation was introduced by Bogoliubov and Mitropolsky (1961, 1963) and extended by Hale (1969). Tori can emerge in a different way by Hopf bifurcation of a periodic solution; this is called Neimark-Sacker bifurcation. Usually, numerical bifurcation path-following programs are used to pinpoint such bifurcations, but the demonstration by averaging is for instance shown in Bakri et al. (2004).

Invariant tori in the Hamiltonian context is a very large subject with an extensive body of literature and many interesting books; for an introduction, see Arnold (1978) or Verhulst (2000). Lochak et al. (2003) discusses what happens between the tori and gives many references. A survey of normalisation and averaging for Hamiltonian systems is given in Verhulst (1998).

The theory of adiabatic invariants is a classical subject that is tied in both with problems of averaging over angles and bifurcation theory. For the classical theory, see Arnold (1978) and, in particular, the survey by Henrard (1993). Slowly varying coefficients may lead a system to passage through a bifurcation. Neishstadt (1986, 1991), Cary et al. (1986), Haberman (1978) and Bourland and Haberman (1990) analysed the slow passage through a separatrix. See also Diminnie and Haberman (2002) for changes of the adiabatic invariant in such a setting. Adiabatic changes in a Hamiltonian system with two degrees of freedom are discussed in Verhulst and Huveneers (1998).

Second-order averaging and longer timescales are studied in Sanders and Verhulst (1985). Van der Burgh (1975) produced the first estimates on the timescale $1/\varepsilon^2$. An extension to $O(\varepsilon^2)$ on the timescale $1/\varepsilon^2$ is given in Verhulst (1988).

Multiple-timescale methods are discussed for instance in Hinch (1991), Kevorkian and Cole (1996), and Holmes (1998). For a comparison of multiple timing and averaging, see Perko (1969) and Kevorkian (1987).

12.9 Exercises

Exercise 12.1

$$\ddot{x} + 2\varepsilon\dot{x} + \varepsilon\dot{x}^3 + x = 0.$$

Use the amplitude-angle transformation 12.1 to obtain a system that can be averaged over the angle ϕ; give the result for the approximation of the amplitude.

Exercise 12.2

$$\ddot{x} + \varepsilon\mu\dot{x} + \omega^2(\varepsilon t)x + \varepsilon x^3 = 0.$$

Use the amplitude-angle transformation (12.1) to obtain a system that can be averaged over the angle ϕ; give the result (with additional assumptions) for the approximation of the amplitude.

Exercise 12.3 Consider the system

$$\dot{x} = \varepsilon + \varepsilon\sin(\phi_1 - \phi_2),$$
$$\dot{\phi}_1 = x,$$
$$\dot{\phi}_2 = x^2.$$

Determine the location(s) of the resonance manifold(s) and an approximation away from these location(s).

Exercise 12.4

$$\dot{x} = \varepsilon \cos(t - \phi_1 + \phi_2) + \varepsilon \sin(t + \phi_1 + \phi_2),$$
$$\dot{\phi}_1 = 2x,$$
$$\dot{\phi}_2 = x^2.$$

Determine the location(s) of the resonance manifold(s) and an approximation away from these location(s).

Exercise 12.5 In Example 12.9, we stated that in the two resonance zones the flow is described by two pendulum equations. Verify this statement.

Exercise 12.6 Consider the Duffing equation with slowly varying coefficients in the form

$$\ddot{x} + \omega^2(\varepsilon t)x = \varepsilon a(\varepsilon t)x^3.$$

Compute an adiabatic invariant for the equation with suitable assumptions on the coefficients $a(\varepsilon t)$ and $\omega(\varepsilon t)$.

Exercise 12.7 In the beginning of Section 12.5, we stated that the equation

$$\ddot{x} + \varepsilon \mu \dot{x} + x = \varepsilon^2 \dot{x}(1 - x^2)$$

with positive damping constant μ does not contain a periodic solution. Ignoring the theory of periodic solutions behind this, verify this statement by discussing the second-order approximation.

Exercise 12.8 Determine a second-order approximation for the solutions of

$$\ddot{x} + x - \varepsilon a x^2 = \varepsilon^2 \dot{x}(1 - x^2)$$

with a a suitable constant independent of ε. Can we identify a periodic solution? Replace the $O(\varepsilon)$ term by εx^m with m an even number and repeat the analysis.

Exercise 12.9 A start for the calculation of higher-order Floquet tongues in the case of the Mathieu equation from Example 12.19 with $m \neq 2n$ and $m \neq n$ is the determination of the second-order approximation. Show that the equations are

$$\dot{v}_1 = \varepsilon^2 \frac{v_2}{2n^2}\left(\frac{1}{2m^2 - 8n^2} + b\right), \quad \dot{v}_2 = -\varepsilon^2 \frac{v_1}{2}\left(\frac{1}{2m^2 - 8n^2} + b\right).$$

What can we conclude at this stage for the size of the higher-order Floquet tongues? Note in this context that the first tongue has a separation between the boundaries that are to first order straight lines, the second tongue is bounded by parabolas, and the higher-order tongues have boundaries that are tangent as ε tends to zero; see Fig. 10.1.

Exercise 12.10 Consider the system

$$\dot{x} = y + \varepsilon(x^2 \sin 2t - \sin 2t),$$
$$\dot{y} = -4x.$$

Find equilibria and corresponding periodic solutions of the associated averaged system. For which initial conditions can we extend the timescale of validity beyond $1/\varepsilon$?

Exercise 12.11 Consider a Hamiltonian system with two degrees of freedom, the so-called Hénon-Heiles family:

$$\ddot{x} + x = \varepsilon(a_1 x^2 + a_2 y^2),$$
$$\ddot{y} + \omega^2 y = \varepsilon 2 a_2 xy, \ a_2 \neq 0.$$

a. Show that for $\omega = 2$ first-order averaging produces a nontrivial result.
b. Consider $\omega = 1$ and determine the equations for the second-order approximation.
c. Determine in the case $\omega = 1$ two integrals of motion of the averaged equations and indicate their geometrical meaning.
d. Determine in the case $\omega = 1$ the conditions for the existence of short-periodic solutions in a general position away from the normal modes.

Exercise 12.12 Consider a Hamiltonian system with two degrees of freedom to which we have added damping and parametric excitation:

$$\ddot{x} + 2\varepsilon\dot{x} + (4 + \varepsilon a \cos t)x = \varepsilon y^2,$$
$$\ddot{y} + 2\varepsilon\dot{y} + (1 + \varepsilon b \cos 2t)y = 2\varepsilon xy.$$

a. If $\varepsilon = 0$, we have two normal modes. Can we continue them for $\varepsilon > 0$?
b. Can you find other periodic solutions?

Averaging for Evolution Equations

13.1 Introduction

The analysis of weakly nonlinear partial differential equations with evolution in time is an exciting field of investigation. In this chapter, we consider specific results related to averaging and do not aim at completeness; see also the guide to the literature, in particular the book by Kevorkian and Cole (1996), for many other problem formulations and techniques. As the analysis of evolution equations involves rather subtle problems, we will in this chapter discuss in more detail some theoretical results of averaging.

Note that the examples discussed in this chapter are really research problems and involve many open questions. Some of our examples will concern conservative systems. In the theory of finite-dimensional Hamiltonian systems, we have for nearly integrable systems the celebrated KAM theorem, which, under certain nondegeneracy conditions, guarantees the persistence of many tori in the nonintegrable system. For infinite-dimensional, conservative systems, we now have the KKAM theorems developed by Kuksin (1991). Finite-dimensional invariant manifolds obtained in this way are densely filled with quasiperiodic orbits; these are the kind of solutions we obtain by approximation methods that involve projection on finite-dimensional subspaces. It is stressed, however, that identification of approximate solutions with solutions covering invariant manifolds is only possible if the validity of the approximation has been demonstrated.

13.2 Operators with a Continuous Spectrum

Various forms of averaging techniques are used in the literature. They are sometimes indicated by terms such as "homogenisation" or "regularisation" methods, and their main purpose is to stabilise numerical integration schemes for partial differential equations. However, apart from numerical improvements, we are also interested in qualitative characteristics of the solutions. This will be the subject of the subsequent sections.

13.2.1 Averaging of Operators

A typical problem formulation would be to consider the Cauchy problem (or later an initial boundary value problem) for equations such as

$$u_t + Lu = \varepsilon f(u), \ t > 0, u(0) = u_0, \tag{13.1}$$

where L is a linear operator, u is an element of a suitable function space, and $f(u)$ represents the nonlinear terms.

To obtain a standard form for averaging in the case of a partial differential equation can already pose a formidable technical problem, even in the case of simple geometries. However, it is reasonable to suppose that one can solve the "unperturbed" ($\varepsilon = 0$) problem in sufficient explicit form before proceeding to the nonlinear equation.

A number of authors, in particular in the former Soviet Union, have addressed problem (13.1). For a survey of results, see Mitropolsky, Khoma, and Gromyak (1997); see also Shtaras (1989). There still does not exist a unified mathematical theory with a satisfactory approach for higher-order approximations (normalisation to arbitrary order) and enough convincing examples. Here we shall follow the theory developed by Krol (1991), which has some interesting applications. Consider the problem (13.1) with two spatial variables x, y and time t; assume that, after solving the unperturbed problem, by a variation of constants procedure we can write the problem in the form of the initial value problem

$$\frac{\partial F}{\partial t} = \varepsilon L(t)F, \ F(x, y, 0) = \gamma(x, y). \tag{13.2}$$

We have

$$L(t) = L_2(t) + L_1(t),$$

where

$$L_2(t) = b_1(x, y, t)\frac{\partial^2}{\partial x^2} + b_2(x, y, t)\frac{\partial^2}{\partial x \partial y} + b_3(x, y, t)\frac{\partial^2}{\partial y^2},$$

$$L_1(t) = a_1(x, y, t)\frac{\partial}{\partial x} + a_2(x, y, t)\frac{\partial}{\partial y},$$

in which $L_2(t)$ is a uniformly elliptic operator on the domain and L_1, L_2 and thus L are T-periodic in t; the coefficients a_i, b_i and the initial value γ are C^∞ and bounded with bounded derivatives.

We average the operator L by averaging the coefficients a_i, b_i over t

$$\bar{a}_i(x, y) = \frac{1}{T}\int_0^T a_i(x, y, t)dt, \ \bar{b}_i(x, y) = \frac{1}{T}\int_0^T b_i(x, y, t)dt,$$

producing the averaged operator $\bar{L}$. As an approximating problem for Eq. (13.2), we now take

$$\frac{\partial \bar{F}}{\partial t} = \varepsilon \bar{L} \bar{F}, \quad \bar{F}(x, y, 0) = \gamma(x, y). \tag{13.3}$$

A rather straightforward analysis shows the existence and uniqueness of the solutions of problems (13.2) and (13.3) on the timescale $1/\varepsilon$. Krol (1991) proves the following result.

Theorem 13.1
Let F be the solution of initial value problem (13.2) and $\bar{F}$ the solution of initial value problem (13.3). Then we have the estimate $\|F - \bar{F}\| = O(\varepsilon)$ on the timescale $1/\varepsilon$. The norm $\|.\|$ is the supnorm on the spatial domain and on the timescale $1/\varepsilon$.

13.2.2 Time-Periodic Advection-Diffusion

As an application, Krol (1991) considers the transport of material (chemicals or sediment) by advection and diffusion in a tidal basin. In this case, the advective flow is nearly periodic and diffusive effects are small.

Example 13.1
The problem can be formulated as

$$\frac{\partial C}{\partial t} + \nabla.(uC) - \varepsilon \Delta C = 0, \quad C(x, y, 0) = \gamma(x, y), \tag{13.4}$$

where $C(x, y, t)$ is the concentration of the transported material, the flow $u = u_0(x, y, t) + \varepsilon u_1(x, y)$ is given, u_0 is T-periodic in time and represents the tidal flow, and εu_1 is a small rest stream. As the diffusion process of chemicals and sediment in a tidal basin is slow, we are interested in a long-timescale approximation.

If the flow is divergence-free, the unperturbed ($\varepsilon = 0$) problem is given by

$$\frac{\partial C_0}{\partial t} + u_0 \nabla C_0 = 0, \quad C_0(x, y, 0) = \gamma(x, y),$$

a first-order equation that can be integrated along the characteristics or streamlines. C_0 is constant along the characteristics, which are the solutions of

$$\frac{d}{dt}(P(t)(x, y)) = u_0(P(t)(x, y), t).$$

$C_0 = \gamma(Q(t)(x, y))$ is the solution with $Q(t)$ the inverse of $P(t)$. In the spirit of variation of constants, we introduce the change of variables

$$C(x, y, t) = F(Q(t)(x, y), t). \tag{13.5}$$

Note that the technical construction depends very much on the geometry of the tidal basin. We expect F to be slowly time-dependent when introducing Eq. (13.5) into the original equation (13.4). By differentiation, we find explicitly

$$
\frac{\partial}{\partial t} F(x, y, t) = \frac{\partial}{\partial t}(C(P(t)(x, y), t))
$$
$$
= \frac{\partial C}{\partial t} P(t)(x, y), t) \nabla C(P(t)(x, y), t)
$$
$$
= \varepsilon \triangle C(P(t)(x, y), t) - \varepsilon u_1(P(t)(x, y)) \nabla C(P(t)(x, y), t).
$$

We have found a slowly varying equation of the form (13.2). It is not essential to use the assumption that the flow $u_0 + \varepsilon u_1$ is divergence-free, it only facilitates the calculations.

Averaging produces a parabolic equation for $\bar{F}$, the solutions of which tend to the equilibrium solution obtained by putting the right-hand side equal to zero. Equilibrium is described by an elliptic equation that can for instance be solved by a numerical package.

Krol (1991) presents some extensions of the theory and explicit examples where the slowly varying equation is averaged to obtain a time-independent parabolic problem. Quite often such a problem still has to be solved numerically and one may wonder what then is the use of this technique. The answer is that one needs solutions on a long timescale and that numerical integration of an equation where the fast periodic oscillations have been eliminated has been shown to be a much safer procedure.

The analysis presented thus far can be used for bounded and unbounded domains. To study the equation on spatially bounded domains while adding boundary conditions, does not present serious obstacles to the techniques and the proofs.

13.3 Operators with a Discrete Spectrum

In this section, we shall be concerned with theory and examples of weakly nonlinear hyperbolic equations. Such equations can be studied in various ways. Krol (1989) and Buitelaar (1993) consider semilinear wave equations with a discrete spectrum to prove asymptotic estimates on the $1/\varepsilon$ timescale.

The procedure of the averaging theorem by Buitelaar involves solving an infinite number of ordinary differential equations. In most interesting cases, resonance will make this virtually impossible and we have to take recourse to truncation techniques; we discuss results by Krol (1989) on the asymptotic validity of truncation methods that at the same time yield information on the timescale of interaction of modes.

13.3.1 A General Averaging Theorem

We would like to study semilinear initial value problems of hyperbolic type

$$u_{tt} + Au = \varepsilon g(u, u_t, t, \varepsilon), \quad u(0) = u_0, u_t(0) = v_0,$$

where A is a positive self-adjoint linear operator on a separable Hilbert space like the Laplacian with boundary conditions. Examples are the wave equation

$$u_{tt} - u_{xx} = \varepsilon f(u, u_x, u_t, t, x, \varepsilon), \quad t \geq 0, 0 < x < \pi,$$

where

$$u(0,t) = u(\pi,t) = 0, u(x,0) = \phi(x), u_t(x,0) = \psi(x), 0 \leq x \leq \pi,$$

and the Klein-Gordon equation

$$u_{tt} - u_{xx} + a^2 u = \varepsilon u^3, \quad t \geq 0, 0 < x < \pi, a > 0,$$

with similar initial boundary conditions. More generally, consider the semilinear initial value problem

$$\frac{dw}{dt} + \mathcal{A}w = \varepsilon f(w, t, \varepsilon), \quad w(0) = w_0,$$

where $-\mathcal{A}$ generates a uniformly bounded C_0-group $H(t), -\infty < t < +\infty$, on the separable Hilbert space X, and f satisfies certain regularity conditions and can be expanded with respect to ε in a Taylor series, at least to some order. A generalised solution is defined as a solution of the integral equation

$$w(t) = H(t)w_0 + \varepsilon \int_0^t H(t-s)f(w(s), s, \varepsilon)ds.$$

Using the variation of constants transformation $w(t) = H(t)z(t)$, we find the integral equation corresponding with the standard form

$$z(t) = w_0 + \varepsilon \int_0^t F(z(s), s, \varepsilon)ds, \quad F(z, s, \varepsilon) = H(-s)f(H(s)z, s, \varepsilon).$$

Introduce the average F^0 of F by

$$F^0(z) = \lim_{T \to \infty} \frac{1}{T} \int_0^T F(z, s, 0)ds$$

and the averaging approximation $\bar{z}(t)$ of $z(t)$ by

$$\bar{z}(t) = w_0 + \varepsilon \int_0^t F_0(\bar{z}(s))ds.$$

Under rather general conditions, Buitelaar (1993) proves that $z(t)-\bar{z}(t) = o(1)$ on the timescale $1/\varepsilon$; see Section 15.9.

In the case where $F(z,t,\varepsilon)$ is T-periodic in t or quasiperiodic, we have the estimate $z(t) - \bar{z}(t) = O(\varepsilon)$ on the timescale $1/\varepsilon$.

Remark

The procedure described here is important for the formulation of the theory. In practice, the procedure of producing an integral equation and the subsequent operations are usually replaced by expanding the solution in a series of orthogonal functions derived from the unperturbed ($\varepsilon = 0$) equation. Substitution of this series in the equation and taking inner products yields an infinite set of coupled ordinary differential equations that is equivalent with the original equation. This will be demonstrated in the following subsections.

13.3.2 Nonlinear Dispersive Waves

As a prototype of a nonlinear wave equation with dispersion, one often considers the equation

$$u_{tt} - u_{xx} + u = \varepsilon f(u),\ t \geq 0, 0 < x < \pi.$$

In the case $f(u) = \sin u$, this is called the sine-Gordon equation, and if $f(u) = u^3$ it is called a nonlinear, cubic Klein-Gordon equation. The first one is "completely integrable", the second one is not but behaves approximately like an integrable equation.

Example 13.2

Consider the nonlinear Klein-Gordon equation

$$u_{tt} - u_{xx} + u = \varepsilon u^3,\ t \geq 0, 0 < x < \pi, \tag{13.6}$$

with boundary conditions $u(0,t) = u(\pi,t) = 0$ and initial values $u(x,0) = \phi(x), u_t(x,0) = \psi(x)$, which are supposed to be sufficiently smooth. The problem has been studied by many authors; for an introduction to formal approximation procedures, see Kevorkian and Cole (1996).

What do we know qualitatively? We have the existence and uniqueness of solutions on the timescale $1/\varepsilon$ and for all time if we add a minus sign on the right-hand side. Kuksin (1991) and Bobenko and Kuksin (1995) consider Klein-Gordon equations as a perturbation of the (integrable) sine-Gordon equation and prove, in an infinite-dimensional version of the KAM theorem, the persistence of many finite-dimensional invariant manifolds in system (13.6). See also the subsequent discussion of results by Bourgain (1996) and Bambusi (1999).

The equivalent integral equation given above is suitable for proving results. For a quantitative analysis, a Fourier analysis producing an infinite number of ordinary differential equations looks more convenient. Putting $\varepsilon = 0$, we have for the eigenfunctions and eigenvalues

$$v_n(x) = \sin(nx), \lambda_n = \omega_n^2 = n^2 + 1, n = 1, 2, \cdots.$$

We propose to expand the solution of the initial boundary value problem for Eq. (13.6) in a Fourier series with respect to these orthogonal eigenfunctions of the form

$$u(t, x) = \sum_{n=1}^{\infty} u_n(t) v_n(x). \tag{13.7}$$

Substitution of the expansion into Eq. (13.6) produces an infinite series with linear terms on the left-hand side and a series with mixed cubic terms on the right-hand side. At this point, we will use the orthogonality of the eigenfunctions $v_n(x)$. By taking inner products for $n = 1, 2, \cdots$ (i.e., multiplying the equation with $v_m(x)$ and integrating over the segment $[0, \pi]$), most of the terms drop out. Performing this for $m = 1, 2, \cdots$, we find an infinite system of ordinary differential equations that is equivalent to the original problem:

$$\ddot{u}_n + \omega_n^2 u_n = \varepsilon f_n(u_1, u_2, \cdots), \quad n = 1, 2, \cdots.$$

We have not only obtained an infinite number of equations, but the right-hand sides also still contain an infinite number of terms. However, the spectrum given by $\lambda_n, n = 1, 2, \cdots$ is such that for Eq. (13.6) nearly all terms vanish by averaging. Introducing the standard form

$$u_n(t) = a_n(t) \cos \omega_n t + b_n(t) \sin \omega_n t, \tag{13.8}$$

$$\dot{u}_n(t) = -\omega_n a_n(t) \sin \omega_n t + \omega_n b_n(t) \cos \omega_n t, \tag{13.9}$$

we find after averaging (a tilde denotes approximation)

$$\dot{\tilde{a}}_n = -\varepsilon \sigma_n \tilde{b}_n, \quad \dot{\tilde{b}}_n = \varepsilon \sigma_n \tilde{a}_n.$$

We have that the energy is conserved in each mode,

$$\tilde{a}_n^2 + \tilde{b}_n^2 = E_n, \quad n = 1, 2, \cdots,$$

and we have for the constants σ_n

$$\sigma_n = \frac{1}{\lambda_n} \left(\frac{3}{8} \sum_{k=1}^{\infty} E_k - \frac{3}{32} E_n \right).$$

The averaging theorem yields that this approximation has precision $o(\varepsilon)$ on the timescale $1/\varepsilon$; if we start with initial conditions in a finite number of modes, the error is $O(\varepsilon)$.

The result is that the actions or amplitudes are constant to this order of approximation, and the angles are varying slowly as a function of the energy level of the modes. Clearly, more interesting things may happen on a longer timescale or at higher-order approximations.

In a second-order calculation, Stroucken and Verhulst (1987) find inter-action between modes n and $3n$, which leads them to conjecture that on a timescale, longer than $1/\varepsilon$ this interaction will be more prominent.

An example with no conservation of energy is studied by Keller and Ko-gelman (1970), who consider a Rayleigh type of excitation. The authors use multiple timing to first order, which yields the same results as averaging.

Example 13.3
Consider the equation

$$u_{tt} - u_{xx} + u = \varepsilon\left(u_t - \frac{1}{3}u_t^3\right), \ t \geq 0, 0 < x < \pi,$$

with boundary conditions $u(0,t) = u(\pi,t) = 0$ and initial values $u(x,0) = \phi(x), u_t(x,0) = \psi(x)$ that are supposed to be sufficiently smooth. As before, putting $\varepsilon = 0$, we have for the eigenfunctions and eigenvalues

$$v_n(x) = \sin(nx), \ \lambda_n = \omega_n^2 = n^2 + 1, \ n = 1, 2, \cdots,$$

and again we propose to expand the solution of the initial boundary value problem for the equation in a Fourier series with respect to these eigenfunctions of the form (13.7). Substituting the expansion into the differential equation we have

$$\sum_{n=1}^{\infty} \ddot{u}_n \sin nx + \sum_{n=1}^{\infty}(n^2+1)u_n \sin nx = \varepsilon\sum_{n=1}^{\infty} \dot{u}_n \sin nx - \frac{\varepsilon}{3}(\sum_{n=1}^{\infty} \dot{u}_n \sin nx)^3.$$

When taking inner products with $\sin mx, m = 1, 2, \cdots$, the linear terms are easy to obtain; we have to Fourier analyse the cubic term

$$(\sum_{n=1}^{\infty} \dot{u}_n \sin nx)^3 =$$

$$\sum_{1}^{\infty} \dot{u}_n^3 \sin^3 nx + 3\sum_{i \neq j} \dot{u}_i^2\dot{u}_j \sin^2 ix \sin jx + 6\sum_{i \neq j \neq k} \dot{u}_i\dot{u}_j\dot{u}_k \sin ix \sin jx \sin kx.$$

This produces many terms, as the equation for u_m is produced by the terms $\sin mx$ that arise if $n = m$, $3n = m$, $j - 2i = m$, and so on. At this stage, it is clear that we will not have exact normal mode solutions, as for instance mode m will excite mode $3m$.

At this point we can start averaging and it becomes important that the spectrum not be resonant. In particular, we have in the averaged equation for u_m only terms arising from $\dot{u}_m^3$ and $\sum_{i \neq m}^{\infty} \dot{u}_i^2\dot{u}_m$. The other cubic terms do not survive the averaging process; the part of the equation for $m = 1, 2, \cdots$ that produces nontrivial terms is

$$\ddot{u}_m + \omega_m^2 u_m = \varepsilon \left(\dot{u}_m - \frac{1}{4} \dot{u}_m^3 - \frac{1}{2} \sum_{i \neq m}^{\infty} \dot{u}_i^2 \dot{u}_m \right) + \cdots ,$$

where the dots stand for nonresonant terms. This is an infinite system of ordinary differential equations that, apart from the nonresonant terms, is equivalent to the original problem.

Note that later, in Example 13.5, we will have an operator that produces many more complicating resonances.

We can now perform the actual averaging in a notation that contains only minor differences with that of Keller and Kogelman (1970). Introducing the standard form (13.8) as before, we find after averaging the approximations given by (a tilde denotes approximation)

$$2\dot{\tilde{a}}_n = \varepsilon \tilde{a}_n \left(1 + \frac{n^2 + 1}{16} (\tilde{a}_n^2 + \tilde{b}_n^2) - \frac{1}{4} \sum_{k=1}^{\infty} (k^2 + 1)(\tilde{a}_k^2 + \tilde{b}_k^2) \right),$$

$$2\dot{\tilde{b}}_n = \varepsilon \tilde{b}_n \left(1 + \frac{n^2 + 1}{16} (\tilde{a}_n^2 + \tilde{b}_n^2) - \frac{1}{4} \sum_{k=1}^{\infty} (k^2 + 1)(\tilde{a}_k^2 + \tilde{b}_k^2) \right).$$

This system shows fairly strong (although not complete) decoupling because of the nonresonant character of the spectrum. Because of the self-excitation, we have no conservation of energy. Putting $\tilde{a}_n^2 + \tilde{b}_n^2 = E_n$, $n = 1, 2, \cdots$, multiplying the first equation with $\tilde{a}_n$ and the second equation with $\tilde{b}_n$, and adding the equations, we have

$$\dot{E}_n = \varepsilon E_n \left(1 + \frac{n^2 + 1}{16} E_n - \frac{1}{4} \sum_{k=1}^{\infty} (k^2 + 1) E_k \right).$$

We have immediately a nontrivial result: starting in a mode with zero energy, this mode will not be excited on a timescale $1/\varepsilon$. Another observation is that if we have initially only one nonzero mode, say for $n = m$, the equation for E_m becomes

$$\dot{E}_m = \varepsilon E_m \left(1 - \frac{3}{16} (m^2 + 1) E_m \right).$$

We conclude that we have stable equilibrium at the value

$$E_m = \frac{16}{3(m^2 + 1)}.$$

More generally, the averaging theorem yields that the approximate solutions have precision $o(\varepsilon)$ on the timescale $1/\varepsilon$; if we start with initial conditions in a finite number of modes the error is $O(\varepsilon)$.

We consider now a conservative system with a few more complications. The example is based on Buitelaar (1994).

Example 13.4

Consider a homogeneous rod that is distorted by external forces. The deformations are described by the extension $u(x,t)$ and the torsion angle $\theta(x,t)$. The rod has fixed ends and without external forces is located along the x-axis, $0 \leq x \leq \pi$. The equations we will consider are

$$u_{tt} = b_1 u_{xx} + b_2 \theta_{xx} - a_1 u - a_2 \theta + \varepsilon(c_0 u^2 + c_1 u \theta + c_2 \theta^2),$$

$$\theta_{tt} = b_2 u_{xx} + b_3 \theta_{xx} - a_2 u - a_3 \theta + \varepsilon\left(\frac{1}{2}c_1 u^2 + 2c_2 u\theta + d_2\theta^2\right).$$

We assume smooth initial values and zero boundary conditions; the external force is derived from a potential. The coefficients are constants that are determined by the physical properties of the rod.

Analysis of the linear system obtained for $\varepsilon = 0$ produces the eigenfunctions for u and θ,

$$U_n(x) = \sin nx, \ V_n(x) = \sin nx, \ n = 1, 2, \cdots,$$

so we propose to expand the solution of the initial boundary value problem for the system in two Fourier series with respect to these eigenfunctions of the form (13.7):

$$u(x,t) = \sum_{n=1}^{\infty} u_n(t) \sin nx, \ \theta(x,t) = \sum_{n=1}^{\infty} \Theta_n(t) \sin nx.$$

Substituting into the system and, after taking the inner products with $\sin nx$, we find for $n = 1, 2, \cdots$

$$\ddot{u}_n + (a_1 + b_1 n^2) u_n + (a_2 + b_2 n^2)\Theta_n = \varepsilon\ldots,$$

$$\ddot{\Theta}_n + (a_2 + b_2 n^2) u_n + (a_3 + b_3 n^2)\Theta_n = \varepsilon\ldots,$$

where the dots stand for an infinite series of quadratic terms in u_n and Θ_n. The basic frequencies of the unperturbed system can be obtained in the usual way as the eigenvalues of a matrix of coefficients, in this case

$$\begin{pmatrix} 0 & 1 & 0 & 0 \\ -(a_1 + b_1 n^2) & 0 & -(a_2 + b_2 n^2) & 0 \\ 0 & 0 & 0 & 1 \\ -(a_2 + b_2 n^2) & 0 & -(a_3 + b_3 n^2) & 0 \end{pmatrix}$$

with characteristic equation for the eigenvalues

$$(\lambda^2 + a_1 + b_1 n^2)(\lambda^2 + a_3 + b_3 n^2) = (a_2 + b_2 n^2)^2.$$

It is possible, by linear transformation, to write the system as two coupled harmonic equations. We will not do that, but we use the well-known fact that when coupling harmonic equations by nonlinear polynomials, when averaging to first order only the 2 : 1 resonance appears; for more details, see Sanders and Verhulst (1985) or Verhulst (1998). Other resonances may appear at higher order. As an example (following Buitelaar, 1994), we choose

$$a_1 = b_1 = \frac{5}{2}, \ a_2 = b_2 = \frac{3}{2}, \ a_3 = b_3 = \frac{5}{2},$$

leading to the frequencies $\sqrt{n^2 + 1}$ and $2\sqrt{n^2 + 1}, n = 1, 2, \cdots$. The implication is that in this case the distorted rod can be described by an infinite set of decoupled systems of two degrees of freedom in 2 : 1 resonance. Whether this resonance is effective (i.e., there is coupling between the modes of the systems with two degrees of freedom) still depends on the nonlinearities and on the initial conditions. Buitelaar (1994) shows that in this example the even modes (n even) average to zero and the odd modes produce resonance.

Another remark is that we can also choose the coefficients such that there is no resonance at first order. In this case, resonances may appear on timescales longer than $1/\varepsilon$ and between more than two modes.

13.3.3 Averaging and Truncation

The averaging result by Buitelaar is of importance in its generality; in many interesting cases, however, because of resonances, the resulting averaged system is difficult to analyse and we need additional theorems. One of the most important techniques involves truncation, which has been validated for a number of wave equations by Krol (1989).

To fix the idea, consider the initial boundary value problem for the nonlinear wave equation

$$u_{tt} - u_{xx} = \varepsilon f(u, u_x, u_t, t, x, \varepsilon), \quad t \geq 0, 0 < x < \pi, \qquad (13.10)$$

with

$$u(0,t) = u(\pi,t) = 0, u(x,0) = \phi(x), u_t(x,0) = \psi(x), 0 \leq x \leq \pi.$$

The eigenfunctions of the unperturbed ($\varepsilon = 0$) problem are $v_n(x) = \sin(nx)$, $n = 1, 2, \cdots$ (normalisation is not necessary), and we propose as before to expand the solution of the initial boundary value problem for Eq. (13.10) in a Fourier series with respect to these eigenfunctions of the form

$$u(t,x) = \sum_{n=1}^{\infty} u_n(t)v_n(x).$$

After taking inner products to obtain an infinite system of ordinary differential equations, the next step is then to truncate this infinite-dimensional system to N modes. Subsequently we apply averaging to the truncated system. The truncation, also called projection on a finite-dimensional subspace, is known as Galerkin's method, and one has to estimate the combined error of truncation and averaging.

The first step is that Eq. (13.10) with its initial boundary values has exactly one solution in a suitably chosen Hilbert space $\mathcal{H}_k = H_0^k \times H_0^{k-1}$, where H_0^k are the well-known Sobolev spaces consisting of functions u with derivatives $U^{(k)} \in L_2[0,\pi]$ and $u^{(2l)}$ zero on the boundary whenever $2l < k$. It is not trivial but rather standard to establish existence and uniqueness of solutions *on the timescale* $1/\varepsilon$ under certain mild conditions on f; examples are right-hand sides f such as $u^3, uu_t^2, \sin u, \sinh u_t$, etc. Moreover, we note that:

1. If $k \geq 3, u$ is a classical solution of Eq. (13.10).
2. If $f = f(u)$ is an odd function of u, one can find an even energy integral. If such an integral represents a positive definite energy integral, we are able to prove existence and uniqueness for all time.

To find u_N, the projection or truncation of the solution u, we have to solve a $2N$-dimensional system of ordinary differential equations for the expansion coefficients $u_n(t)$ with appropriate (projected) initial values. The estimates for the error $\|u - u_N\|$ depend strongly on the smoothness of the right-hand side f of Eq. (13.10) and the initial values $\phi(x), \psi(x)$ but, remarkably enough, not on ε. The truncated system is in general difficult to solve. Averaging in the periodic case produces an approximation $\bar{u}_N$ of u_N and finally the following theorem.

Theorem 13.2
Galerkin averaging theorem (Krol, 1989)
Consider the initial-boundary value problem for Eq. (13.10)

$$u_{tt} - u_{xx} = \varepsilon f(u, u_x, u_t, t, x, \varepsilon), \quad t \geq 0, 0 < x < \pi,$$

where

$$u(0,t) = u(\pi,t) = 0, u(x,0) = \phi(x), u_t(x,0) = \psi(x), 0 \leq x \leq \pi.$$

Suppose that f is k times continuously differentiable and satisfies the existence and uniqueness conditions on the timescale $1/\varepsilon, (\phi, \psi) \in \mathcal{H}_k$; if the solution of the initial boundary problem is (u, u_t) and the approximation obtained by the Galerkin averaging procedure $(\bar{u}_N, \bar{u}_{Nt})$, we have on the timescale $1/\varepsilon$

$$\|u - \bar{u}_N\|_{\sup} = O(N^{\frac{1}{2}-k}) + O(\varepsilon), \quad N \to \infty, \varepsilon \to 0,$$
$$\|u_t - \bar{u}_{Nt}\|_{\sup} = O(N^{\frac{3}{2}-k}) + O(\varepsilon), \quad N \to \infty, \varepsilon \to 0.$$

There are a number of remarks:

- Taking $N = O(\varepsilon^{-\frac{2}{2k-1}})$, we obtain an $O(\varepsilon)$-approximation on the timescale $1/\varepsilon$, so the required number of modes decreases when the regularity of the data and the order up to which they satisfy the boundary conditions increases.

- However, this decrease in the number of required modes is not uniform in k, so it is not obvious for which choice of k the estimates are optimal at a given value of ε.

- An interesting case arises if the nonlinearity f satisfies the regularity conditions for all k. This happens for instance if f *is an odd polynomial in u* and with analytic initial values. In such cases, the results are

$$\|u - \bar{u}_N\|_{\sup} = O(N^{-1}a^{-N}) + O(\varepsilon), \ N \to \infty, \varepsilon \to 0,$$
$$\|u_t - \bar{u}_{Nt}\|_{\sup} = O(a^{-N}) + O(\varepsilon), \ N \to \infty, \varepsilon \to 0,$$

where the constant a arises from the bound one has to impose on the size of the strip around the real axis on which analytic continuation is permitted in the initial boundary value problem. The important implication is that, because of the a^{-N} term we need only $N = O(|\ln \varepsilon|)$ terms to obtain an $O(\varepsilon)$-approximation on the timescale $1/\varepsilon$.

- Here and in the following we have chosen Dirichlet boundary conditions. It is stressed that this is by way of example and not a restriction. In Krol's method, we can also include Neumann conditions, periodic boundary conditions, etc.

- It is possible to generalise these results to higher-dimensional (spatial) problems; see Krol (1989) for remarks and see the example of Pals (1996) below for an analysis of a two-dimensional nonlinear Klein-Gordon equation with Dirichlet boundary conditions. It is possible to include dispersion, although not without some additional difficulties.

It is not difficult to apply the Galerkin averaging method to the examples of the preceding subsection; this is left to the reader.

The following example of the Rayleigh wave equation is used sometimes as a model for wind-induced vibration phenomena of overhead power lines. Formal approximations obtained by different techniques can be found in Chikwendu and Kevorkian (1972) and Lardner (1977); see also the discussion and estimates in Van Horssen (1988).

Example 13.5
Consider the Rayleigh wave equation with initial boundary conditions

$$u_{tt} - u_{xx} = \varepsilon(u_t - u_t^3), \ t \geq 0, 0 < x < \pi,$$

where

$$u(0,t) = u(\pi, t) = 0, u(x,0) = \phi(x), u_t(x,0) = \psi(x), 0 \leq x \leq \pi.$$

In most papers, one uses $\frac{1}{3}u_t^3$ instead of u_t^3; replacing u by $\sqrt{3}u$ changes the coefficient to 1. For the eigenfunctions, we take $v_n(x) = \sin nx$, and after substitution of the eigenfunction expansion (13.7) we obtain an equation of the form

$$\sum_{n=1}^{\infty} \ddot{u}_n \sin nx + \sum_{n=1}^{\infty} n^2 u_n \sin nx = \varepsilon \sum_{n=1}^{\infty} \dot{u}_n \sin nx - \varepsilon (\sum_{n=1}^{\infty} \dot{u}_n \sin nx)^3.$$

Multiplying with $\sin mx$ and integrating over $[0, \pi]$, we find for $m = 1, 2, \cdots$ an infinite system of nonlinearly coupled equations.

Are (approximate) normal modes or, more generally, periodic solutions possible? As a simple case, we choose $N = 3$, a three-mode expansion, so that

$$u_N(x, t) = u_1(t) \sin x + u_2(t) \sin 2x + u_3(t) \sin 3x.$$

This makes sense if we restrict the initial values to the first three modes. Substitution into the Rayleigh wave equation and taking inner products with v_1, v_2, v_3 produces

$$\ddot{u}_1 + u_1 = \varepsilon \left(\dot{u}_1 - \frac{3}{4}\dot{u}_1^3 - \frac{3}{2}\dot{u}_1\dot{u}_2^2 - \frac{3}{2}\dot{u}_1\dot{u}_3^2 + \frac{3}{4}\dot{u}_1^2\dot{u}_3 - \frac{3}{4}\dot{u}_2^2\dot{u}_3 \right),$$

$$\ddot{u}_2 + 4u_2 = \varepsilon \left(\dot{u}_2 - \frac{3}{4}\dot{u}_2^3 - \frac{3}{2}\dot{u}_1^2\dot{u}_2 - \frac{3}{2}\dot{u}_2\dot{u}_3^2 - \frac{3}{2}\dot{u}_1\dot{u}_2\dot{u}_3 \right),$$

$$\ddot{u}_3 + 9u_3 = \varepsilon \left(\dot{u}_3 - \frac{3}{4}\dot{u}_3^3 - \frac{3}{2}\dot{u}_1^2\dot{u}_3 - \frac{3}{2}\dot{u}_2^2\dot{u}_3 - \frac{3}{4}\dot{u}_1\dot{u}_2^2 + \frac{1}{4}\dot{u}_1^3 \right).$$

Because of the resonant spectrum of the reduced ($\varepsilon = 0$) equation, there are many interacting terms. On the right-hand side of each equation, the first four terms are generic (in the case of N modes, this would be $N + 1$ terms). The other nonlinearities are caused by special spectral resonances.

It is clear that, taking in this three-mode expansion the second mode zero, we have a four-dimensional invariant manifold. Consider as an example this manifold (i.e., $u_2 = \dot{u}_2 = 0$).

To illustrate the phenomena, we shall use amplitude-phase variables. Introducing the transformation to a slowly varying system by $u_i = r_i \cos(\omega t + \phi_i)$, $\dot{u}_i = -r_i \omega \sin(\omega t + \phi_i)$, $i = 1, 3$ and first-order averaging produces

$$\dot{\tilde{r}}_1 = \frac{\varepsilon}{2}\tilde{r}_1 \left(1 - \frac{9}{16}\tilde{r}_1^2 - \frac{27}{4}\tilde{r}_3^2 - \frac{9}{16}\tilde{r}_1\tilde{r}_3 \cos(3\tilde{\phi}_1 - \tilde{\phi}_3) \right),$$

$$\dot{\tilde{\phi}}_1 = \frac{9\varepsilon}{32}\tilde{r}_1\tilde{r}_3 \sin(3\tilde{\phi}_1 - \tilde{\phi}_3),$$

$$\dot{\tilde{r}}_3 = \frac{\varepsilon}{2}\tilde{r}_3 \left(1 - \frac{81}{16}\tilde{r}_3^2 - \frac{3}{4}\tilde{r}_1^2 \right) - \frac{\varepsilon}{96}\tilde{r}_1^3 \cos(3\tilde{\phi}_1 - \tilde{\phi}_3),$$

$$\dot{\tilde{\phi}}_3 = -\frac{\varepsilon\tilde{r}_1^3}{96\tilde{r}_3} \sin(3\tilde{\phi}_1 - \tilde{\phi}_3).$$

Looking for periodic solutions outside the normal modes, we have the requirement of phase-locking,

$$\sin(3\tilde{\phi}_1 - \tilde{\phi}_3) = 0,$$

and simultaneously

$$1 - \frac{9}{16}\tilde{r}_1^2 - \frac{27}{4}\tilde{r}_3^2 \pm \tilde{r}_1\tilde{r}_3 = 0, \ \tilde{r}_3 - \frac{81}{16}\tilde{r}_3^3 - \frac{3}{4}\tilde{r}_1^2\tilde{r}_3 \pm \frac{1}{48}\tilde{r}_1^3 = 0.$$

Combining the requirements, we find that there is one positive solution $\tilde{r}_1, \tilde{r}_3$ corresponding with a periodic solution of the projected three-mode system. On a timescale longer than $1/\varepsilon$, the interactions with higher-order modes have to be taken into account. It cannot be expected that such a periodic solution survives these interactions.

Considering the nonlinear (cubic) Klein-Gordon equation with one spatial variable, we found rather simple behaviour by first-order averaging. Pals (1996) found that the situation is different and much more complicated in the case of more spatial variables.

Example 13.6
Consider the cubic Klein-Gordon equation with initial boundary conditions

$$u_{tt} - u_{xx} - \alpha u_{yy} + \beta u = \varepsilon u^3, \ t \ge 0, 0 < x, y < \pi,$$

with $\alpha, \beta > 0$ and where on the boundaries

$$u(0, y, t) = u(\pi, y, t) = u(x, 0, t) = u(x, \pi, t) = 0,$$

and initially

$$u(x, y, 0) = \phi(x, y), u_t(x, y, 0) = \psi(x, y), \ 0 \le x, y \le \pi.$$

The parameter α originates from rescaling a rectangle to a square.
Pals (1996) gives a fairly complete quantitative analysis of the problem together with proofs of asymptotic validity of the approximations. We have the orthogonal eigenfunctions $v_{kl} = \sin kx \sin ly$ and eigenvalues $\lambda_{kl} = k^2 + \alpha l^2 + \beta$. Introducing the eigenfunction expansion

$$u(x, y, t) = \sum_{k,l=1}^{\infty} u_{kl}(t) v_{kl}(x, y)$$

and taking inner products, we find the infinite system of coupled equations

$$\ddot{u}_{kl} + \lambda_{kl} u_{kl} = \varepsilon \sum_{k_i, l_i=1}^{\infty} c_{k_i l_i} u_{k_1 l_1} u_{k_2 l_2} u_{k_3 l_3},$$

where

$$c_{k_i l_i} = \int_{x=0}^{\pi} \int_{y=0}^{\pi} \sin k_1 x \sin k_2 x \sin k_3 x \sin kx \sin l_1 y \sin l_2 y \sin l_3 y \sin ly \, dx \, dy.$$

We can write the product of four sines as the sum of eight cosines. The arguments are of the form

$$(k \pm k_1 \pm k_2 \pm k_3)x, \ (l \pm l_1 \pm l_2 \pm l_3)y,$$

and $c_{k_i l_i}$ does not vanish if simultaneously

$$k = \pm k_1 \pm k_2 \pm k_3, \ l = \pm l_1 \pm l_2 \pm l_3,$$

for some combinations of k_i, l_i. Pals (1996) shows that the normal modes are exact solutions of the infinite system and there are many other invariant manifolds with dimension larger than two. He also shows that the key resonance is 1 : 1. This resonance may be isolated and its dynamics takes place on a four-dimensional invariant manifold, or it can be connected to other 1 : 1 resonances, as in the $1 : 1 : a : a$-resonance $(a > 0)$.

As an example of the results, we consider the isolated 1 : 1 resonance. Suppose that $\lambda_{ij} = \lambda_{pq} = \lambda$, whereas there exists no $(k, l) \neq (i, j)$ and $(k, l) \neq (p, q)$ such that $\lambda_{kl'} = \lambda$. If the initial conditions are only nonzero in the (i, j) and (p, q) modes, the other modes are not excited, and there is no energy transfer out of this system with two degrees of freedom in 1 : 1 resonance. Averaging to first order in amplitude-phase variables $r_{ij}, \phi_{ij}, r_{pq}, \phi_{pq}$, rescaling time, and putting for the averaged quantities $\tilde{r}_{ij} = r_1, \tilde{\phi}_{ij} = \phi_1, \tilde{r}_{pq} = r_2, \tilde{\phi}_{pq} = \phi_2$, we find

$$\dot{r}_1 = r_1 r_2^2 \sin 2(\phi_1 - \phi_2),$$
$$\dot{\phi}_1 = r_2^2 \cos 2(\phi_1 - \phi_2) + \frac{9}{4}r_1^2 + 2r_2^2,$$
$$\dot{r}_2 = -r_1^2 r_2 \sin 2(\phi_1 - \phi_2),$$
$$\dot{\phi}_2 = r_1^2 \cos 2(\phi_1 - \phi_2) + \frac{9}{4}r_2^2 + 2r_1^2.$$

The Klein-Gordon equation is conservative, it can be put in a Hamiltonian framework, and this also holds for the truncated subsystem. The integral $r_1^2 + r_2^2 = $ constant of the averaged system is an approximate energy integral; as usual when normalising Hamiltonian systems, there exists a second integral. Apart from the (unstable) normal modes, the averaged system has two critical points, which are both stable. They correspond with stable periodic solutions of the equations for u_{ij} and u_{pq}. In turn, these periodic solutions correspond with two packets of standing waves, parametrised by the energy, of the original two-dimensional Klein-Gordon equation. One should consult Pals (1996) for details and the analysis of other interesting cases.

In the case of Example 13.5 for the Rayleigh wave equation, a fundamental problem is that the spectrum of the (unperturbed) equation is fully resonant. This problem is still poorly understood. We consider now a prototype problem of full resonance to review the results that are available.

Example 13.7

In Kevorkian and Cole (1996) and Stroucken and Verhulst (1987), the following initial boundary value problem is briefly discussed:

$$u_{tt} - u_{xx} = \varepsilon u^3, \quad t \geq 0, 0 < x < \pi,$$

with boundary conditions $u(0,t) = u(\pi,t) = 0$ and initial values $u(x,0) = \phi(x), u_t(x,0) = \psi(x)$, which are supposed to be sufficiently smooth.

Using again the eigenfunction expansion (13.7) with $v_n(x) = \sin(nx), \lambda_n = n^2, n = 1, 2, \cdots$, we have to solve an infinite system of coupled ordinary differential equations for the coefficients $u_n(t)$. In fact, the problem is reminiscent of the famous Fermi-Pasta-Ulam problem (see for references and a discussion Jackson (1978, 1991)), and it displays similar interaction between the modes and recurrence.

Apart from numerical approximation, Galerkin averaging seems to be a possible approach, and we state here the application of Krol's (1989) Galerkin averaging theorem to this problem with the cubic term. Suppose that for the initial values ϕ, ψ we have a finite-mode expansion of M modes only, where of course we take $N \geq M$ in the eigenfunction expansion. Now the initial values ϕ, ψ are analytic, and Krol (1989) optimises the way in which the analytic continuation of the initial values takes place. The analysis leads to the estimate for the approximation $\bar{u}_N$ obtained by Galerkin averaging:

$$\|u - \bar{u}_N\|_\infty = O(\varepsilon^{\frac{N+1-M}{N+1+2M}}), \quad 0 \leq \varepsilon^{\frac{N+1}{N+1+2M}} t \leq 1.$$

It is clear that if $N \gg M$ the error estimate tends to $O(\varepsilon)$ and the timescale to $1/\varepsilon$. The result can be interpreted as an upper bound for the speed of energy transfer from the first M modes to higher-order modes.

Important progress has been achieved by Van der Aa and Krol (1990), who apply Birkhoff normalisation, which is asymptotically equivalent to averaging, to the Hamiltonian system generated by H^N, which is the Hamiltonian of the truncated system for arbitrary N; the normalised Hamiltonian is indicated by $\bar{H}^N$. Remarkably enough, the flow generated by $\bar{H}^N$ for arbitrary N contains many invariant manifolds.

Consider the "odd" manifold M_1, which is characterised by the fact that only odd-numbered modes are involved in M_1. Inspection of $\bar{H}^N$ reveals that M_1 is an invariant manifold.

In the same way the "even" manifold M_2 is characterised by the fact that only even-numbered modes are involved; this is again an invariant manifold of $\bar{H}^N$. In Stroucken and Verhulst (1987), this was noted for $N = 3$, which is rather restricted; moreover, it can be extended to manifolds M_m with $m = 2^k q, q$ an odd natural number, and k a natural number. It turns out that projections to two modes yield little interaction, so this motivates looking at projections with at least $N = 6$ involving the odd modes $1, 3, 5$ on M_1 and $2, 4, 6$ on M_2.

Van der Aa and Krol (1990) analyse $\bar{H}^6$, in particular the periodic solutions on M_1. For each value of the energy, this Hamiltonian produces three normal mode (periodic) solutions that are stable on M_1. Analysing the stability in the full system generated by $\bar{H}^6$ Van der Aa and Krol again find stability.

An open question is whether there exist periodic solutions in the flow generated by $\bar{H}^6$ that are not contained in either M_1 or M_2.

What is the relation between the periodic solutions found by averaging and periodic solutions of the original nonlinear wave problem? Van der Aa and Krol (1990) compare these solutions with results obtained by Fink et al. (1974), who employ the Poincaré-Lindstedt continuation method, to prove existence and to approximate periodic solutions. Related results employing elliptic functions have been derived by Lidskii and Shulman (1988). It turns out that there is very good agreement between the various results, but the calculation by the Galerkin averaging method is technically simpler.

We conclude with a problem that still contains many interesting open questions. It was introduced by Rand et al. (1995) and concerns a nonlinear partial differential equation with parametric excitation.

Example 13.8

Consider the equation

$$u_{tt} - c^2 u_{xx} + \varepsilon\beta u_t + (\omega_0^2 + \varepsilon\gamma\cos t)u = \varepsilon\alpha u^3, \quad t \geq 0, 0 < x < \pi,$$

with $\beta \geq 0$, the other constants positive, and Neumann boundary conditions $u_x(0,t) = u_x(\pi,t) = 0$.

Rand et al. (1995) consider a three-mode truncation that is complicated and contains many phenomena. As predicted by Krol's (1989) analysis, the results are confirmed by numerical analysis on a timescale $1/\varepsilon$. It is easy to see that when truncating to an arbitrary number of modes N, all the normal modes are solutions of the truncated system and of course this carries over to the averaged system. Interestingly, on a timescale longer than $1/\varepsilon$, other mode interactions may come up, and the treatment of this is still an open problem. We shall choose $\beta = 0, \alpha = 1$ in this example. Rand (1996) considers a two-mode truncation, and we shall discuss some aspects of this analysis. Putting

$$u_{(N=2)}(x,t) = u_1(t) + u_2(t)\cos x,$$

substituting this into the differential equation and taking inner products produces the system

$$\ddot{u}_1 + \omega_0^2 u_1 + \varepsilon\gamma u_1 \cos t = \varepsilon\left(u_1^3 + \frac{3}{2}u_1 u_2^2\right),$$

$$\ddot{u}_2 + (\omega_0^2 + c^2)u_2 + \varepsilon\gamma u_2 \cos t = \varepsilon\left(3u_1^2 u_2 + \frac{3}{4}u_2^3\right).$$

Even with a two-mode expansion, several low-order resonances are possible. A first choice could be $\omega_0^2 = 1/4, c^2 = \delta\varepsilon$ (with δ a positive constant independent of ε) corresponding with a low wave speed. A second interesting choice could be $\omega_0^2 = 1/4, c^2 = 3/4$. In the second case, we have a first-order Flocquet tongue in the first equation and a second order tongue in the second equation (see Example 10.9). The nonlinear interaction might be interesting.

Consider now the first choice and the question of whether the trivial solution $u_1(t) = u_2(t) = 0$ and the normal modes $u_1(t) \neq 0, u_2(t) = 0$ and $u_1(t) = 0, u_2(t) \neq 0$ are stable. As we have seen in Chapter 11, the trivial solution of the Mathieu equation at this resonance value is unstable. We conclude that this also holds for the trivial solution $u_1(t) = u_2(t) = 0$ and the normal modes; the nonlinear terms do not change this. A more complete analysis shows many interesting bifurcations, see Rand (1996).

13.4 Guide to the Literature

Formal approximation methods have been nicely presented by Kevorkian and Cole (1996). A number of results of the Kiev school of mathematics can be found in Mitropolsky, Khoma, and Gromyak (1997).

An interesting formal approach of averaging Lagrangians by Whitham (1974) has been influential; see also Luke (1966), Bourland and Haberman (1989), and Haberman (1991). Some of these formal methods for nonlinear hyperbolic equations on unbounded domains have been analysed with respect to the question of asymptotic validity by Van der Burgh (1979). See also the survey paper by Verhulst (1999).

An adaptation of the Poincaré-Lindstedt method for periodic solutions of weakly nonlinear hyperbolic equations was given by Hale (1967) based on the implicit function theorem.

An extension of the advection-diffusion problem discussed above has been obtained by Heijnekamp et al. (1995). They considered the problem with initial and boundary values on a two-dimensional domain with small nonlinear reaction terms and a periodic source term.

An early version of Krol's (1989) Galerkin averaging method can be found in Rafel (1983), who considers vibrations of bars. For interesting extensions of Galerkin averaging see Fečkan (2000, 2001).

Applying Buitelaar's (1993) averaging theorem (Section 15.9), we often have the case of averaging of an almost periodic infinite-dimensional vector field that yields an $o(1)$-approximation on the timescale $1/\varepsilon$ in the case of general, smooth initial values. As mentioned earlier, it is not difficult to improve this result in the case of finite-mode initial values (i.e., the initial values can be expressed in a finite number of eigenfunctions $v_n(x)$). In this case, the error becomes $O(\varepsilon)$ on the timescale $1/\varepsilon$ if N is taken large enough.

Using the method of two timescales, Van Horssen and Van der Burgh (1988) construct an asymptotic approximation of a wave equation with estimate $O(\varepsilon)$ on the timescale $1/\sqrt{\varepsilon}$. Van Horssen (1992) develops a method to prove an $O(\varepsilon)$-approximation on the timescale $1/\varepsilon$ that is applied to the nonlinear Klein-Gordon equation with a quadratic nonlinearity $(-\varepsilon u^2)$.

Qualitative insight is added by Bourgain (1996), who considers the nonlinear Klein-Gordon equation in the rather general form

$$u_{tt} - u_{xx} + V(x)u = \varepsilon f(u), \quad t \geq 0, 0 < x < \pi,$$

with V a periodic, even function and $f(u)$ an odd polynomial in u. Assuming a rapid decrease of the amplitudes in the eigenfunction expansion (13.7) and Diophantine (nonresonance) conditions on the spectrum, it is proved that *infinite*-dimensional invariant tori persist in this nonlinear wave equation corresponding with almost-periodic solutions. The proof involves a perturbation expansion that is valid on a timescale $1/\varepsilon^M$ with $M > 0$ a fixed number. Bambusi (1999) considers the nonlinear Klein-Gordon equation in the general form

$$u_{tt} - u_{xx} + mu = \varepsilon \phi(x, u), \quad t \geq 0, 0 < x < \pi,$$

with boundary conditions. The function $\phi(x, u)$ is polynomial in u, analytic and periodic in x, and odd in the sense that $\phi(x, u) = -\phi(-x, -u)$. Under a certain nonresonance condition on the spectrum, Bambusi (1999) shows that the solutions remain close to finite-dimensional invariant tori corresponding with quasiperiodic motion on timescales longer than $1/\varepsilon$. These statements add to the understanding and interpretation of the averaging results.

13.5 Exercises

Exercise 13.1 Consider again the equation with a Rayleigh-type nonlinearity from Example 13.3, studied by Keller and Kogelman (1970). Formulate the problem with initial values in two modes only. Try to draw conclusions from the averaged equations.

Exercise 13.2 As an example of averaging for advection-diffusion problems, consider the problem

$$C_t + e^{-y} \cos t C_x - \varepsilon \Delta C = \varepsilon \delta(0, 0)$$

for $-\infty \leq x \leq \infty, y \geq 0, t \geq 0$, where δ is the Dirac delta function. We assume that on the boundary $C(x, 0, t) = 0$ and suitable initial values. This can be seen as a simple model of advection diffusion in a tidal motion along the coast where at $(0, 0)$ a constant inflow of polluted material takes place.

Derive the averaged equation and show that the flow tends to a stationary solution described by an elliptic equation. Can you visualise this final state? For more details and other examples, see Krol (1991).

Exercise 13.3 Consider again the Rayleigh wave equation of Example 13.5. Formulate the truncated system in the case of a general two-mode expansion (mode m and mode n). Choose $n \neq 3m, m \neq 3n$. The system reminds us of example 12.13 (Hale); compare the results.

Exercise 13.4 Another interesting possibility is to consider Example 13.3 for two spatial variables. Formulate the problem on a square with Dirichlet conditions (u vanishes on the boundary) and study a two-mode Galerkin projection. There is of course some freedom in choosing which two modes one considers.

Exercise 13.5 As a research problem, consider again from Example 13.8 the equation

$$u_{tt} - c^2 u_{xx} + \varepsilon \beta u_t + (\omega_0^2 + \varepsilon \gamma \cos t)u = \varepsilon \alpha u^3, \quad t \geq 0, 0 < x < \pi,$$

with Neumann boundary conditions $u_x(0, t) = u_x(\pi, t) = 0$.

We allow for damping to this wave equation, so $\beta > 0$. How does this affect the stability of the trivial solution and the two normal modes in a two-mode expansion as in Example 13.8? Note that there are still many unsolved questions regarding this equation.

Wave Equations on Unbounded Domains

In the study of evolution equations describing wave phenomena on unbounded domains, one is confronted with a great many concepts and methods but usually only formal results. This is not a good reason to avoid the subject, as there are many interesting mathematical questions and physical phenomena in this field. Also, many parts of physics and engineering require a practical approach to real-life problems that cannot wait until rigorous mathematical methods are available.

So, this chapter will be different from the preceding ones because in some of the results discussed here a mathematical justification is lacking. This holds in particular when we are discussing perturbations of strongly nonlinear partial differential equations. The interest of the problems and the elegance of the methods will hopefully make up for this. Also, it may inspire exploration of the mathematical foundations of the methods discussed in this chapter.

14.1 The Linear Wave Equation with Dissipation

Consider as a simple example the wave equation with weak energy dissipation (damping)

$$\frac{\partial^2 u}{\partial x^2} = \frac{\partial^2 u}{\partial t^2} + \varepsilon \frac{\partial u}{\partial t}, \quad -\infty < x < \infty, \ t > 0,$$

with initial values $u(x,0) = f(x)$, $u_t(x,0) = 0$.

It is not difficult to see that a regular expansion of the form $u(x,t) = u_0(x,t) + \varepsilon u_1(x,t) + \cdots$ produces secular terms (Exercise 14.1). If $\varepsilon = 0$, it is useful to use characteristic coordinates

$$\xi = x - t, \ \eta = x + t.$$

From earlier experiences, we expect that the perturbation will also involve a timescale $\tau = \varepsilon t$ and maybe a spatial scale εx. For simplicity, we will consider a multiple-timescale expansion in three variables: ξ, η, and τ. Assuming that the solutions are C^2, we have

$$\frac{\partial}{\partial x} = \frac{\partial}{\partial \xi} + \frac{\partial}{\partial \eta}, \ \frac{\partial}{\partial t} = -\frac{\partial}{\partial \xi} + \frac{\partial}{\partial \eta} + \varepsilon \frac{\partial}{\partial \tau},$$

$$\frac{\partial^2}{\partial x^2} = \frac{\partial^2}{\partial \xi^2} + 2\frac{\partial^2}{\partial \xi \partial \eta} + \frac{\partial^2}{\partial \eta^2},$$

$$\frac{\partial^2}{\partial t^2} = \frac{\partial^2}{\partial \xi^2} - 2\frac{\partial^2}{\partial \xi \partial \eta} + \frac{\partial^2}{\partial \eta^2} + 2\varepsilon \left(-\frac{\partial^2}{\partial \xi \partial \tau} + \frac{\partial^2}{\partial \eta \partial \tau} \right) + \varepsilon^2 \frac{\partial^2}{\partial \tau^2}.$$

The wave equation transforms to

$$\frac{\partial^2 u}{\partial \xi \partial \eta} = \frac{\varepsilon}{2} \left(-\frac{\partial^2 u}{\partial \xi \partial \tau} + \frac{\partial^2 u}{\partial \eta \partial \tau} \right) + \frac{\varepsilon}{4} \left(-\frac{\partial u}{\partial \xi} + \frac{\partial u}{\partial \eta} \right) + \varepsilon^2 \cdots.$$

The multiple-timescale expansion is of the form

$$u = u_0(\xi, \eta, \tau) + \varepsilon u_1(\xi, \eta, \tau) + \varepsilon^2 \cdots.$$

Substitution in the transformed wave equation yields equations for u_0 and u_1:

$$\frac{\partial^2 u_0}{\partial \xi \partial \eta} = 0, \ \frac{\partial^2 u_1}{\partial \xi \partial \eta} = \frac{1}{2} \left(-\frac{\partial^2 u_0}{\partial \xi \partial \tau} + \frac{\partial^2 u_0}{\partial \eta \partial \tau} \right) + \frac{1}{4} \left(-\frac{\partial u_0}{\partial \xi} + \frac{\partial u_0}{\partial \eta} \right).$$

From the first equation, we obtain

$$u_0(\xi, \eta, \tau) = F(\xi, \tau) + G(\eta, \tau)$$

with F and G arbitrary C^2 functions. Integration of the equation for u_1 yields

$$u_1(\xi, \eta, \tau) = \frac{1}{2} \left(-\eta \frac{\partial F}{\partial \tau} + \xi \frac{\partial G}{\partial \tau} \right) + \frac{1}{4}(-F\eta + G\xi) + A(\xi) + B(\eta)$$

with A, B arbitrary C^2 functions. There are secular terms that can be eliminated by putting

$$-\frac{\partial F}{\partial \tau} - \frac{1}{2}F = 0, \ \frac{\partial G}{\partial \tau} + \frac{1}{2}G = 0.$$

In the next section, we shall see that the secularity conditions are obtained from an averaging process. The initial conditions require that

$$F(x, 0) + G(x, 0) = f(x), \ -\frac{\partial F}{\partial x}(x, 0) + \frac{\partial G}{\partial x}(x, 0) = 0,$$

so that

$$u_0(\xi, \eta, \tau) = \frac{1}{2}(f(\xi) + f(\eta))e^{-\tau/2}.$$

If $f(x)$ is localised (compact support), this corresponds with two waves, initially half the size of $f(x)$, moving respectively to the right and to the left but slowly damped in time.

14.2 Averaging over the Characteristics

The multiple-scale expansion of the preceding section is a simple example of a more general method developed by Chikwendu and Kevorkian (1972) that can also be called "averaging over the characteristics". They consider problems of the form

$$\frac{\partial^2 u}{\partial x^2} = \frac{\partial^2 u}{\partial t^2} + \varepsilon H\left(\frac{\partial u}{\partial t}, \frac{\partial u}{\partial x}\right), \quad -\infty < x < \infty, \ t > 0,$$

with initial values $u(x,0) = f(x)$, $u_t(x,0) = g(x)$.

The nonlinearity H is chosen such that the solutions of the nonlinear wave equation are bounded. However, with the actual constructions, one can consider various useful generalisations of H and other wave equations. We return to this later.

We assume again that the characteristic coordinates $\xi = x - t$, $\eta = x + t$ play a part and that the perturbation will also involve the timescale $\tau = \varepsilon t$. In fact, Chikwendu and Kevorkian (1972) introduce a more general fast timescale T by putting $dT/dt = 1 + \omega_1(\tau)\varepsilon + \omega_2(\tau)\varepsilon^2 + \varepsilon^3 \cdots$. They show that $\omega_1(\tau) = 0$ so that, restricting ourselves to first order and $O(\varepsilon)$ expansions, we may as well use t. Transforming the equation as in Section 14.1, we have

$$\frac{\partial^2 u}{\partial \xi \partial \eta} = \frac{\varepsilon}{2}\left(-\frac{\partial^2 u}{\partial \xi \partial \tau} + \frac{\partial^2 u}{\partial \eta \partial \tau}\right) + \frac{\varepsilon}{4}H\left(-\frac{\partial u}{\partial \xi} + \frac{\partial u}{\partial \eta} + \varepsilon\frac{\partial u}{\partial \tau}, \frac{\partial u}{\partial \xi} + \frac{\partial u}{\partial \eta}\right).$$

The multiple-timescale expansion is again of the form

$$u = u_0(\xi, \eta, \tau) + \varepsilon u_1(\xi, \eta, \tau) + \varepsilon^2 \cdots,$$

and substitution in the wave equation yields equations for u_0 and u_1:

$$\frac{\partial^2 u_0}{\partial \xi \partial \eta} = 0,$$

$$\frac{\partial^2 u_1}{\partial \xi \partial \eta} = \frac{1}{2}\left(-\frac{\partial^2 u_0}{\partial \xi \partial \tau} + \frac{\partial^2 u_0}{\partial \eta \partial \tau}\right) + \frac{1}{4}H\left(-\frac{\partial u_0}{\partial \xi} + \frac{\partial u_0}{\partial \eta}, \frac{\partial u_0}{\partial \xi} + \frac{\partial u_0}{\partial \eta}\right).$$

As before, we obtain from the first equation

$$u_0(\xi, \eta, \tau) = F(\xi, \tau) + G(\eta, \tau)$$

with F and G arbitrary C^2 functions. The equation for u_1 can then be written as

$$\frac{\partial^2 u_1}{\partial \xi \partial \eta} = \frac{1}{2}\left(-\frac{\partial^2 F}{\partial \xi \partial \tau} + \frac{\partial^2 G}{\partial \eta \partial \tau}\right) + \frac{1}{4}H\left(-\frac{\partial F}{\partial \xi} + \frac{\partial G}{\partial \eta}, \frac{\partial F}{\partial \xi} + \frac{\partial G}{\partial \eta}\right).$$

Integration gives for the first derivatives

$$\frac{\partial u_1}{\partial \xi} = \frac{1}{2}\left(-\eta \frac{\partial^2 F}{\partial \xi \partial \tau} + \frac{\partial G}{\partial \tau}\right) + \frac{1}{4}\int^{\eta} H(\cdots , \cdots)d\eta,$$

$$\frac{\partial u_1}{\partial \eta} = \frac{1}{2}\left(-\frac{\partial F}{\partial \tau} + \xi \frac{\partial^2 G}{\partial \eta \partial \tau}\right) + \frac{1}{4}\int^{\xi} H(\cdots , \cdots)d\xi.$$

As $\xi, \eta \to \infty$, the derivatives must be bounded, which results in the conditions

$$\frac{\partial^2 F}{\partial \xi \partial \tau} = \lim_{\eta \to \infty} \frac{1}{2\eta}\int^{\eta} H\left(-\frac{\partial F}{\partial \xi} + \frac{\partial G}{\partial \eta}, \frac{\partial F}{\partial \xi} + \frac{\partial G}{\partial \eta}\right)d\eta,$$

$$\frac{\partial^2 G}{\partial \eta \partial \tau} = -\lim_{\xi \to \infty} \frac{1}{2\xi}\int^{\xi} H\left(-\frac{\partial F}{\partial \xi} + \frac{\partial G}{\partial \eta}, \frac{\partial F}{\partial \xi} + \frac{\partial G}{\partial \eta}\right)d\xi.$$

These secularity conditions are partial differential equations for F and G corresponding with (general) averaging over ξ and η. The averaged equations have to be solved while applying the initial conditions. The next step is to solve the equation for u_1. This again produces arbitrary functions that are determined by considering the equation for u_2 and again applying secularity conditions. If we have no a priori estimates for boundedness of the solutions, we still have to check whether the resulting approximation for $u_0 + \varepsilon u_1$ is bounded, as the secularity conditions on the derivatives are necessary but not sufficient.

Remark
The actual boundedness of the solutions is not essential for the constructions as long as the solutions are bounded on a timescale of the order $1/\varepsilon$. In this respect, the secularity conditions are misleading. As we have seen for ordinary differential equations in Chapters 11 and 12, the averaging process is the basic technique producing a normal form for the original equation. After obtaining a normal form, one can usually estimate the error introduced by normalisation on a long timescale.

Following Chikwendu and Kevorkian (1972), we give some examples.

Example 14.1   •
Suppose that we have nonlinear damping $H = u_t^3$ and so

$$\frac{\partial^2 u}{\partial x^2} = \frac{\partial^2 u}{\partial t^2} + \varepsilon\left(\frac{\partial u}{\partial t}\right)^3, \quad -\infty < x < \infty, \ t > 0.$$

Assume that if $\varepsilon = 0$ we have a progressive wave, initially $u(x,0) = f(x), u_t(x,0) = -f_x(x)$. This gives a drastic simplification, as in this case

$$u_0 = F(\xi, \tau).$$

The secularity conditions reduce to

$$\frac{\partial^2 F}{\partial \xi \partial \tau} = \lim_{\eta \to \infty} \frac{1}{2\eta} \int^\eta \left(-\frac{\partial F}{\partial \xi}\right)^3 d\eta = -\frac{1}{2} \left(\frac{\partial F}{\partial \xi}\right)^3.$$

Considering this as an ordinary differential equation of the form $w_\tau = -\frac{1}{2}w^3, w = F_\xi$, we find

$$\frac{\partial F}{\partial \xi} = \frac{1}{(\tau + A(\xi))^{1/2}}$$

with $A(\xi)$ still arbitrary. As $\xi = x$ when $t = \tau = 0$, we can apply the initial condition so that

$$A(\xi) = (f_\xi(\xi))^{-2}.$$

The approximation to first order becomes finally

$$u_0(\xi, \tau) = \int_0^\xi \frac{f_s(s)}{(1 + f_s^2(s)\tau)^{1/2}} ds + f(0).$$

For a number of elementary functions $f(x)$, we can evaluate the integral explicitly.

Remark
We have started with an initial progressive wave, which simplifies the calculation. For more general initial conditions, we have to put $u_0 = F(\xi, \tau) + G(\eta, \tau)$, which enables the presence of another wave moving to the left.

Another classical example is the Rayleigh wave equation.

Example 14.2
Choosing $H = -u_t + \frac{1}{3}u_t^3$, we have

$$\frac{\partial^2 u}{\partial x^2} = \frac{\partial^2 u}{\partial t^2} + \varepsilon \left(-\frac{\partial u}{\partial t} + \frac{1}{3}\left(\frac{\partial u}{\partial t}\right)^3\right), \quad -\infty < x < \infty, \ t > 0.$$

Starting again with a progressive wave $u(x, 0) = f(x), u_t(x, 0) = -f_x(x)$, we find with $u_0 = F(\xi, \tau)$ from the secularity condition

$$\frac{\partial^2 F}{\partial \xi \partial \tau} = \lim_{\eta \to \infty} \frac{1}{2\eta} \int^\eta \left(\frac{\partial F}{\partial \xi} - \frac{1}{3}\left(\frac{\partial F}{\partial \xi}\right)^3\right) d\eta = \frac{1}{2}\left(\frac{\partial F}{\partial \xi} - \frac{1}{3}\left(\frac{\partial F}{\partial \xi}\right)^3\right).$$

We consider this as an ordinary differential equation of the form $w_\tau = \frac{1}{2}(w - \frac{1}{3}w^3), w = F_\xi$, with solution

$$w(\tau) = \left(\frac{C(\xi)e^\tau}{1 + \frac{1}{3}C(\xi)e^\tau}\right)^{\frac{1}{2}},$$

where $C(\xi)$ will be determined by the initial conditions so that

$$\frac{\partial F}{\partial \xi} = \frac{|f_\xi(\xi)|}{(\frac{1}{3}f_\xi^2(\xi) + (1 - \frac{1}{3}f_\xi^2(\xi))e^{-\tau})^{\frac{1}{2}}}.$$

Choosing certain elementary functions $f(x)$, we can explicitly integrate the equation for $F(\xi, \tau)$.

Chikwendu and Kevorkian (1972) note that an interesting result is obtained by letting τ tend to infinity. We find

$$\lim_{\tau \to \infty} \frac{\partial F}{\partial \xi} = \sqrt{3}\frac{f_\xi(\xi)}{|f_\xi(\xi)|},$$

which corresponds with a sawtooth oscillation. Choosing for instance that initially $u(x,0) = A\sin px$, $u_t(x,0) = -Ap\cos px$, we have a sawtooth limit with amplitude $\sqrt{3}/p$ (depending on the wave number only) with spatial oscillation period (wavelength) $2\pi/p$. This limiting behaviour where an initially smooth wave train evolves towards a nonsmooth, generalised solution is confirmed by numerical analysis. Also, it is easy to see that there exist an infinite number of exact sawtooth solutions with slopes $\pm\sqrt{3}$. Their stability for general initial conditions is still an open question.

14.3 A Weakly Nonlinear Klein-Gordon Equation

We return again to the cubic Klein-Gordon equation discussed earlier with boundary values (Chapter 13) but now on an unbounded domain,

$$\frac{\partial^2 u}{\partial t^2} - \frac{\partial^2 u}{\partial x^2} + u = \varepsilon u^3, \quad -\infty < x < \infty, \; t > 0.$$

This is an example of a nonlinear dispersive wave equation displaying slowly varying wave trains. It is well-known that if $\varepsilon = 0$ we can substitute functions of the form $f(kx \pm \omega t)$, with k and ω constants, to obtain Fourier (trigonometric) wave trains satisfying the equation

$$(\omega^2 - k^2)f'' + f = 0.$$

With dispersion relation $\omega^2 - k^2 = 1$, this produces for instance wave solutions of the form
$$A\cos(kx - \omega t) + B\sin(kx - \omega t).$$

In this nonlinear case, we shall take a more restricted approach than in Section 14.1. We want to investigate for instance what happens to the wave trains found moving to the right for $\varepsilon = 0$ when the nonlinearity is turned on. We put $\theta = kx - \omega t$, with k and ω constants, and assume the dispersion relation $\omega^2 - k^2 = 1$ for $\varepsilon > 0$ and moreover that the modulated wave train, at least to first order, only depends on θ and $\tau = \varepsilon t$. With these assumptions, transforming the equation, we find

$$\frac{\partial^2 u}{\partial \theta^2} - 2\omega\varepsilon\frac{\partial^2 u}{\partial\theta\partial\tau} + \varepsilon^2\frac{\partial^2 u}{\partial\tau^2} + u = \varepsilon u^3.$$

We assume that we may substitute the expansion

$$u = u_0(\theta,\tau) + \varepsilon u_1(\theta,\tau) + \varepsilon^2 \cdots.$$

To order 1, we find

$$\frac{\partial^2 u_0}{\partial\theta^2} + u_0 = 0$$

with solution $u_0 = a(\tau)\cos(\theta + \phi(\tau))$. To $O(\varepsilon)$, we find

$$\frac{\partial^2 u_1}{\partial\theta^2} + u_1 = 2\omega\frac{\partial^2 u_0}{\partial\theta\partial\tau} + u_0^3$$

or

$$\frac{\partial^2 u_1}{\partial\theta^2} + u_1 = -2\omega\left(\frac{da}{d\tau}\sin(\theta + \phi) + a(\tau)\cos(\theta + \phi)\frac{d\phi}{d\tau}\right)$$
$$+ a^3(\tau)\left(\frac{3}{4}\cos(\theta + \phi) + \frac{1}{4}\cos(3\theta + 3\phi)\right).$$

To avoid secular terms, we put

$$\frac{da}{d\tau} = 0, \quad -2\omega a(\tau)\frac{d\phi}{d\tau} + \frac{3}{4}a^3(\tau) = 0,$$

with solutions $a(\tau) = a_0, \phi(\tau) = \frac{3}{8\omega}a_0^2\tau$, where $a(0) = a_0$. We conclude that

$$u(x,t) = a_0\cos\left(kx - \omega t + \frac{3}{8\omega}a_0^2\varepsilon t\right) + \cdots$$

represents the first-order (formal) approximation of the solution. Note that in this approximation the amplitude is still constant but there is a modulation of the phase speed.

Remark
In this problem, we have fixed k and ω and allowed for slow variations of amplitude and phase. Another classical approach is to look for solutions of θ, or explicitly $U = U(kx - \omega t)$. Again putting $\omega^2 - k^2 = 1$, we find after substitution

$$\frac{d^2 U}{d\theta^2} + U = \varepsilon U^3.$$

We can solve this equation in terms of elliptic functions or alternatively we can approximate U. In the latter case, the perturbation scheme again must allow for variations of amplitude and phase.

14.4 Multiple Scaling and Variational Principles

A large number of equations, in particular conservative ones, can be derived from a variational principle. Consider for instance the function $u : \mathbb{R}^n \times \mathbb{R} \mapsto \mathbb{R}$ characterised by the Lagrangian

$$L = L(u_t, u_x, u)$$

and the variational principle

$$\delta \int \int L(u_t, u_x, u) dt dx = 0.$$

This so-called first-order variation leads to a Euler equation for $u(x, t)$ of the form

$$\frac{\partial}{\partial t} \frac{\partial L}{\partial u_t} + \frac{\partial}{\partial x} \frac{\partial L}{\partial u_x} - \frac{\partial L}{\partial u} = 0.$$

Note that u_x and $\partial L / \partial u_x$ are vectors with components $\partial u / \partial x_i$ and $\partial L / \partial u_{x_i}$, $i = 1, \cdots, n$. Assuming that the Euler equation corresponds with a dispersive wave problem, we can look for special solutions of the form

$$u = U(\theta), \ \theta = k_i x_i - \omega t,$$

where $k_i, i = 1, \cdots, n$ and ω are constants, respectively called wave numbers and frequency. Substitution in the Euler equation produces a second-order ordinary differential equation, in general nonlinear, which upon integration will contain two free constants, the amplitude and the phase. The free parameters k_i, ω, amplitude and phase, are not arbitrary but must satisfy a so-called dispersion relation. We shall see examples later on. This approach to Euler (wave) equations has been applied by many scientists since the end of the nineteenth century.

In a number of papers starting in 1965, Whitham gave a new perturbation approach; in the description we will follow Whitham (1970, 1974) and Luke (1966). To fix the idea, consider the strongly nonlinear, one-dimensional Klein-Gordon equation

$$\frac{\partial^2 u}{\partial t^2} - \frac{\partial^2 u}{\partial x^2} + V'(u) = 0,$$

which can be derived as the Euler equation generated by the Lagrangian

$$L = \frac{1}{2} u_t^2 - \frac{1}{2} u_x^2 - V(u).$$

Exact periodic wave trains can be produced by substituting $u = U(\theta)$ as discussed above. We will study the slowly varying behaviour of the wave train over large distances and for large times by introducing the slow variables $X = \varepsilon x$ and $\tau = \varepsilon t$. The quantity θ will depend on X and τ; moreover the rescaling

$$\theta = \frac{\Theta(X,\tau)}{\varepsilon}$$

is sometimes used. The wave number and the frequency will be slowly varying:

$$k = \theta_x \to k(X,\tau) = \Theta_X, \quad \omega = -\theta_t \to \omega(X,\tau) = -\Theta_\tau.$$

Consistency then requires that

$$k_\tau + \omega_X = 0.$$

The solution describing the slowly varying wave train is supposed to have the expansion

$$u = U(\theta, X, \tau) + \varepsilon U_1(\theta, X, \tau) + \varepsilon^2 \cdots.$$

Transformation of the differential operators as in the previous sections yields

$$u_{xx} \to U_{\theta\theta}k^2 + \varepsilon(U_\theta k_X + 2U_{\theta X}k + U_{1\theta\theta}k^2) + \varepsilon^2 \cdots,$$
$$u_{tt} \to U_{\theta\theta}\omega^2 + \varepsilon(-U_\theta\omega_\tau - 2U_{\theta\tau}\omega + U_{1\theta\theta}\omega^2) + \varepsilon^2 \cdots,$$
$$V' \to V'(U) + \varepsilon U_1 V''(U) + \varepsilon^2 \cdots.$$

Substitution into the Klein-Gordon equation produces at lowest order

$$(\omega^2 - k^2)U_{\theta\theta} + V'(U) = 0$$

and to $O(\varepsilon)$

$$(\omega^2 - k^2)U_{1\theta\theta} + V''(U)U_1 = 2\omega U_{\theta\tau} + 2kU_{\theta X} + \omega_\tau U_\theta + k_X U_\theta.$$

As before, we have obtained a system of ordinary differential equations, only the first one is nonlinear. The constants of integration may depend on τ and X. In general, the nonlinear equation will have an infinite number of solutions periodic in θ (see, for instance, Verhulst, 2000). We choose one, $U_0(\theta)$, and normalise the period to 2π. The lowest order equation has the integral

$$\frac{1}{2}(\omega^2 - k^2)U_\theta^2 + V(U) = E(X,\tau),$$

where the parameter E ("energy") still depends on τ and X. From the energy integral, we can extract U_θ and integrate

$$\theta = \sqrt{\frac{1}{2}(\omega^2 - k^2)} \int \frac{dU_0}{\sqrt{E - V(U_0)}}.$$

Integration for $U_0(\theta)$ over the whole period in the phase-plane yields

$$2\pi = \sqrt{\frac{1}{2}(\omega^2 - k^2)} \oint \frac{dU_0}{\sqrt{E - V(U_0)}}.$$

This is a relation between ω, k, and E that we will again call the *dispersion relation*. Whitham (1970, 1974) introduces the averaged Lagrangian $\bar{L}$ by substituting $U_0(\theta)$ into the expression for the Lagrangian and averaging over the period

$$\bar{L} = \frac{1}{2\pi} \int_0^{2\pi} \left(\frac{1}{2}(\omega^2 - k^2)U_0'^2 - V(U_0) \right) d\theta.$$

We eliminate $V(U_0)$ using the energy integral so that

$$\bar{L} = \frac{1}{2\pi} \int_0^{2\pi} (\omega^2 - k^2)U_0'^2 d\theta - E = \frac{1}{2\pi} \oint (\omega^2 - k^2)U_0' dU_0 - E,$$

which can also be written as

$$\bar{L} = \frac{1}{2\pi} \sqrt{2(\omega^2 - k^2)} \oint \sqrt{E - V(U_0)} dU_0 - E.$$

The last expression for the averaged Lagrangian $\bar{L}$ depends for a given potential V on ω, k, and E only; for the integration, U_0 is a dummy variable.

   Whitham proposes to use this averaged Lagrangian to obtain appropriate Euler equations for the unknown quantities ω, k, and E. So far, we did not apply the secularity conditions to the equation for U_1. Remarkably enough, these are tied in with the variations of the averaged Lagrangian. For technical details, see the literature cited.

Example 14.3

Consider the strongly nonlinear, one-dimensional Klein-Gordon equation

$$\frac{\partial^2 u}{\partial t^2} - \frac{\partial^2 u}{\partial x^2} + u + au^3 = 0,$$

corresponding with $V(u) = \frac{1}{2}u^2 + \frac{a}{4}u^4$ with a a constant; the equation can be derived as a Euler equation by variation of the Lagrangian

$$L = \frac{1}{2}u_t^2 - \frac{1}{2}u_x^2 - \frac{1}{2}u^2 - \frac{a}{4}u^4.$$

With the multiple-timescale expansion $u = U + \varepsilon U_1 + \varepsilon^2 \cdots$, we have

$$(\omega^2 - k^2)U_{\theta\theta} + U + aU^3 = 0$$

and to $O(\varepsilon)$

$$(\omega^2 - k^2)U_{1\theta\theta} + (1 + 3aU^2)U_1 = 2\omega U_{\theta\tau} + 2kU_{\theta X} + \omega_\tau U_\theta + k_X U_\theta.$$

The solutions for U can be obtained as elliptic functions that are periodic in θ. They oscillate between the two zeros of $E(X, \tau) - \frac{1}{2}U^2 - \frac{a}{4}U^4$. The dispersion relation among ω, k, and E takes the form

$$2\pi = \sqrt{\frac{1}{2}(\omega^2 - k^2)} \oint \frac{dU_0}{\sqrt{E - \frac{1}{2}U_0^2 - \frac{a}{4}U_0^4}}.$$

The averaged Lagrangian becomes

$$\bar{L} = \frac{1}{2\pi} \sqrt{2(\omega^2 - k^2)} \oint \sqrt{E - \frac{1}{2}U_0^2 - \frac{a}{4}U_0^4} \, dU_0 - E,$$

for which various series expansions are available.

In the case of weakly nonlinear problems, it is easier to obtain explicit expressions. We show this with another example that is used quite often in the literature.

Example 14.4
Consider Bretherton's model equation

$$\frac{\partial^2 u}{\partial t^2} + \frac{\partial^4 u}{\partial x^4} + \frac{\partial^2 u}{\partial x^2} + u = \varepsilon u^3,$$

which can be derived from the Lagrangian

$$L = \frac{1}{2}u_t^2 - \frac{1}{2}u_{xx}^2 + \frac{1}{2}u_x^2 - \frac{1}{2}u^2 + \frac{1}{4}\varepsilon u^4.$$

If $\varepsilon = 0$, substitution of $u = a\cos(kx - \omega t)$ produces the dispersion relation

$$\omega^2 - k^4 + k^2 = 1.$$

For $\varepsilon > 0$, we assume that the solution is slowly varying in $\theta = k(X, \tau)x - \omega(X, \tau)t$, $X = \varepsilon x$, and $\tau = \varepsilon t$:

$$u = a(X, \tau)\cos(k(X, \tau)x - \omega(X, \tau)t) + \varepsilon \cdots.$$

For the averaged Lagrangian at lowest order, we find

$$\bar{L} = \frac{1}{2\pi}\int_0^{2\pi} L\,d\theta = \frac{1}{4}a^2(\omega^2 - k^4 + k^2 - 1) + \frac{1}{32}\varepsilon a^4 + \cdots.$$

The Euler-Lagrange equation with respect to the amplitude a is $\partial \bar{L}/da = 0$; this produces the dispersion relation

$$\omega^2 - k^4 + k^2 = 1 + \frac{3}{4}\varepsilon a^2,$$

which is an extension of the "linear" dispersion relation. Other variations of $\bar{L}$ will produce relations among amplitude a, frequency ω, and wave number k that play a part in wave mechanics. For explicit calculations, see for instance Shivamoggi (2003).

14.5 Adiabatic Invariants and Energy Changes

In Chapter 12, we looked at adiabatic invariants that can be seen as asymptotic integrals or asymptotic conservation laws of a system. In contrast with "classical integrals of motion," these adiabatic invariants represent a relation between phase variables and time that is conserved with a certain precision on a certain timescale. This is of particular interest when we want to characterise changes of energy or angular momentum without having to integrate the complete equations of motion.

In this section, we explore the ideas for ordinary differential equations, after which, in the next section, we give an application to the Korteweg-de Vries equation.

Example 14.5
Consider initial value problems for the equation

$$\ddot{x} + x = \varepsilon f(x, \dot{x}, \varepsilon t)$$

with sufficiently smooth right-hand side. Applying averaging as in Chapter 11, we can introduce amplitude-phase variables of the form (11.5) by putting $x(t) = r(t)\cos(t + \phi(t)), \dot{x}(t) = -r(t)\sin(t + \phi(t))$. We will add two variables,

$$E(t) = \frac{1}{2}\dot{x}^2(t) + \frac{1}{2}x^2(t), \ \tau = \varepsilon t,$$

and, after differentiation of $E(t)$, the two equations

$$\frac{dE}{dt} = \varepsilon\dot{x}f(x, \dot{x}, \tau), \ \dot{\tau} = \varepsilon.$$

Note that $E(t) = \frac{1}{2}r^2(t)$. The four equations (for r, ϕ, E, τ) that we can derive, are all slowly varying; we only average the equation for E:

$$\frac{dE_a}{dt} = -\frac{\varepsilon}{2\pi}\int_0^{2\pi} r\sin(t + \phi)f(r\cos(t + \phi), -r\sin(t + \phi), \tau)dt.$$

As in Chapter 11, we can put $s = t + \phi$ with the result that ϕ does not occur in the averaged equation. We find after averaging an expression of the form

$$\frac{dE_a}{dt} = \varepsilon F(r_a(t), \tau) = \varepsilon F(\sqrt{2E_a}, \tau).$$

This is a first-order differential equation for $E_a(t)$ that we can study without solving the averaged equations of motion. $E_a(t)$ is an adiabatic invariant with $E_a(t) - E(t) = O(\varepsilon)$ on the timescale $1/\varepsilon$.

Consider as an example the Rayleigh equation

$$\ddot{x} + x = \varepsilon\dot{x}\left(1 - \frac{1}{3}\dot{x}^2\right)$$

with

$$\frac{dE}{dt} = \varepsilon \dot{x}^2 \left(1 - \frac{1}{3}\dot{x}^2\right).$$

Amplitude-phase variables produce

$$\frac{dE}{dt} = \varepsilon r^2(t) \sin^2(t + \phi(t)) \left(1 - \frac{1}{3}r^2(t) \sin^2(t + \phi(t))\right)$$

and, after averaging over t,

$$\frac{dE_a}{dt} = \frac{1}{2}\varepsilon r_a^2(t) \left(1 - \frac{1}{4}r_a^2(t)\right) = \varepsilon E_a \left(1 - \frac{1}{2}E_a\right).$$

If we choose $E_a(t) = 2$, the energy does not change in time. This is the energy value of the well-known limit cycle of the Rayleigh equation. If we start with $E(0) < 2$, the energy grows to the value 2 as the solution tends to the limit cycle, and if we start with $E(0) > 2$, the energy decreases to this value.

Integrating the equation for E_a, we obtain an expression that we can interpret as an adiabatic invariant for the Rayleigh equation.

We consider now a strongly nonlinear problem based on Huveneers and Verhulst (1997).

Example 14.6
Consider the equation

$$\ddot{x} + x = a(\varepsilon t)x^2$$

with $a(0) = 1$ and $a(\varepsilon t)$ a smooth, positive function decreasing towards zero. This is a simple model exemplifying a Hamiltonian system with asymmetric potential that by some evolution process tends towards a symmetric one. Transforming $y = a(\varepsilon t)x$, we obtain the equation

$$\ddot{y} + y = y^2 + 2\varepsilon\frac{a'(\varepsilon t)}{a(\varepsilon t)}\dot{y} + \varepsilon^2 \cdots .$$

A prime denotes differentiation with respect to its argument. The result is surprising. The $O(\varepsilon)$ term represents a dissipative term, which means that our system in evolution towards symmetry is characterised by an autonomous Hamiltonian system with dissipation added. To see what happens, we consider a special choice: $a(\varepsilon t) = e^{-\varepsilon t}$. The equation becomes

$$\ddot{y} + y = y^2 - 2\varepsilon\dot{y} + \varepsilon^2 \cdots .$$

If $\varepsilon = 0$, we have a centre point at $(0,0)$ and a saddle at $(1,0)$ with a homoclinic loop emerging from the saddle and intersecting the y-axis at $(-\frac{1}{2},0)$. If $\varepsilon > 0$, the loop will break up but the saddle still has two stable and two unstable one-dimensional manifolds. If $\varepsilon = 0$, we can associate with the equation the energy

$$E = \frac{1}{2}\dot{y}^2 + \frac{1}{2}y^2 - \frac{1}{3}y^3$$

and by differentiation and using the equation for $\varepsilon > 0$

$$\frac{dE}{dt} = -2\varepsilon\dot{y}^2 + \varepsilon^2 \cdots.$$

To see what happens to the homoclinic loop, we approximate E by E_a, omitting the ε^2 terms and using in the equation the unperturbed homoclinic loop behaviour of $\dot{y}$. Integrating from y_0 to y_1, we have

$$E_a = -2\varepsilon \int_{t(y_0)}^{t(y_1)} \dot{y}^2(t)dt = -2\varepsilon \int_{y_0}^{y_1} \dot{y}dy.$$

We now use that for $\varepsilon = 0$ the homoclinic loop is given by

$$\frac{1}{2}\dot{y}^2 + \frac{1}{2}y^2 - \frac{1}{3}y^3 = \frac{1}{6}$$

and that the loop is symmetric with respect to the y-axis; we have

$$E_a = \frac{1}{6} - 4\varepsilon \int_{-\frac{1}{2}}^{1} \sqrt{\frac{2}{3}y^3 - y^2 + \frac{1}{3}}dy.$$

Fortunately, this is an elementary integral. (Use MATHEMATICA or an integral table.) We find for the first-order changed energy

$$E_a = \frac{1}{6} - \frac{12}{5}\varepsilon.$$

It is interesting to deduce from this the position where the stable manifold of the unstable equilibrium intersects the y-axis. The top half of the homoclinic loop is bent inwards with an energy change of $\frac{6}{5}\varepsilon$, so the stable manifold has energy (to a first approximation) $\frac{1}{6} + \frac{6}{5}\varepsilon$ and is approximately described by

$$\frac{1}{2}\dot{y}^2 + \frac{1}{2}y^2 - \frac{1}{3}y^3 = \frac{1}{6} + \frac{6}{5}\varepsilon.$$

Introducing into this equation $\dot{y} = 0, y = -\frac{1}{2} + \varepsilon\alpha + \cdots$, we find $\alpha = -\frac{8}{5}$, so the intersection takes place at approximately $(-\frac{1}{2} - \varepsilon\frac{8}{5}, 0)$.

Remark

This type of energy change for one homoclinic loop can be found in Guckenheimer and Holmes (1997). In the interior of the homoclinic loop "ordinary" averaging is valid, but this is not the case in a boundary layer near the loop. In Huveneers and Verhulst (1997), the computations use elliptic functions and cover both the interior of the homoclinic loop and the boundary layer near the homoclinic loop and the saddle. The error analysis is subtle and involves various domains and different expressions for the adiabatic invariants; for the passage of the saddle, the analysis follows Bourland and Haberman (1990).

14.6 The Perturbed Korteweg-de Vries Equation

In the spirit of example 14.6 and following Scott (1999), we discuss perturbations of solitons in the Korteweg-de Vries (KdV) equation. The equation is

$$u_t + uu_x + u_{xxx} = \varepsilon f(\cdots), \quad -\infty < x < \infty, t > 0,$$

where we have the KdV equation if $\varepsilon = 0$, and f is a perturbation that may depend on x, t, u and the derivatives of u. The well-known single soliton solution of the KdV equation is

$$u(x,t) = 3v \operatorname{sech}^2\left(\frac{\sqrt{v}}{2}(x - vt)\right)$$

with v the constant soliton velocity. The perturbation f can have many consequences, but we will study the case with the assumption that we have only small variations of the velocity v, $v = v(\tau), \tau = \varepsilon t$.

As in the preceding section, we can directly derive an equation for the behaviour of the energy with time. It is convenient to introduce the function w by $w_x = u$. The KdV equation can be derived from the Lagrangian density

$$\mathcal{L} = \frac{1}{2}w_x w_t + \frac{1}{6}w_x^3 - \frac{1}{2}w_{xx}^2.$$

Instead of the Lagrangian, we will use the associated Hamiltonian density

$$\mathcal{H} = w_t \frac{\partial \mathcal{L}}{\partial w_t} - \mathcal{L} = -\frac{1}{6}w_x^3 + \frac{1}{2}w_{xx}^2.$$

The total energy (which we will later specify for a soliton) is

$$H = \int_{-\infty}^{\infty} \mathcal{H} dx.$$

From this energy functional, we find by differentiation

$$\frac{dH}{dt} = \int_{-\infty}^{\infty} \left(-\frac{1}{2}w_x^2 w_{xt} + w_{xx}w_{xxt}\right) dx.$$

The first term is partially integrated once and the second term twice; we also assume that the first three derivatives of w vanish as $x \to \pm\infty$. We find

$$\frac{dH}{dt} = \int_{-\infty}^{\infty} (w_{xxx} + w_{xxxx})w_t dx$$

and finally, using that w_x satisfies the perturbed KdV equation,

$$\frac{dH}{dt} = \int_{-\infty}^{\infty} (-w_{xt} + \varepsilon f)w_t dx = \varepsilon \int_{-\infty}^{\infty} f w_t dx.$$

We can explicitly compute the total soliton energy by substituting the expression for the soliton while neglecting variations of $v(\tau)$; they are of higher order. We find after integration

$$H = -\frac{36v^{\frac{5}{2}}(\tau)}{5}$$

and so derive variations of the energy from long-term variations of $v(\tau)$:

$$\frac{dH}{dt} = -18v^{\frac{3}{2}}\frac{dv}{dt}.$$

Combining the general expression for dH/dt with this specific one, we find for the velocity variation

$$\frac{dv}{dt} = -\varepsilon\frac{1}{18v^{\frac{3}{2}}}\int_{-\infty}^{\infty} fw_t dx.$$

For the soliton ($\varepsilon = 0$), we have the relation

$$w_t = -vw_x = -vu,$$

and assuming that for $\varepsilon > 0$ we have to first order $w_t = -v(\tau)u$, we obtain

$$\frac{dv}{dt} = \varepsilon\frac{1}{18v^{\frac{1}{2}}}\int_{-\infty}^{\infty} fw_t dx.$$

One of the simplest examples is the choice $f = -u$; inserting this and using the expression for a single soliton yields after integration

$$\frac{dv}{dt} = -\varepsilon\frac{4}{3}v$$

with solution

$$v(\varepsilon t) = v(0)e^{-\frac{4}{3}\varepsilon t}.$$

As we can observe, in the expression for the single soliton, the amplitude obeys the same variation with time. In an interesting discussion, Scott (1999) notes that this result is confirmed by numerical calculations. On the other hand, when analysing other conservation laws of the KdV equation with the same technique, the results are not always correct. The implication is that mathematical analysis of these approximation techniques is much needed.

14.7 Guide to the Literature

An early reference is Benney and Newell (1967), where multiple-scale methods are developed to study wave envelopes and interacting nonlinear waves. Around the same time, Luke (1966) gave a multiple-scale analysis for some prominent wave equations; this is tied in with work started in 1965 and extensively described in Whitham (1970, 1974). In Whitham's work, multiple-scale analysis is imbedded in variational principles, employing averaged Lagrangians, which puts the computations in a more fundamental although still formal framework. If the medium is not homogeneous, Whitham shows that the wave action is conserved for strongly nonlinear dispersive waves analogous to the action being an adiabatic invariant for strongly nonlinear Hamiltonian oscillations.

Until the papers of Haberman and Bourland (1988) and Bourland and Haberman (1989), modulations of the phase shift for strongly dispersive waves got little attention. Using the equation for the wave action, they developed a modification of the approximation scheme to include higher-order effects and characterise at what order of the perturbation scheme such effects play a part.

Chikwendu and Kevorkian (1972) developed a general multiple-scale approach for wave equations that we have called "averaging over the characteristics". See also Kevorkian and Cole (1996), where wave equations and conservation laws are discussed. Chikwendu and Easwaran (1992) extended this for semi-infinite domains.

Van der Burgh (1979) has shown that in a large number of weakly nonlinear equations the multiple-scale analysis yields asymptotically valid results. This is based on the usual integral inequality estimates, and the asymptotic estimates may even be improved. See also the discussion in De Jager and Jiang Furu (1996). We are not aware of other proofs of asymptotic validity for unbounded domains.

Important extensions and applications have been found and are being implemented. Ablowitz and Benney (1970) extended Whitham's averaged variational principle to multiphase dispersive nonlinear waves. McLaughlin and Scott (1978) did this extension for solitary waves and solitons. In Scott's (1999) book, used in the previous section, applications to the sine-Gordon equation, the nonlinear Schrödinger equation, and other interesting examples can be found.

14.8 Exercises

Exercise 14.1 Consider the weakly damped Klein-Gordon equation in the form

$$\frac{\partial^2 u}{\partial x^2} - \frac{\partial^2 u}{\partial t^2} = \varepsilon \frac{\partial u}{\partial t}, \quad -\infty < x < \infty, \ t > 0,$$

with $u(x,0) = f(x)$, $u_t(x,0) = 0$.

a. Introduce a regular expansion $u(x,t) = u_0(x,t) + \varepsilon u_1(x,t) + \cdots$ and formulate the initial value problems for u_0 and u_1.

b. Reformulate the initial value problems for u_0 and u_1 in characteristic coordinates ξ, η.

c. Solve the initial value problems for u_0 and u_1 to show that secular terms arise.

Exercise 14.2 Compare the first-order multiple-timescale expansion of Section 14.1 with the exact solution. For comparison, see Chikwendu and Kevorkian (1972).

Exercise 14.3 Consider the weakly nonlinear wave equation from Section 14.3 with damping added:

$$\frac{\partial^2 u}{\partial t^2} - \frac{\partial^2 u}{\partial x^2} + u + \varepsilon \frac{\partial u}{\partial t} = \varepsilon u^3, \quad -\infty < x < \infty, \ t > 0.$$

Consider a wave train moving to the right. Determine the change in the secularity conditions.

Exercise 14.4 Consider the wave equation with a Van der Pol perturbation

$$\frac{\partial^2 u}{\partial x^2} = \frac{\partial^2 u}{\partial t^2} - \varepsilon \frac{\partial u}{\partial t}(1 - u^2), \quad -\infty < x < \infty, \ t > 0.$$

Initially we have a progressive wave $u(x,0) = f(x), u_t(x,0) = -f_x(x)$. Compute a first-order approximation as in Section 14.2. Consider a suitable initial $f(x)$ explicitly.

15

Appendices

15.1 The du Bois-Reymond Theorem

We present the proof of the theorem in the version of Eckhaus (1979).

Theorem 15.1
(du Bois-Reymond)
For any asymptotic sequence $\delta_n(\varepsilon), n = 0, 1, 2, \cdots$, there exist order functions $\delta^0(\varepsilon)$ such that for all n

$$\delta^0 = o(\delta_n).$$

Proof
Assume that $\delta_n(\varepsilon) = o(1), n = 1, 2, \cdots$; if not, the statement of the theorem is trivial. As the δ_n sequence is asymptotic, we have for $0 < \varepsilon < \varepsilon_n \leq \varepsilon_0$

$$\delta_{n+1}(\varepsilon) < \delta_n(\varepsilon).$$

If the graphs of the functions $\delta_n(\varepsilon)$ intersect, we may take ε_n from the intersection points; if not, we may take more arbitrarily a sequence $\varepsilon_n \to 0$. We construct the order function $\delta^0(\varepsilon)$ as follows. At $\varepsilon = \varepsilon_n$, we define $\delta^0(\varepsilon_n) = \delta_{n+1}(\varepsilon_n)$.
Repeating this process for $n = 1, 2, \cdots$, we obtain after connecting the $\delta^0(\varepsilon_n)$ a function $\delta^0(\varepsilon)$ with

$$0 < \delta^0(\varepsilon) \leq \delta_{n+1}(\varepsilon_n) \text{ for } \varepsilon_{n+1} \leq \varepsilon \leq \varepsilon_n.$$

The estimate holds clearly for $0 < \varepsilon \leq \varepsilon_n$ so that

$$\frac{\delta^0(\varepsilon)}{\delta_n(\varepsilon)} \leq \frac{\delta_{n+1}(\varepsilon)}{\delta_n(\varepsilon)}$$

for $0 < \varepsilon \leq \varepsilon_n$ and all n. Taking the limit $\varepsilon \to 0$, we have

$$\lim_{\varepsilon \to 0} \frac{\delta^0(\varepsilon)}{\delta_n(\varepsilon)} = 0 \text{ for all } n.$$

$\square$

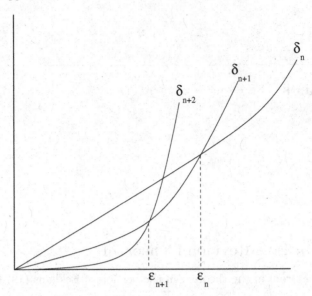

Fig. 15.1. Construction of $\delta^0(\varepsilon)$ in the Eckhaus (1979) version of the du Bois-Reymond theorem.

15.2 Approximation of Integrals

The proofs of the theorems for the approximation of integrals that we presented in Chapter 3 are based on partial integration and quite elementary.

Theorem 15.2
Consider the Laplace integral

$$\phi(x) = \int_0^\infty e^{-xt} f(t)dt$$

with $f(t)$ uniformly bounded for $t \geq 0$; the derivatives $f^{(1)}(t) \cdots f^{(m+1)}(t)$ exist and are bounded on $[0, a], a > 0$. Then

$$\phi(x) = \sum_{n=0}^{m-1} \frac{1}{x^{n+1}} f^{(n)}(0) + 0\left(\frac{1}{x^{m+1}}\right) + 0\left(\frac{1}{x}e^{-ax}\right)$$

for $x \to \infty$.

Proof
The idea is that, because of the exponent, the largest contribution arises near $t = 0$ so that only the values of $f^{(n)}(t)$ at $t = 0$ play a part. Introducing the positive constant a, we have

$$\phi(x) = \int_0^a e^{-xt} f(t)dt + \int_a^\infty e^{-xt} f(t)dt.$$

Assuming that $|f(t)| \leq M$, we have for the second integral

$$\left| \int_a^\infty e^{-xt} f(t)dt \right| \leq M \int_a^\infty e^{-xt} dt = \frac{M}{x} e^{-ax}.$$

Partial integration of the first integral produces

$$\int_0^a e^{-xt} f(t)dt = \frac{1}{x}(f(0) - e^{-ax} f(a)) + \frac{1}{x} \int_0^a e^{-xt} f^{(1)}(t)dt,$$

which after repeated integration yields the result of the theorem.

$\square$

Theorem 15.3

Consider the Fourier integral $\phi(x) = \int_\alpha^\beta e^{ixt} f(t)dt$ with $f(t)$ $(m+1)$ times continuously differentiable in $[\alpha, \beta]$; then

$$\phi(x) = -\sum_{n=0}^{m-1} i^{n-1} \frac{f^{(n)}(\alpha)}{x^{n+1}} e^{ix\alpha} + \sum_{n=0}^{m-1} i^{n-1} \frac{f^{(n)}(\beta)}{x^{n+1}} e^{ix\beta} + 0(x^{-m-1}).$$

Proof

Partial integration produces

$$\int_\alpha^\beta e^{ixt} f(t)dt = \frac{1}{ix} e^{ixt} f(t)|_\alpha^\beta - \frac{1}{ix} \int_\alpha^\beta e^{ixt} f^{(1)}(t)dt$$

$$= \frac{i^{-1}}{x}(f(\beta)e^{ix\beta} - f(\alpha)e^{ix\alpha}) + \cdots .$$

Repeated partial integration produces the expansion with remainder

$$R_m = \frac{1}{(-ix)^m} \int_\alpha^\beta e^{ixt} f^{(m)}(t)dt.$$

Partial integration of the integral on the right-hand side yields the estimate of the theorem.

$\square$

15.3 Perturbations of Constant Matrices

In studying differential equations, one often has to analyse square matrices $A(\varepsilon)$ that depend smoothly on a small parameter ε and are otherwise constant. As usual, our quantities are real, but one should be aware that many of the following results also hold for complex matrix entries, for instance in the case of the general expansion results 1 and 2 in the next subsection.

Perturbations of matrices arise for instance when considering the stability of individual solutions and when analysing bifurcations. In the latter case, the "unperturbed" matrix $A(0)$ will have eigenvalues zero or purely imaginary. In all of these cases, we wish to know how the eigenvalues, and sometimes also the eigenvectors, depend on the parameter ε.

We will assume that each element of the matrix $A(\varepsilon)$ can be expanded in a Taylor series with respect to ε so that we can write

$$A(\varepsilon) = A_0 + \varepsilon A_1 + \varepsilon^2 A_2 + \varepsilon^3 \cdots ,$$

where $A(0) = A_0$. $A_1, A_2, \cdots$ are square matrices with the same dimension as A_0 and do not depend on ε. One immediate problem is that often we know A_0, a first-order perturbation analysis, will give us A_1, but to obtain A_2, etc., usually requires a substantial computational effort. So one of the first questions is what conclusions we can draw from our knowledge of A_0 and A_1. In this respect, we will be interested in several different questions.

- Can the eigenvalues be expanded in a convergent series of the form

$$\lambda = \lambda_0 + \varepsilon \lambda_1 + \varepsilon^2 \cdots ,$$

  where λ_0 is an eigenvalue of the matrix A_0?
- If we are in the critical case where λ_0 is zero or purely imaginary, how do the perturbations affect the eigenvalues and thus the qualitative behaviour of the solutions of the corresponding differential equations?

The first question is a classical one and was solved a long time ago. As we shall see, in the second case these questions are not always easy to answer. First we will summarise some results for *real* $n \times n$ matrices. An important part will be played by whether the eigenvalues of $A_0 + \varepsilon A_1$ or in general $A(\varepsilon)$ are distinct or not. However, we have to clarify what we mean by "distinct" in this perturbation context:
The eigenvalues $\lambda_a(\varepsilon)$ and $\lambda_b(\varepsilon)$ are distinct if $\lambda_a(\varepsilon) - \lambda_b(\varepsilon) \neq 0$ and moreover not $o(1)$ as ε tends to zero.

In fact we may relax this condition somewhat. In many cases, $\lambda_a - \lambda_b$ may be $o(1)$ if the difference tends to zero slowly enough. We omit such refinements.

15.3.1 General Results

We start with some basic results.

1. If an eigenvalue λ_0 of A_0 is single and thus distinct from the other eigenvalues, the perturbed eigenvalue $\lambda(\varepsilon)$ of $A(\varepsilon)$ can be expanded in a convergent power series (a Taylor series) of the form

$$\lambda(\varepsilon) = \lambda_0 + \varepsilon \lambda_1 + \varepsilon^2 \cdots .$$

On realising that the characteristic equation is algebraic in λ, this is a straightforward application of the implicit function theorem.

If one of the single eigenvalues of A_0 vanishes, $A(\varepsilon)$ has one small eigenvalue of size $O(\varepsilon)$.

Note that here and in the following confusion can arise from a slightly different terminology in numerical analysis. A numerical analyst might say: what happens when λ_1 is a large number? The answer is that in perturbation theory the starting point will be a rescaling of all quantities so that we can identify the size of the parameters with respect to ε. It follows that in this perturbation context λ_1 is a fixed constant of size $O(1)$ as ε tends to zero.

2. If some of the eigenvalues of A_0 are multiple, the expansion of the eigenvalues of $A(\varepsilon)$ is modified by the inclusion of fractional powers of ε in the convergent series. This is called a Puisieux series. A proof, in the more general context of functions of two variables, can be found in Hörmander (1983, Vol. 2, Appendix A); basically, arguments from complex function theory are used.

The implication of the convergent power series result and the Puisieux series is that if the matrix A_0 is nonsingular and if all the eigenvalues have a nonzero real part, there is little qualitative change by adding higher-order perturbations. In this case, A_0 is called *structurally stable*. An extensive discussion and many applications can be found in Vainberg and Trenogin (1974).

3. If the matrix $A(\varepsilon)$ is *symmetric* with respect to the diagonal (in another context also called *Hermitian* or *self-adjoint*), all eigenvalues are real. Coincident eigenvalues do not affect this statement. A large number of operators in physical problems are symmetric; for an extensive treatment, see Parlett (1998), and for perturbations see Wilkinson (1965) or Stewart and Sun (1990).

In this symmetric case, the eigenvalues can again be expanded in a Taylor series with respect to ε. An important consequence is that if A_0 is nonsingular, stability statements will be completely decided by the eigenvalues of A_0. Adding the perturbations will only move the eigenvalues an $O(\varepsilon)$ distance on the real axis.

An interesting special case arises if A_0 is symmetric and A_1 is not. If the matrix A_0 is nonsingular, stability is again completely determined by the eigenvalues of A_0 but the eigenvalues may move into the complex plane, which may trigger other interesting dynamics.

In general, for bifurcation problems where eigenvalues may be zero or purely imaginary, these results for symmetric matrices can seldom be applied.

4. If the matrix $A(\varepsilon)$ is derived by linearisation near an equilibrium or periodic solution of an autonomous Hamiltonian system, we know that the eigenvalues in the complex plane are located symmetrically with respect to the imaginary and real axes. Moreover, in the case of a periodic solution, there are two eigenvalues zero (one arising from the family of energy manifolds and one from the usual translation property in autonomous sys-

tems); see for instance Verhulst (2000). In the following, a and b will be real, positive numbers. Note that in a Hamiltonian system with n degrees of freedom, the dimension of $A(\varepsilon)$ is $2n$.

- Systems with two degrees of freedom

  Suppose that A_0 has rank 2. The possible eigenvalues for A_0 near a periodic solution are $(0, 0, -a, +a)$, $(0, 0, -bi, +bi)$. Because of the symmetry requirement, changes take place along the axes so this location cannot be changed by higher-order perturbations.

  Suppose that A_0 has rank 4 near an equilibrium. The possible eigenvalues for A_0 are $(-a, +a, -b, +b)$, $(-a, +a, -bi, +bi)$, $(-ai, +ai, -bi, +bi)$, $(-a + bi, -a - bi, +a + bi, +a - bi)$. As far as A_0 determines the outcome, the first, second, and fourth cases are unstable and the third case stable. If $a \neq b$, higher-order perturbations cannot change this picture because of the symmetry requirement; if $a = b$, the third case may also lead to instability if the eigenvalues move from the imaginary axis into the complex plane.

  The Hamiltonian Hopf bifurcation arises when the system has a parameter that, when varied, moves eigenvalues on the imaginary axis until at the bifurcation value of the parameter the eigenvalues are coincident, after which they move away from the imaginary axis in opposite directions.

- Systems with three degrees of freedom

  We consider periodic solutions only. On perturbing around such solutions, assuming that A_0 has rank 4, we have for the possible eigenvalues $(0, 0, -a, +a, -b, +b)$, $(0, 0, -ai, +ai, -bi, +bi)$, $(0, 0, -a, +a, -bi, +bi)$, $(0, 0, -a + bi, -a - bi, +a + bi, +a - bi)$. Such sets of eigenvalues allow for interesting bifurcations.

5. Another interesting case with many applications, for instance for incompressible fluids, arises if the matrix $A(\varepsilon)$ is derived by linearisation near an equilibrium or periodic solution of an autonomous system of differential equations in which the phaseflow is measure-preserving. In such a case, the trace of the matrix $A(\varepsilon)$ and so the sum of the eigenvalues equals zero. This also holds for the traces of the matrices $A_0, A_1, A_2, \cdots$ etc. If we have for instance a three-dimensional system, the eigenvalues of A_0 can be $(0, -a, +a)$, $(0, -ai, +ai)$, $(2a, -a + bi, -a - bi)$. In the first and third cases, we have instability; in the second case, we have stability from A_0 but instability may be caused by higher-order perturbations.

   The last two cases are associated with the possibility of the so-called Šhilnikov bifurcation (see Guckenheimer and Holmes, 1997).

Although we have explicit results for the dependence of the eigenvalues on the parameter ε mentioned above, we are left with the basic problem of how far we have to go in calculating the perturbation matrices A_1, A_2, etc., in a number of problems. A few simple examples will illustrate this.

15.3.2 Some Examples

In the first example, we show a result of the Newton-Puisieux expansion of eigenvalues.

Example 15.1
Consider the matrix $A(\varepsilon)$ with the expansion

$$A(\varepsilon) = \begin{pmatrix} 0 & 0 & 1 \\ 0 & 0 & 0 \\ 0 & 0 & 0 \end{pmatrix} + \varepsilon \begin{pmatrix} 0 & 0 & 0 \\ 1 & 0 & 0 \\ 3 & 0 & 0 \end{pmatrix} + \varepsilon^2 \begin{pmatrix} 0 & 0 & 0 \\ 0 & 0 & 0 \\ 0 & -1 & 0 \end{pmatrix}$$

in which A_0 is very degenerate. Including the $O(\varepsilon)$ term, we find the approximate characteristic equation

$$\lambda^3 - 3\varepsilon\lambda = 0$$

and eigenvalues $\lambda_1 = 0, \lambda_{2,3} = \pm 3\sqrt{\varepsilon}$. Adding the $O(\varepsilon^2)$ term, the full characteristic equation becomes

$$\lambda^3 - 3\varepsilon\lambda + \varepsilon^3 = 0.$$

Allowing for fractional powers, the Newton-Puisieux expansion produces

$$\lambda_1 = \frac{1}{3}\varepsilon^2 + \frac{1}{81}\varepsilon^5 + \cdots, \lambda_{2,3} = \pm 3\sqrt{\varepsilon} - \frac{1}{6}\varepsilon^2 + \cdots.$$

Note that if we have to solve the differential equation $\dot{x} = A(\varepsilon)x$ time-like variables (timescales) $\sqrt{\varepsilon}t, \varepsilon^2 t, \varepsilon^5 t$ and longer timescales clearly play a part. Such problems concerning timescales are discussed in Chapters 11 and 12.

Above we introduced the concept of a structurally stable matrix, and one might think that if at a certain order of approximation the eigenvalues have real part nonzero, higher-order approximations will have no qualitative influence. This is too optimistic, as the following example shows (see also Sanders and Verhulst, 1985, and Murdock, 2003).

Example 15.2
Consider the matrix $A(\varepsilon)$ with the expansion

$$A(\varepsilon) = \begin{pmatrix} 0 & 1 \\ 0 & 0 \end{pmatrix} + \varepsilon \begin{pmatrix} -1 & 0 \\ 0 & -2 \end{pmatrix} + \varepsilon^2 \begin{pmatrix} 0 & 0 \\ 6 & 0 \end{pmatrix}.$$

Including the $O(\varepsilon)$ term, we find the approximate characteristic equation

$$(-\varepsilon - \lambda)(-2\varepsilon - \lambda) = 0$$

with eigenvalues $\lambda_1 = -\varepsilon, \lambda_2 = -2\varepsilon$. The full characteristic equation, including the $O(\varepsilon^2)$ term, is

$$\lambda^2 + 3\varepsilon\lambda - 4\varepsilon^2 = 0$$

with solutions $\lambda_1 = \varepsilon, \lambda_2 = -4\varepsilon$.

This is a disturbing example. As mentioned before, in differential equations higher-order approximations require a considerable computational effort. We would like to stop the calculation if going on does not change the qualitative picture but only adds some very small quantitative corrections. From the preceding subsection, we know that we can expand the eigenvalues with respect to the small parameter ε but, in this case of a singular matrix A_0, it does not tell us how many terms of the expansion of the matrix $A(\varepsilon)$ we need. A rough *upper limit* is obtained as follows.

An upper limit for the matrix expansion
Suppose that for the expansion of $A(\varepsilon)$ to order m

$$A_0 + \varepsilon A_1 + \varepsilon^2 A_2 + \cdots + \varepsilon^m A_m,$$

we have that all real parts of the eigenvalues are nonzero and that the smallest of the real parts has (or have) size $O_s(\varepsilon^m)$. We assume that A_0 has at least one nonzero element; if not, we rescale by extracting ε. Suppose we compute the eigenvalues of

$$A_0 + \varepsilon A_1 + \varepsilon^2 A_2 + \cdots + \varepsilon^m A_m + \cdots + \varepsilon^M A_M$$

with $M > m$. If the dimension of $A(\varepsilon)$ is n, we may have eigenvalues of size $O(\varepsilon^{M/n})$, so in our computations we have to include the matrix expansion as far as A_M with $M/n = m$ or $M = mn$.

This agrees with our earlier observation that if A_0 has eigenvalues with real part nonzero, A_0 is structurally stable, and also with our Example 15.2 where $n = 2, m = 1$.

An improvement of the estimate can be realised when using Jordan normal forms. Suppose that the eigenvalue we are analysing is located in a Jordan block of size n^*. Then we may use n^* instead of n for our estimate. We omit the technical details.

The relation with numerical analysis
Perturbation bounds for matrix eigenvalues are studied extensively in numerical linear algebra. Our upper limit estimate is related to general estimates by Ostrowski and Elsner; see Stewart and Sun (1990, Section IV.1).

Example 15.3
In Chapters 11 and 12, we considered the Mathieu equation

$$\ddot{x} + (1 + \varepsilon a + \varepsilon^2 b + \varepsilon \cos 2t)x = 0$$

with free parameters a and b. An (averaging) normal form approach produces $A_0 = 0$ (because of the use of co-moving variables) and to first order the matrix

$$A(\varepsilon) = +\varepsilon \begin{pmatrix} 0 & \frac{1}{2}(a - \frac{1}{2}) \\ -\frac{1}{2}(a + \frac{1}{2}) & 0 \end{pmatrix} + O(\varepsilon^2).$$

The eigenvalues are

$$\lambda_{1,2} = \pm \frac{1}{2} \sqrt{\frac{1}{4} - a^2}.$$

This leads to the well-known result (see Chapter 11 for details) that for $a^2 > \frac{1}{4}$ the solutions of the Mathieu equation are stable and for $a^2 < \frac{1}{4}$ they are unstable. What happens at the transition values, for instance at $a = \frac{1}{2}$? In this case, we have

$$A_1 = \begin{pmatrix} 0 & 0 \\ -\frac{1}{2} & 0 \end{pmatrix}$$

and on performing a second-order averaging calculation

$$A_2 = \begin{pmatrix} 0 & \frac{1}{64} + \frac{1}{2}b \\ \frac{7}{64} - \frac{1}{2}b & 0 \end{pmatrix}.$$

We find for the eigenvalues of $A(\varepsilon)$ to this order of approximation

$$\lambda^2 = -\frac{1}{4}\left(b + \frac{1}{32}\right)\varepsilon^3 + \frac{1}{4}\left(b + \frac{1}{32}\right)\left(\frac{7}{32} - b\right)\varepsilon^4,$$

so if $b > -\frac{1}{32}$ we again have stability, and if $b < -\frac{3}{32}$ we have instability. Note that the timescale $\varepsilon^{\frac{3}{2}}t$ plays a part in this problem. The boundary case of $b = \frac{1}{32}$ again requires a higher-order calculation or different reasoning; see again Chapter 12.

Example 15.4
Consider the (real) matrix $A(\varepsilon)$, where $n = 2$ and

$$A_0 = \begin{pmatrix} 0 & 1 \\ -1 & 0 \end{pmatrix},$$

which, derived from a differential equation, corresponds with the harmonic oscillator. The eigenvalues of A_0 are $\pm i$, and the eigenvalues of $A(\varepsilon)$ are more generally complex. What can be the influence of perturbations? If t_k is the trace of A_k, $k = 1, 2, \cdots$, the eigenvalues of $A(\varepsilon)$ can be written as

$$\lambda_{1,2} = \frac{1}{2}\sum_{k \geq 1} \varepsilon^k t_k \pm i(1 + O(\varepsilon)).$$

The implication is that if for a certain value m we have $t_m \neq 0$, the structural stability is not affected by perturbations higher than m. Do we have such a nice result for coupled oscillators where $n = 4, 6$, or higher? Unfortunately not, as is shown by the following example. Consider the matrix

$$A(\varepsilon) = \begin{pmatrix} 0 & 1 & 0 & 0 \\ -1 & -2\varepsilon & 0 & 1 \\ 0 & 0 & 0 & 1 \\ 0 & a^2\varepsilon^2 & -1 & -2\varepsilon \end{pmatrix}.$$

The characteristic equation can be written as

$$(\lambda^2 + 2\varepsilon\lambda + 1)^2 = a^2\varepsilon^2\lambda^2$$

or

$$\lambda^2 + 2\varepsilon\lambda + 1 = \pm a\varepsilon\lambda,$$

so that A_0 has double eigenvalues $\pm i$, and $A_0 + \varepsilon A_1$ has double eigenvalues with negative real parts $-\varepsilon \pm \sqrt{1 - \varepsilon^2}i$. However, if we take $a > 2$, the full matrix A_ε has eigenvalues with positive real parts, so calculations to $O(\varepsilon^2)$ are necessary in this case of coupled oscillator equations with equal frequencies.

In Chapter 11, we also considered bifurcation problems that are treated by averaging normal form methods. Perturbations of matrices in the context of normal form methods are discussed extensively in Bogaevsky and Povzner (1991) and Murdock (2003). Theory, examples, and references for infinite-dimensional problems can be found in Chow and Hale (1982).

15.4 Intermediate Matching

As we have seen in many problems, the approximation of solutions of differential equations with a small parameter usually involves several local approximations (local in space, time, or both). To obtain a global approximation, one has to match the local solutions, assuming that there is an overlap of domains.

There is some freedom in the choice of the way the solutions are matched, the so-called *matching principle*. Both mathematical analysis and personal taste play a part here. In this context, the basic ideas of Van Dyke, Lagerstrom, and Eckhaus should be mentioned. Discussions can be found in Fraenkel (1969) and Eckhaus (1979), historical notes in Eckhaus (1994). As mentioned before, a different approach to matching is by blow-up techniques; see Krupa and Szmolyan (2001) or Popović and Szmolyan (2004).

The most elementary way of matching, used extensively in Chapters 4–8, is "matching against zero". To fix the idea, consider an interval with left endpoint $x = 0$ where we want to determine an approximation of the function $\phi_\varepsilon(x)$. Suppose that, away from $x = 0$, in the interior of the interval, we have a regular expansion of the form $\phi_0(x) + \varepsilon\phi_1(x) + \cdots$. Using the subtraction trick by putting

$$\psi_\varepsilon(x) = \phi_\varepsilon(x) - (\phi_0(x) + \varepsilon\phi_1(x) + \cdots),$$

we obtain an equation for $\psi_\varepsilon(x)$. For this equation, the regular expansion vanishes. Near $x = 0$, we assume that we have a boundary layer expansion $\psi_0(\xi) + \kappa_1(\varepsilon)\psi_1(\xi) + \cdots$, $\kappa_1(\varepsilon) = o(1)$, in the local (or boundary layer) variable

$$\xi = \frac{x}{\delta(\varepsilon)}, \ \delta(\varepsilon) = o(1),$$

which plays a part near the left endpoint of the interval. "Matching against zero" takes the form

$$\lim_{\xi \to +\infty} \psi_0(\xi) = 0.$$

This is justified by stating that if we move outside an $O(\delta(\varepsilon))$ neighbourhood of $x = 0$, the boundary layer terms have to vanish.

As we have mentioned before, the overlap hypothesis assumes that there exists a subdomain near $x = 0$ where both the regular and the boundary layer expansions are valid and where they can be matched. A local variable ξ_0 corresponding with such a subdomain can be of the form

$$\xi_0 = \frac{x}{\delta_0(\varepsilon)}, \ \delta_0(\varepsilon) = o(1), \delta(\varepsilon) = o(\delta_0(\varepsilon)).$$

In the overlap domain, the first terms of the regular and the boundary layer expansions are transformed:

$$\psi_0(\xi) \to \psi_0\left(\frac{\delta_0(\varepsilon)}{\delta(\varepsilon)}\xi_0\right), \ \phi_0(x) \to \phi_0(\delta_0(\varepsilon)\xi_0).$$

For $\psi_0(\xi)$ and $\phi_0(x)$ to be matched in the overlap domain as $\varepsilon \to 0$, we have upon expanding from the equality of the leading terms

$$\lim_{\xi \to +\infty} \psi_0(\xi) = \lim_{x \to 0} \phi_0(x).$$

This is a slightly more general matching rule and the simplest form of *intermediate matching*. As Eckhaus (1994) formulates it: "it may baffle serious students because it says: the regular approximation when extended to values where it is no longer valid equals the local approximation when extended to values where it is no longer valid".

In a number of problems, this matching rule is still too simple and one has to use a natural generalisation. This consists in rewriting the boundary layer expansion to some order and the regular expansion to (in general) some other order in the intermediate variable and then, after expanding both of them, asking for equality. Explicitly, consider the regular expansion to order m re-written in the intermediate variable ξ_0,

$$\phi_0(\delta_0(\varepsilon)\xi_0) + \varepsilon\phi_1(\delta_0(\varepsilon)\xi_0) + \cdots + \varepsilon^m\phi_m(\delta_0(\varepsilon)\xi_0),$$

and the local (boundary layer) expansion to order p,

$$\psi_0\left(\frac{\delta_0(\varepsilon)}{\delta(\varepsilon)}\xi_0\right) + \kappa_1(\varepsilon)\psi_1\left(\frac{\delta_0(\varepsilon)}{\delta(\varepsilon)}\xi_0\right) + \cdots + \kappa_p(\varepsilon)\psi_p\left(\frac{\delta_0(\varepsilon)}{\delta(\varepsilon)}\xi_0\right).$$

The generalised matching rule is now that both expansions, re-expanded and truncated at some suitable order, should be equal for certain m and p.

We have seen the results of intermediate matching between a local and a regular expansion in Chapter 6. The matching principle can also be applied to two local expansions on nearby domains, as we shall demonstrate below.

Example 15.5

To facilitate the calculations, we study an explicitly given function

$$\phi_\varepsilon(x) = \frac{\varepsilon^2}{x + \varepsilon^2}e^{-\frac{x}{\varepsilon}}, \ x \geq 0.$$

We note that $\phi_\varepsilon(0) = 1$, and the function is exponentially small outside a boundary layer of size $O(\varepsilon)$. There is, however, a smaller boundary layer imbedded of size $O(\varepsilon^2)$, and we identify the local variables

$$\xi_1 = \frac{x}{\varepsilon}, \ \xi_2 = \frac{x}{\varepsilon^2}.$$

Transforming $x \to \xi_1$, we find that $\phi_\varepsilon(x)$ takes the form

$$\psi_1(\xi_1) = \frac{\varepsilon}{\xi_1 + \varepsilon}e^{-\xi_1} = e^{-\xi_1}\left(\frac{\varepsilon}{\xi_1} - \frac{\varepsilon^2}{\xi_1^2} + \cdots\right).$$

Transforming $x \to \xi_2$, we find

$$\psi_2(\xi_2) = \frac{1}{\xi_2 + 1}e^{-\varepsilon\xi_2} = \frac{1}{\xi_2 + 1}\left(1 - \varepsilon\xi_2 + \frac{1}{2}\varepsilon^2\xi_2^2 + \cdots\right).$$

Using the overlap hypothesis, we assume that an intermediate local variable $\xi_0(\varepsilon) = x/\delta_0(\varepsilon)$ exists with the properties

$$\delta_0(\varepsilon) = o(\varepsilon), \ \varepsilon^2 = o(\delta_0(\varepsilon)).$$

Rewriting ψ_1 in the intermediate variable ξ_0 and expanding, we have

$$\psi_1(\xi_1) \to e^{-\frac{\delta_0\xi_0}{\varepsilon}}\left(\frac{\varepsilon^2}{\delta_0\xi_0} - \varepsilon^4\cdots\right)$$

$$= \frac{\varepsilon^2}{\delta_0\xi_0} + \cdots.$$

Rewriting ψ_2 in the intermediate variable ξ_0 and expanding, we have

$$\psi_2(\xi_2) \to \frac{\varepsilon^2}{\delta_0\xi_0 + \varepsilon^2}\left(1 - \frac{\delta_0\xi_0}{\varepsilon} + \cdots\right)$$

$$= \frac{\varepsilon^2}{\delta_0\xi_0} + \cdots.$$

This completes the matching of the two local expansions.

There still exist a number of basic analysis problems with respect to the overlap hypothesis and the justification of matching rules. The reader is referred to the literature cited above. In most if not all real-life applications, matching in one of its different forms often meets computational obstacles but seldom fundamental ones.

15.5 Quadratic Boundary Value Problems

Consider the boundary value problem

$$\varepsilon\frac{d^2\phi}{dx^2} = a(x,\phi)\left(\frac{d\phi}{dx}\right)^2 + b(x,\phi)\frac{d\phi}{dx} + c(x,\phi), \; \phi_\varepsilon(0) = \alpha, \phi_\varepsilon(1) = \beta.$$

The coefficients a, b, c are supposed to be sufficiently smooth, and from the outset we assume that $a(x,\phi) \geq a_0 > 0$ for $x \in [0,1], \phi \in \mathbb{R}$, where a_0 is a positive constant. The analysis will follow in large part Van Harten (1975), and we have added some examples.

15.5.1 No Roots

As we have seen in Example 6.1 of Chapter 6, the construction and the solvability of the problem may fail if the right-hand side has no roots. This can be formulated as follows:

Theorem 15.4
(unsolvability)
If $b^2(x,\phi) - 4a(x,\phi)c(x,\phi) \leq -d_0 < 0$ for $x \in [0,1], \phi \in \mathbb{R}$, and d_0 a positive constant, then for ε sufficiently small no solution of the boundary value problem exists.

Proof
It follows from the theory of quadratic forms that we have

$$\varepsilon\frac{d^2\phi}{dx^2} \geq \frac{d_0}{4a_0} + c_0\left(\frac{d\phi}{dx}\right)^2$$

with c_0 a positive constant. By integration, we have that $d\phi/dx$ will become unbounded in an ε-neighbourhood of each starting point in the interval.

$\square$

15.5.2 One Root

In Exercise 6.4, we have seen a quadratic equation with one root of the right-hand side.

Example 15.6

$$\varepsilon \frac{d^2\phi}{dx^2} = \left(\frac{d\phi}{dx}\right)^2, \quad \phi_\varepsilon(0) = \alpha, \phi_\varepsilon(1) = \beta.$$

It is instructive to analyse this problem without using the exact solution. Substituting a regular expansion of the form $\phi_0(x) + \varepsilon\phi_1(x) + \cdots$, we find

$$\frac{d\phi_0}{dx} = 0, \frac{d^2\phi_1}{dx^2} - \left(\frac{d\phi_1}{dx}\right)^2 = 0.$$

We can satisfy one boundary condition. Assuming that $\phi_0(1) = \beta$, we expect a boundary layer near $x = 0$. For $\phi_1(x)$, we find

$$\phi_1(x) = -\ln\frac{A+x}{A+1},$$

where A is a suitable constant that should be determined by the matching process. Near $x = 0$, we introduce the boundary layer variable

$$\xi = \frac{x}{\delta(\varepsilon)}, \quad \delta(\varepsilon) = o(1).$$

The equation takes the form

$$\frac{\varepsilon}{\delta^2(\varepsilon)}\frac{d^2\psi}{d\xi^2} = \frac{1}{\delta^2(\varepsilon)}\left(\frac{d\psi}{d\xi}\right)^2.$$

No significant degeneration is found, and we can only hope that $\delta(\varepsilon)$ will be determined by the subsequent steps. Although we cannot justify this at present, we try an expansion of the form $\psi(\xi) = \psi_0(\xi) + \varepsilon\psi_1(\xi) + \cdots$, which leads to

$$\frac{d\psi_0}{d\xi} = 0, \frac{d^2\psi_1}{d\xi^2} - \left(\frac{d\psi_1}{d\xi}\right)^2 = 0.$$

We conclude that $\psi(\xi) = \alpha$. Adding the condition $\psi_1(0) = 0$, we find

$$\psi_1(\xi) = -\ln(1 + B\xi)$$

with B a positive constant. Without knowing $\delta(\varepsilon)$, we introduce an intermediate variable (see Section 15.4 for intermediate matching)

$$\xi_0 = \frac{x}{\delta_0(\varepsilon)}, \quad \delta(\varepsilon) = o(\delta_0(\varepsilon)), \delta_0(\varepsilon) = o(1).$$

We have to expand in the intermediate variable:

$$\phi_0(x) + \varepsilon\phi_1(x) \rightarrow$$
$$\beta - \varepsilon\ln\frac{A + \delta_0\xi_0}{A+1} = \beta - \varepsilon\ln\frac{A}{A+1} + \cdots, \quad A > 0, A < -1,$$
$$\beta - \varepsilon\ln(\delta_0\xi_0), \quad A = 0,$$
$$\psi_0(\xi) + \varepsilon\psi_1(\xi) \rightarrow$$
$$\alpha - \varepsilon\ln\left(1 + B\frac{\delta_0}{\delta}\xi_0\right) = \alpha + \varepsilon\ln\delta - \varepsilon\ln B - \varepsilon\ln(\delta_0\xi_0) + \cdots.$$

The expansions match if $A = 0$ and $\beta = \alpha + \varepsilon\delta - \varepsilon \ln B$, so

$$\beta - \alpha = \varepsilon \ln \frac{\delta(\varepsilon)}{B}.$$

The constant B can be taken equal to 1 as it only scales $\delta(\varepsilon)$. We find for the scaling of the local variable near $x = 0$

$$\delta(\varepsilon) = e^{-\frac{\alpha - \beta}{\varepsilon}}.$$

This is correct if $\alpha > \beta$. If the reverse holds, the boundary layer is shifted to $x = 1$ and the regular expansion extends to $x = 0$.

It is easy to check from the exact solution that the expansions represent valid approximations. We could not determine the boundary layer variable from a significant degeneration of the operator, but it turns out that one boundary layer variable exists corresponding with an exponentially small boundary layer. Remarkably enough, this local variable is derived from the matching process.

Generalisation to the boundary problem for $\varepsilon\phi'' = a(x)(\phi')^2$ is straightforward except that there can be technical difficulties in computing the regular expansion.

The analysis becomes very different if we assume that $a(x, \phi)$ depends on ϕ explicitly. Van Harten (1975) gives a general result, but without a detailed example. It is instructive to consider such a special case first.

Example 15.7
Consider the problem

$$\varepsilon\frac{d^2\phi}{dx^2} = \frac{1}{1 + |\phi|}\left(\frac{d\phi}{dx}\right)^2, \ \phi_\varepsilon(0) = \alpha, \phi_\varepsilon(1) = \beta, \alpha > \beta > 0.$$

The term $|\phi|$ need not worry us as, the solution with these boundary values turns out to be positive. Substituting a regular expansion of the form $\phi_0(x) + \varepsilon\phi_1(x) + \ldots$, we find

$$\frac{d\phi_0}{dx} = 0, \frac{d^2\phi_1}{dx^2} - \frac{1}{1 + |\phi_0(x)|}\left(\frac{d\phi_1}{dx}\right)^2 = 0.$$

We assume that the regular expansion satisfies the boundary condition at $x = 1$. This produces

$$\phi_0(x) = \beta, \phi_1(x) = -(1 + \beta)\ln\frac{A + x}{A + 1},$$

with A a suitable constant to be determined later. Near $x = 0$, we introduce the boundary layer variable

$$\xi = \frac{x}{\delta(\varepsilon)}, \ \delta(\varepsilon) = o(1).$$

As before, no significant degeneration is found. We try an expansion of the form

$$\psi(\xi) = \psi_0(\xi) + \varepsilon\psi_1(\xi) + \cdots,$$

leading to

$$\frac{d\psi_0}{d\xi} = 0, \ \frac{d^2\psi_1}{d\xi^2} - \frac{1}{1 + |\psi_0(\xi)|}\left(\frac{d\psi_1}{d\xi}\right)^2 = 0.$$

Applying the boundary value at $x = 0$, we have

$$\psi_0(\xi) = \alpha, \ \psi_1(\xi) = -(1+\alpha)\ln(1+B\xi),$$

with B a suitable constant to be determined by the matching process. We introduce an intermediate variable

$$\xi_0 = \frac{x}{\delta_0(\varepsilon)}, \delta(\varepsilon) = o(\delta_0(\varepsilon), \delta_0(\varepsilon) = o(1),$$

and reexpand the regular and the boundary layer expansions:

$$\phi_0(x) + \varepsilon\phi_1(x) \rightarrow$$
$$\beta - \varepsilon(1+\beta)\ln\frac{A}{A+1}, \ A > 0, A < -1,$$
$$\beta - \varepsilon(1+\beta)\ln(\delta_0\xi_0), A = 0,$$
$$\psi_0(\xi) + \varepsilon\psi_1(\xi) \rightarrow$$
$$\alpha - \varepsilon(1+\alpha)(-\ln\delta + \varepsilon\ln B + \varepsilon\ln(\delta_0\xi_0)).$$

We cannot match these expansions unless $\alpha = \beta$, which produces the exact, constant solution. Of course, we made a number of assumptions, but it turns out that if we expand with respect to order functions more general than ε^n or include higher-order terms, the expansions still do not match. The reason for this is very surprising; it will follow from an analysis of the exact solution

$$\phi_\varepsilon(x) = (1+\alpha)\left(1 - x + \left(\frac{1+\beta}{1+\alpha}\right)^{\frac{\varepsilon-1}{\varepsilon}} x\right)^{\frac{\varepsilon-1}{\varepsilon}} - 1.$$

Putting $(1+\beta)/(1+\alpha) = q$, we have from our assumptions $0 < q < 1$; this makes $q^{1/\varepsilon}$ a (very) small parameter. So, outside a neighbourhood of $x = 0$, we have the regular expansion

$$\phi_\varepsilon(x) = (1+\alpha)(q^{\frac{\varepsilon-1}{\varepsilon}} x)^{\frac{\varepsilon}{\varepsilon-1}} - 1 + \cdots = \beta - \varepsilon(1+\beta)\ln x + \cdots.$$

Near $x = 0$, it is natural to introduce the local variable

$$\xi = \frac{x}{q^{1/\varepsilon}},$$

which produces the expansion

$$\phi_\varepsilon(x) = (1+\alpha)(1 - q^{\frac{1}{\varepsilon}}\xi + q^\varepsilon \xi)^{\frac{\varepsilon}{\varepsilon-1}} - 1 = \alpha - \varepsilon(1+\alpha)\ln(1+\xi) + \cdots .$$

However, we can define many more local variables by taking

$$\eta = \frac{x}{\delta(\varepsilon)}, \quad q^{\frac{1}{\varepsilon}} = o(\delta(\varepsilon)), \quad \delta(\varepsilon) = o(1),$$

which produces on expanding $\phi_\varepsilon(x)$ with respect to η

$$(1+\alpha)(1 - \delta\eta + q^{\frac{\varepsilon-1}{\varepsilon}}\delta\eta)^{\frac{\varepsilon}{\varepsilon-1}} - 1 =$$
$$(1+\alpha)q((1 - \delta\eta)q^{\frac{1-\varepsilon}{\varepsilon}} + \delta\eta)^{\frac{\varepsilon}{\varepsilon-1}} - 1 =$$
$$(1+\beta)(\delta\eta)^{\frac{\varepsilon}{\varepsilon-1}} - 1 + \cdots =$$
$$\beta - \varepsilon\ln\delta(\varepsilon)(1+\beta) - \varepsilon(1+\beta)\ln\eta + \cdots .$$

We conclude that we have a *continuum* of local variables and significant degenerations. Our matching procedures do not allow for this.

Van Harten (1975) analyses the more general problem

$$\varepsilon\frac{d^2\phi}{dx^2} = a(\phi)\left(\frac{d\phi}{dx}\right)^2, \quad \phi_\varepsilon(0) = \alpha, \phi_\varepsilon(1) = \beta, \alpha > \beta > 0.$$

Dividing by $d\phi/dx$ and integrating yields

$$\varepsilon\ln\left|\frac{d\phi}{dx}\right| = \int_0^\phi a(t)dt + c = A(\phi) + c,$$

with c a constant of integration. Separation of variables, integrating and applying the boundary values yields the implicitly defined solution

$$\int_\phi^\alpha e^{-A(t)/\varepsilon}dt = \int_\beta^\alpha e^{-A(t)/\varepsilon}dt \; x.$$

The integral on the right-hand side is exponentially small, and by introducing local variables near $x = 0$, one can show again that a continuum of local variables and significant degenerations exists in this problem.

Example 15.8

As a final case of a right-hand side with one root, we consider the problem

$$\varepsilon\frac{d^2\phi}{dx^2} = a(x)\left(\frac{d\phi}{dx} - f(x)\right)^2, \quad \phi_\varepsilon(0) = \alpha, \phi_\varepsilon(1) = \beta,$$

with again $a(x) \geq a_0 > 0$. We do not expect anything new from this equation, but surprisingly there are new aspects. It is convenient to transform

$$\phi = u + \int_0^x f(t)dt,$$

which leads to the problem

$$\varepsilon \frac{d^2u}{dx^2} = a(x)\left(\frac{du}{dx}\right)^2 - \varepsilon f'(x), \; u_\varepsilon(0) = \alpha, u_\varepsilon(1) = \beta - \int_0^1 f(t)dt = \gamma.$$

Substituting a regular expansion $u_\varepsilon(x) = u_0(x) + \varepsilon u_1(x) + \cdots$ does not produce a solution for $u_1(x)$ unless $f(x)$ is constant. We try a regular expansion of the form $u_\varepsilon(x) = u_0(x) + \varepsilon^\nu u_1(x) + \cdots$:

$$\frac{du_0}{dx} = 0, \; \varepsilon^{1+\nu} \frac{d^2u_1}{dx^2} = \varepsilon^{2\nu} a(x)\left(\frac{du_1}{dx}\right)^2 - \varepsilon f'(x) + \cdots.$$

A significant degeneration arises if $\nu = 1/2$, so that

$$a(x)\left(\frac{du_1}{dx}\right)^2 - f'(x) = 0 \rightarrow \frac{du_1}{dx} = \pm\sqrt{f'(x)/a(x)}.$$

There will be solutions only if $f'(x) \geq 0, x \in [0,1]$.

What happens if $f'(x) < 0$ in some part of the interval, say in a neighbourhood of $x = x_0$? With nearly the same elementary estimate as before (the case "no roots"), we find that du/dx becomes unbounded in a $\sqrt{\varepsilon}$-neighbourhood of x_0.

We shall assume now that $f'(x) > 0$ and also that the regular expansion extends to $x = 1$ with $u_\varepsilon(x) = u_0(x) + \sqrt{\varepsilon}u_1(x) + \cdots$ and

$$u_0(x) = \gamma, \; u_1(x) = \pm\int_x^1 \sqrt{f'(t)/a(t)}dt.$$

For a boundary layer expansion near $x = 0$, we find, using the local variable $\xi = x/\delta(\varepsilon)$,

$$\frac{\varepsilon}{\delta^2(\varepsilon)} \frac{d^2\psi}{d\xi^2} = \frac{a(\delta(\varepsilon)\xi)}{\delta^2(\varepsilon)}\left(\frac{d\psi}{d\xi}\right)^2 - \varepsilon f'(\delta(\varepsilon)\xi)$$

and a significant degeneration if $\delta(\varepsilon) = \sqrt{\varepsilon}$. Trying various expansions, it turns out that we can take $\psi(\xi) = \psi_0(\xi) + \varepsilon\psi_1(\xi) + \cdots$ with

$$\frac{d\psi_0}{d\xi} = 0, \; \frac{d^2\psi_1}{d\xi^2} = a(0)\left(\frac{d\psi_1}{d\xi}\right)^2 - f'(0).$$

Both equations can be integrated and produce, when applying the initial values $\psi_0(0) = \alpha, \psi_1(0) = 0$,

$$\psi_0(\xi) = \alpha, \; \psi_1(\xi) = \ln\left|\frac{1+C}{e^{-\xi} + Ce^\xi}\right|,$$

where C is a suitable constant to be determined by the matching process; we have also chosen $a(0) = f'(0) = 1$ to avoid too many non-essential parameters. Intermediate matching requires introducing the intermediate variable

$$\xi_0 = \frac{x}{\delta_0(\varepsilon)}, \quad \varepsilon^{1/2} = o(\delta_0(\varepsilon)), \quad \delta_0(\varepsilon) = o(1).$$

Re-expanding the regular and the boundary layer expansions, we have

$$u_0(x) + \varepsilon^{\frac{1}{2}} u_1(x) \rightarrow \gamma \pm (\varepsilon^{\frac{1}{2}} - \varepsilon^{\frac{1}{2}} \delta_0(\varepsilon)\xi_0),$$

$$\psi_0(\xi) + \varepsilon\psi_1(\xi) \rightarrow \alpha + \ln\left|\frac{1+C}{C}\right| - \varepsilon^{\frac{1}{2}}\delta_0(\varepsilon)\xi_0.$$

Matching is possible if
a. we take the plus sign for $u_1(x)$;
b. we choose C dependent of ε:

$$C = (1 - e^{\frac{\gamma - \alpha + \varepsilon^{1/2}}{\varepsilon}})^{-1}.$$

The result makes sense if $\alpha > \gamma$; the reader may verify that if $\alpha < \gamma$, the boundary layer will be located near $x = 1$.

15.5.3 Two Roots

The case of two roots is in general unsolved. It will be clear from the preceding subsection and from the examples below that many complications are possible. Apart from examples, it is therefore important to have theory available for guidance to make the right choices. Chang and Howes (1984) obtained results that are useful in this respect. The use of maximum principles, can also be helpful, see in particular Dorr, Parter, and Shampine (1973).

Example 15.9
We turn now to a relatively simple case. Consider the problem

$$\varepsilon\frac{d^2\phi}{dx^2} = \left(\frac{d\phi}{dx} - f(x)\right)\left(\frac{d\phi}{dx} - g(x)\right), \quad \phi_\varepsilon(0) = \alpha, \phi_\varepsilon(1) = \beta, \alpha, \beta > 0.$$

In analysing the problem, it turns out that the question of whether there are points x_0 such that $f(x_0) = g(x_0)$ in $[0,1]$ plays a part in the construction. Assuming that there is no such point, it is not a restriction to put

$$f(x) < g(x), \quad x \in [0,1].$$

There are many possibilities to explore, as putting $\varepsilon = 0$ enables us to choose as a first-order part of a regular expansion

$$\int_0^x f(t)dt + \alpha, \quad \int_0^x g(t)dt + \alpha, \quad \int_x^1 f(t)dt + \beta, \quad \int_x^1 g(t)dt + \beta,$$

or combinations of such functions; see Section 6.3, where a similar case with two roots is discussed. Let us assume that the regular expansion $\phi_0(x) + \varepsilon\phi_1(x) + \cdots$ extends to $x = 1$ and that the choice

$$\phi_0(x) = \int_x^1 f(t)dt + \beta$$

is the correct one. For $\phi_1(x)$, we find in the usual way

$$\phi_1(x) = \int_x^1 \frac{f'(t)}{f(t) - g(t)}dt.$$

Our assumptions imply that we have boundary layer behaviour near $x = 0$. Introduce the local variable

$$\xi = \frac{x}{\delta(\varepsilon)}, \quad \delta(\varepsilon) = o(1),$$

to obtain the equation

$$\frac{\varepsilon}{\delta^2}\frac{d^2\psi}{d\xi^2} = \left(\frac{1}{\delta}\frac{d\psi}{d\xi} - f(\delta\xi)\right)\left(\frac{1}{\delta}\frac{d\psi}{d\xi} - g(\delta\xi)\right).$$

An expansion of the form $\psi = \psi_0(\xi) + \varepsilon^\nu\psi_1(\xi) + \cdots$ produces $d\psi_0/d\xi = 0$ or $\psi_0(\xi)$ is constant for any choice of $\delta(\varepsilon)$. A significant degeneration arises on choosing

$$\delta(\varepsilon) = \varepsilon, \quad \nu = 1,$$

with, for ψ_1, the equation

$$\frac{d^2\psi_1}{d\xi^2} = \left(\frac{d\psi_1}{d\xi} - f(0)\right)\left(\frac{d\psi_1}{d\xi} - g(0)\right).$$

Abbreviating $f(0) = f_0, g(0) = g_0$, we find for the solutions

$$\psi_1(\xi) = g_0\xi - \ln|1 - Ae^{(g_0 - f_0)\xi}| + B$$

with A and B suitable constants. Applying the boundary condition at $x = 0$, we find $\psi_0(\xi) = \alpha, B = 0$. The possibility of matching is considered by introducing the intermediate variable

$$\xi_0 = \frac{x}{\delta_0(\varepsilon)}, \quad \varepsilon = o(\delta_0(\varepsilon)) \rightarrow x = \delta_0\xi_0, \quad \xi = \frac{\delta_0}{\varepsilon}\xi_0.$$

Re-expanding $\phi_0(x) + \varepsilon\phi_1(x) + \cdots$ and $\psi_0(\xi) + \varepsilon\psi_1(\xi) + \cdots$ in the intermediate variable shows that matching fails; the reader may wish to check the details as an exercise.

We will repeat the analysis for an equation where we can obtain the exact solution; this is a simple case, and there are other cases solvable by quadrature.

Example 15.10

Consider the problem

$$\varepsilon\frac{d^2\phi}{dx^2} = \frac{d\phi}{dx}\left(\frac{d\phi}{dx} - 1\right), \quad \phi_\varepsilon(0) = \alpha, \phi_\varepsilon(1) = \beta, \alpha, \beta > 0.$$

A regular expansion as proposed earlier takes the form

$$\phi_0(x) = \beta, \phi_1(x) = 0, \cdots.$$

Introducing as before the boundary layer variable $\xi = x/\varepsilon$ and expanding $\psi(\xi) = \psi_0(\xi) + \varepsilon\psi_1(\xi) + \cdots$, we find

$$\psi_0(\xi) = \alpha, \psi_1(\xi) = \xi - \ln|1 - Ae^\xi| = -\ln|A| - \ln\left|1 - \frac{e^{-\xi}}{A}\right|.$$

Re-expanding in the intermediate variable ξ_0 yields

$$\phi_0(x) + \varepsilon\phi_1(x) \to \beta,$$
$$\psi_0(\xi) + \varepsilon\psi_1(\xi) \to \alpha - \varepsilon\ln|A| + \frac{\varepsilon}{A}e^{-\frac{\delta_0}{\varepsilon}\xi_0}.$$

We cannot match these expansions. Instead of changing our assumptions or trying to improve the matching process, we consider the exact solution

$$\phi_\varepsilon(x) = -\varepsilon\ln\left|e^{-\frac{\alpha+x}{\varepsilon}} - e^{-\frac{\beta+x}{\varepsilon}} + e^{-\frac{\beta}{\varepsilon}} - e^{-\frac{\alpha+1}{\varepsilon}}\right| + \varepsilon\ln(1 - e^{-\frac{1}{\varepsilon}}).$$

The behaviour of the solution is very sensitive to the choice of the (positive) boundary values α and β.

Case $0 < \beta < \alpha$

In this case, the term $\exp(-\beta/\varepsilon)$ dominates; extracting this term, we can write

$$\phi_\varepsilon(x) = \beta - \varepsilon\ln\left|1 + e^{-\frac{\alpha-\beta+x}{\varepsilon}} - e^{-\frac{x}{\varepsilon}} - e^{-\frac{\alpha-\beta+1}{\varepsilon}}\right| + \varepsilon\ln(1 - e^{-\frac{1}{\varepsilon}})$$
$$= \beta - \ln|1 + \nu(\varepsilon)e^{-\xi} - e^{-\xi}| + \cdots$$

with $\nu(\varepsilon) = \exp(-(\alpha - \beta)/\varepsilon)$. So the boundary layer variable is as predicted but the expansion involves an unexpected, exponentially small order function.

Case $0 < \alpha < \beta < \alpha + 1$

Extracting again the term $\exp(-\beta/\varepsilon)$, we observe that the exact solution contains two local variables, near $x = 0$ and near $x = \beta - \alpha$:

$$\xi = \frac{x}{\varepsilon}, \quad \eta = \frac{x + \alpha - \beta}{\varepsilon}.$$

We can expand the solution for $0 < x < \beta - \alpha$ away from the boundary layers near 0 and $\beta - \alpha$ to obtain the regular expansion $\alpha + x + \cdots$. In the interior of

$[\beta - \alpha, 1]$, we have the regular expansion found earlier. Both of them should be matched with the boundary layer expansions.

Case $\beta > \alpha + 1$
It is easy to check that in this case the regular expansion extends to $x = 0$ and that we have a boundary layer near $x = 1$.

The analysis of this example inspires handling of the problem of a right-hand side with two roots as formulated initially. More general choices of $f(x)$ and $g(x)$ with $f(x) < g(x)$ show phenomena similar to those calculated here explicitly.

15.6 Application of Maximum Principles

In Chapter 4, we introduced the concept of a formal approximation of a solution. As an illustration, we present a proof of the asymptotic character of a formal approximation obtained in Chapter 7 on elliptic equations.

In the case of the interior Dirichlet problems studied there, the concept of formal approximation has the following meaning. Consider the bounded domain D, boundary Γ, and the elliptic operator L_ε. We have to solve the boundary value problem

$$L_\varepsilon \phi = f(x, y), \quad x \in D; \phi|_\Gamma = \theta(x, y),$$

where $\tilde{\phi}(x, y)$ is a formal approximation of $\phi(x, y)$ if

$$L_\varepsilon \tilde{\phi} = f(x, y) + o(1), \quad x \in D; \tilde{\phi}|_\Gamma = \theta(x, y) + o(1).$$

We can sharpen the definition by specifying the $o(1)$ estimates to be explicit order functions $\delta(\varepsilon)$, provided $\delta(\varepsilon) = o(1)$. For the difference $R = \phi - \tilde{\phi}$, we obtain, because of the linearity of the problem,

$$L_\varepsilon R = L_\varepsilon \phi - L_\varepsilon \tilde{\phi} = o(1),$$
$$R|_\Gamma = o(1).$$

We conclude that to study the difference R we have to consider the same type of boundary value problem but with an asymptotically small right-hand side and small boundary values. In the case of elliptic operators L_ε, which we have studied, maximum principles will tell us that $R = o(1)$. So $\tilde{\phi}$ is indeed an asymptotic approximation of ϕ. The next example illustrates this for the problem studied in Section 7.1.

Example 15.11

$$\varepsilon \Delta \phi - \phi = f(x, y), \quad x^2 + y^2 < 1,$$
$$\phi|_\Gamma = \theta(x, y), \quad \text{with } \Gamma \text{ the boundary of the domain.}$$

We found a formal approximation of the form

$$\tilde{\phi}(x,y) = -f(x,y) + (\theta(x,y) + f(x,y))|_{\Gamma}\left(1 - \frac{1}{2}\rho\right)e^{-\rho/\sqrt{\varepsilon}},$$

in which $\rho = 1 - \sqrt{x^2 + y^2}$. For the difference $R = \phi - \tilde{\phi}$, we have

$$\varepsilon\Delta R - R = r_\varepsilon(x,y), \quad R|_{\Gamma} = 0.$$

It follows from the analysis in Section 6.1 that

$$r_\varepsilon(x,y) = O(\varepsilon), \quad x,y \in D, \quad \varepsilon \to 0.$$

From Protter and Weinberger (1967), we use the following maximum principle: any twice-differentiable function $u(x,y)$ that satisfies

$$\varepsilon\Delta u - u \geq 0, \quad x^2 + y^2 < 1,$$
$$u \leq 0, \quad x^2 + y^2 = 1,$$

satisfies $u \leq 0$ for $x^2 + y^2 \leq 1$. From our estimate for r, we know that there exists a positive constant c such that

$$-c\varepsilon \leq r_\varepsilon(x,y) \leq c\varepsilon, \quad x,y \in D, \, \varepsilon \to 0.$$

We introduce on D the functions $R_+(x,y) = c\varepsilon$, $R_-(x,y) = -c\varepsilon$. Applying the maximum principle to $R - R_+$, we find

$$\varepsilon\Delta(R - R_+) - (R - R_+) = r_\varepsilon(x,y) + c\varepsilon \geq 0,$$
$$(R - R_+)_\Gamma = -c\varepsilon \leq 0,$$

so that $R - R_+ \leq 0$ for $x^2 + y^2 \leq 1$. The principle applied to $R_- - R$ yields

$$R_- - R \leq 0 \quad \text{for} \quad x^2 + y^2 \leq 1.$$

It follows that $R_- \leq R \leq R_+$ or

$$-c\varepsilon \leq (\phi - \tilde{\phi}) \leq c\varepsilon \quad \text{for} \quad x^2 + y^2 \leq 1,$$

so $\tilde{\phi}$ is an asymptotic approximation of ϕ in the domain.

□

The functions R_- and R_+ that we used in the proof are called *barrier functions*. In this problem, they are very simple; in general one needs some ingenuity to find such functions.

15.7 Behaviour near the Slow Manifold

In Section 8.6, we saw that the solutions of certain initial value problems remain in an $O(\varepsilon)$ neighbourhood of the approximate slow manifold M_0 for $0 \leq t \leq L$. As we shall show, the nearby slow manifold M_ε, assuming that it exists, is approached exponentially closely. This has some interesting consequences.

Consider the system

$$\dot{x} = f(x, y, \varepsilon), \ x(0) = x_0, \ x \in D \subset \mathbb{R}^n, \ t \geq 0,$$
$$\varepsilon\dot{y} = g(x, y, \varepsilon), \ y(0) = y_0, \ y \in G \subset \mathbb{R}^m, \ t \geq 0.$$

The vector field f is C^1 and g is C^2 on the compact domains D and G; f and g depend smoothly on the small parameter ε. Suppose that $g(x, y, 0) = 0$ is solved by $\bar{y} = \phi(x)$ and that

$$\text{Re Sp } g_y(x, c, t) \leq -\mu < 0, \ x \in D, 0 \leq t \leq L,$$

where μ and L are independent of ε. Note that the assumptions of compactness of D and G are essential for the existence of a unique slow manifold; see also the discussion in Section 8.5.

- *Step 1* is then to realise that M_0 is normally hyperbolic and that, according to Fenichel (1971) and Hirsch, Pugh, and Shub (1977), for ε small enough an invariant manifold M_ε exists in an $O(\varepsilon)$-neighbourhood of M_0. The dynamics on M_ε is slow (in terms of ε) and the transversal dynamics is fast. The collection of stable manifolds on which this fast dynamics takes place is called the fast foliation.
- *Step 2* consists of choosing suitable coordinates near the slow manifold, usually called Fenichel (1979) coordinates. The smoothness of the vector fields and the slow manifold enables us to write the system of differential equations in the form

$$\dot{x} = f(x, y, \varepsilon), \ x \in D \subset \mathbb{R}^n, \ t \geq 0,$$
$$\varepsilon\dot{y} = A(x, \varepsilon)y + h(x, y, \varepsilon), \ y \in G \subset \mathbb{R}^m, \ t \geq 0,$$

where A is an $m \times m$ matrix, smoothly dependent on ε, with the property

$$\text{Re Sp } A(x, \varepsilon) \leq -\mu < 0, \ x \in D, 0 \leq t \leq L,$$

in which μ and L are independent of ε. The slow manifold M_ε has been shifted to correspond to $y = 0$; for $h(x, y, \varepsilon)$, we have the estimate

$$||h(x, y, \varepsilon)|| \leq k||y||^2$$

with k a positive constant, and $||.||$ indicates the Euclidean norm.
- *Step 3* consists of the actual process of obtaining an exponential estimate based on the steps 1 and 2. First we will present an elegant proof.

15.7.1 The Proof by Jones and Kopell (1994)

Consider a neighbourhood $\mathcal{B}$ of the slow manifold M_ε. Each point $(x, y) \in \mathcal{B}$ has a neighbourhood H, in size independent of ε, such that an exponential estimate for y holds. This follows from the slow variation of x and the classical Poincaré-Lyapunov estimate (see for instance Verhulst, 2000, Chapter 7). In other words, if $\mathcal{B}$ is sufficiently close to M_ε and if $(x(0), y(0)) \in H$, then

$$||y(t)|| \leq C_H e^{-\mu t}$$

as long as $(x(t), y(t)) \in H$ and for some constant C_H. Because of the compactness, there is a finite covering of $\mathcal{B}$ of such neighbourhoods H and it can be arranged that an orbit passes for $0 \leq t \leq L$ through a finite number of these neighbourhoods, independent of ε. If we need m neighbourhoods, we have explicitly the estimate

$$||y(t)|| \leq C_{H_1} C_{H_2} \cdots C_{H_m} e^{-\mu t}, 0 \leq t \leq L.$$

$\square$

Some details have to be filled in such as the analysis of the neighbourhood H.

15.7.2 A Proof by Estimating Solutions

A more explicit (but also more elaborate) step 3 runs as follows:
According to the O'Malley-Vasil'eva expansion theorem (Section 8.3), we have that if $||y(0)|| = O(\varepsilon)$, then $||y(t)|| = O(\varepsilon), 0 \leq t \leq L$. Also, for $h(x, y, \varepsilon)$,

$$||y(0)|| = O(\varepsilon) \Rightarrow ||h(x, y, \varepsilon)|| \leq \varepsilon K ||y||,$$

as long as $0 \leq t \leq L$ and where K is a positive constant independent of ε. Following Flatto and Levinson (1955), the differential equation for y is now written as

$$\varepsilon \frac{dy}{dt} = Q(\tau, \varepsilon)y + Q(t, \varepsilon)y - Q(\tau, \varepsilon)y + h(x, y, \varepsilon),$$

where we put $A(x(t), \varepsilon) = Q(t, \varepsilon)$ smoothly dependent on ε and for the parameter $\tau : 0 \leq \tau \leq t \leq L$. Variation of constants produces the integral equation

$$y(t) = e^{Q(\tau, \varepsilon)t/\varepsilon} y(0) + \frac{1}{\varepsilon} \int_0^t e^{Q(\tau, \varepsilon)(t-s)/\varepsilon} [\, (Q(s, \varepsilon) - Q(\tau, \varepsilon))y(s)$$
$$+ h(x(s), y(s), \varepsilon)] ds.$$

Our assumption on the spectrum of $Q(t, \varepsilon)$ yields

$$||e^{Q(\tau,\varepsilon)t/\varepsilon}|| \leq Ce^{-\mu t/\varepsilon}$$

for some positive constant C, independent of ε. Together with $||y(0)|| = O(\varepsilon)$ and the estimate for h, we find

$$||y(t)|| \leq \varepsilon Ce^{-\mu t/\varepsilon} + \frac{C}{\varepsilon}\int_0^t e^{-\mu(t-s)/\varepsilon}(||Q(s,\varepsilon) - Q(\tau,\varepsilon)|| + \varepsilon K)||y(s)||ds.$$

Now we put $||y(t)||e^{\mu t/2\varepsilon} = z(t)$ and choose $\tau = t$ to obtain

$$z(t) \leq \varepsilon Ce^{-\mu t/2\varepsilon} + \frac{C}{\varepsilon}\int_0^t e^{-\mu(t-s)/2\varepsilon}(||Q(s,\varepsilon) - Q(t,\varepsilon)|| + \varepsilon K)z(s)ds.$$

As g is C^2, we have $||Q(s,\varepsilon) - Q(t,\varepsilon)|| \leq M|s - t|$; we also note that

$$\sup_{0 \leq s \leq t} e^{-\mu(t-s)/2\varepsilon}|s - t| = \frac{2\varepsilon}{\mu e},$$

so that

$$z(t) \leq \varepsilon C + C\int_0^t \left(\frac{2M}{\mu e} + K\right)z(s)ds.$$

Putting $\beta = C(\frac{2M}{\mu e} + K)$, we find with Gronwall's inequality (see Verhulst, 2000, Chapter 1):

$$z(t) \leq \varepsilon Ce^{\beta t}$$

or

$$||y(t)|| \leq \varepsilon Ce^{-(\mu/2 - \varepsilon\beta)t/\varepsilon}.$$

$\square$

Remarks

1. It is possible to relax the differentiability conditions on f and g, but this complicates the proof somewhat.
2. There is no a priori restriction on the interval bound L. The restriction of the time interval arises from the conditions that x and y are in the compacta D and G. If $x(t)$ leaves D as in Examples 8.4 and 8.5 of Section 8.2, this imposes the bound on the time interval of validity of the estimates. In the application to the two body problem with variable mass in Section 8.4, this theorem applies with $L = +\infty$.
3. As discussed in Section 8.6, the exponential closeness of the solutions to the slow manifold M_ε may cause interesting "sticking" phenomena in the cases where the slow manifold becomes unstable.

15.8 An Almost-Periodic Function

As mentioned in Chapter 11, for almost-periodic functions there exists an analogue of Fourier theory so that a function $f(t)$ that is almost-periodic in t can be written as

$$f(t) = \sum_{n=0}^{\infty} (A_n \cos \lambda_n t + B_n \sin \lambda_n t). \tag{15.1}$$

If $f(t)$ is P-periodic, we would have $\lambda_n = n$, but in this more general case the λ_n can be any sequence of real numbers that together form the generalised Fourier spectrum. The Fourier decomposition is always possible, and to prove this is one of the fundamental results of the theory; to simplify our discussion, we assume that the almost-periodic function is already given in this form. For almost-periodic functions $f(t)$, we have the following properties:

1. Almost-periodic functions are bounded for $-\infty < t < +\infty$.
2. The functions can be represented as a uniformly convergent series of the form (15.1) on $-\infty < t < +\infty$.
3. The generalised average

$$\lim_{T \to \infty} \frac{1}{T} \int_0^T f(t)dt$$

always exists.

If $f(t)$ is P-periodic, the generalised average equals the ordinary average. To see this, write $T = mP + r$ with m a natural number and $0 \le r < P$. We have

$$\frac{1}{T} \int_0^T f(t)dt = \frac{1}{T} \int_0^{mP+r} f(t)dt = \frac{m}{T} \int_0^P f(t)dt + \frac{1}{T} \int_0^r f(t)dt.$$

The last integral is bounded, so if T (or m) tends to ∞, this term vanishes. Taking the limit, the ordinary average remains.

If the average vanishes, the primitive of a periodic function is bounded and again periodic. A consequence is that the error arising from averaging is in the periodic case $O(\varepsilon)$. In the almost-periodic case, the primitive of $f(t)$ need not be bounded for all time. This weakens the error estimate, as we shall see in the following example.

Example 15.12
Consider the almost-periodic function

$$f(t) = \sum_{n=0}^{\infty} \frac{1}{(2n+1)^2} \sin\left(\frac{t}{2n+1}\right).$$

The series is absolutely and uniformly convergent, as it is majorised by the series with terms $1/(2n+1)^2, n = 0, 1, \cdots$. Its average is zero. Because of the uniform convergence, the integral of $f(t)$ exists for each value of t and it can be obtained by interchanging summation and integration:

$$\int_0^t f(s)ds = \sum_0^\infty \frac{1}{(2n+1)^2} \int_0^t \sin\left(\frac{s}{2n+1}\right) ds$$

$$= \sum_0^\infty \frac{1}{2n+1}\left(1 - \cos\left(\frac{t}{2n+1}\right)\right)$$

$$= \sum_0^\infty \frac{2}{2n+1} \sin^2\frac{t}{2(2n+1)}.$$

This series consists of terms greater than or equal to 0 and it is split in a finite sum and a tail as

$$\sum_0^\infty \frac{2}{2n+1} \sin^2\frac{t}{2(2n+1)} = \sum_0^{N(\varepsilon)} \cdots + \sum_{N(\varepsilon)}^\infty \cdots,$$

where we define

$$2(2N(\varepsilon)+1) = \text{entier}\left(\frac{1}{\varepsilon}\right).$$

("entier$(1/\varepsilon)$" is the number $1/\varepsilon$ rounded off to the next natural number.) For the tail, we have for $t < 1/\varepsilon$ and as

$$\sin\left(\frac{t}{2(2n+1)}\right) \le \frac{t}{2(2n+1)}$$

the estimate

$$\sum_{N(\varepsilon)}^\infty \frac{2}{2n+1} \sin^2\frac{t}{2(2n+1)} \le \sum_{N(\varepsilon)}^\infty \frac{2}{2n+1} \frac{t^2}{4(2n+1)^2}.$$

As $t \le 2(2n+1)$, this series converges and has size $O(1)$ with respect to ε. The finite sum is estimated as

$$\sum_0^{N(\varepsilon)} \frac{2}{2n+1} \sin^2\frac{t}{2(2n+1)} \le \sum_0^{N(\varepsilon)} \frac{2}{2n+1} \le \ln(2N(\varepsilon)+1).$$

From the definition of $N(\varepsilon)$, we conclude that this sum is $O(|\ln \varepsilon|)$.

The error arising from averaging almost-periodic functions can be found from analysing the growth rate of a (sometimes) divergent series. There is no standard method for this. The reader might try for instance to estimate the integral of the almost-periodic function

$$f(t) = \sum_{n=0}^\infty \frac{1}{2^n} \cos\left(\frac{t}{2^n}\right)$$

which also has average zero.

15.9 Averaging for PDE's

The averaging theorems of Krol (1989, 1991) can be found in the literature. As Buitelaar's (1993) general averaging theorems were published only in his thesis, we will give an abbreviated account of the ideas.

15.9.1 A General Averaging Formulation

Consider the semilinear initial value problem

$$\frac{dw}{dt} + \mathcal{A}w = \varepsilon f(w, t, \varepsilon), \quad w(0) = w_0, \tag{15.2}$$

where $-\mathcal{A}$ generates a uniformly bounded C_0-semigroup $T(t), -\infty < t < +\infty$, on the Banach space X. We assume the following *basic conditions*:

- f is continuously differentiable and uniformly bounded on $\bar{D} \times [0, \infty) \times [0, \varepsilon_0]$, where D is an open, bounded set in X.
- f can be expanded with respect to ε in a Taylor series, at least to some order.

A generalised solution of Eq. (15.2) is defined as a solution of the integral equation:

$$w(t) = T(t)w_0 + \varepsilon \int_0^t T(t - s)f(w(s), s, \varepsilon)ds.$$

It is well-known that under the given conditions for f and with the uniform boundedness of $T(t)$ the integral equation has a unique solution that exists on the timescale $1/\varepsilon$. The proof follows the usual contraction construction in Banach spaces. Examples that can be put in this form are the wave equation

$$u_{tt} - u_{xx} = \varepsilon f(u, u_x, u_t, t, x, \varepsilon), \quad t \geq 0, 0 < x < \pi, \tag{15.3}$$

where

$$u(0, t) = u(\pi, t) = 0, u(x, 0) = \phi(x), u_t(x, 0) = \psi(x), 0 \leq x \leq \pi,$$

and the cubic Klein-Gordon equation

$$u_{tt} - u_{xx} + a^2 u = \varepsilon u^3, \quad t \geq 0, 0 < x < \pi, a > 0.$$

Using the variation of constants transformation $w(t) = T(t)z(t)$ for Eq. (15.2), we find the integral equation corresponding with the so-called standard form

$$z(t) = w_0 + \varepsilon \int_0^t F(z(s), s, \varepsilon)ds, \quad F(z, s, \varepsilon) = T(-s)f(T(s)z, s, \varepsilon). \tag{15.4}$$

Assume the existence of the average F^0 of F given by

$$F^0(z) = \lim_{T \to \infty} \frac{1}{T} \int_0^T F(z, s, 0) ds \tag{15.5}$$

and the averaging approximation $\bar{z}(t)$ of $z(t)$ by

$$\bar{z}(t) = w_0 + \varepsilon \int_0^t F^0(\bar{z}(s)) ds. \tag{15.6}$$

Under these rather general conditions, Buitelaar (1993) proves the following theorems.

Theorem 15.5
(general averaging)
Consider Eq. (15.2) and the corresponding $z(t)$, $\bar{z}(t)$ given by Eqs (15.4) and (15.6) under the basic conditions stated above. If $T(t)\bar{z}(t)$ exists in an interior subset of D on the timescale $1/\varepsilon$, we have

$$z(t) - \bar{z}(t) = o(1)$$

on the timescale $1/\varepsilon$.

Theorem 15.6
(periodic averaging)
If in addition to the assumptions of theorem 15.5 $F(z, t, \varepsilon)$ is T-periodic in t, we have the estimate
$$z(t) - \bar{z}(t) = O(\varepsilon)$$
on the timescale $1/\varepsilon$.

It turns out that in the right framework we can again use the methods of proof as they were developed for averaging in ordinary differential equations. We present some basic results that are preliminary to this.

15.9.2 Averaging in Banach and Hilbert Spaces

The theory of complex-valued almost-periodic functions was created by Harald Bohr; later the theory was extended to functions with values in Banach spaces by Bochner.

Definition (Bochner's criterion)
Let X be a Banach space. Then $h : \mathbb{R} \to X$ is almost-periodic if and only if h belongs to the closure, with respect to the uniform convergence on $\mathbb{R}$, of the set of trigonometric polynomials

$$\left\{ P_n : \mathbb{R} \to X : t \mapsto \sum_{k=1}^n a_k e^{i\lambda_k t} \Big| n \in \mathbb{N}, \lambda_k \in \mathbb{R}, a_k \in X \right\}.$$

The proof of the following basic theorem was pointed out by J.J. Duistermaat.

Theorem 15.7

Let K be a compact metric space, X a Banach space, and h a continuous function: $K \times \mathbb{R} \to X$. Suppose that for every $z \in K, t \mapsto h(z,t)$ is almost-periodic, and assume that the family $z \mapsto h(z,t) : K \to X, t \in \mathbb{R}$ is equicontinuous. Then the average

$$h^0(z) = \lim_{T \to \infty} \frac{1}{T} \int_0^T h(z,s)ds$$

is well-defined and the limit exists uniformly for $z \in K$. Moreover, if $\phi : \mathbb{R} \to K$ is almost-periodic, then $t \mapsto h(\phi(t),t)$ is almost-periodic.

Proof

We start with the proof that $H : t \mapsto (z \mapsto h(z,t))$ is almost-periodic: $\mathbb{R} \to Y$ where $Y \equiv C(K : X)$ is a Banach space, provided with the supnorm. Let $BC(\mathbb{R} : X)$ be the space of bounded continuous mappings: $\mathbb{R} \to X$ (provided with the supnorm). The family $H_\tau : z \mapsto (t \mapsto h(z, t+\tau)), \tau \in \mathbb{R}$ is an equicontinuous family of mappings $K \to BC(\mathbb{R} : X)$. For each $z \in K, \{H_\tau(z) : \tau \in \mathbb{R}\}$ is a relatively compact subset of $BC(\mathbb{R} : X)$. This follows from the almost-periodicity of $t \mapsto h(z,t)$ and Bochner's criterion. According to the lemma of Arzelà-Ascoli, the family $H_\tau, \tau \in \mathbb{R}$ is relatively compact in $C(K : BC(\mathbb{R} : X))$. Now the switch $H \mapsto \hat{H} : (\hat{H}(z))(t) = (H(t))(z)$ is an isometry from $BC(\mathbb{R} : C(K : X))$ onto $C(K : BC(\mathbb{R} : X))$. We have

$$\|\hat{H}\| = \sup_{z \in K} \sup_{t \in \mathbb{R}} \|(\hat{H}(z))(t)\| = \sup_{t \in \mathbb{R}} \sup_{z \in K} \|(H(t))(z)\| = \|H\|.$$

Thus H_τ is equal to the translate $\mathcal{T}_\tau H : t \mapsto H(t+\tau)$ over τ of H, and the set $\{\mathcal{T}_\tau H : \tau \in \mathbb{R}\}$ is relatively compact in $BC(\mathbb{R} : C(K : X))$. It follows that H is almost-periodic (Bochner's criterion). This implies that the average $h_0(z)$ exists uniformly for $z \in K$:

$$\left\| H^0 - \frac{1}{t} \int_0^t H(s)ds \right\| = \sup_{z \in K} \left\| h_0(z) - \frac{1}{t} \int_0^t h(z,s)ds \right\|.$$

To prove next that $t \mapsto h(\phi(t),t)$ is almost-periodic, we proceed as follows. Note that the closure L of $\{\mathcal{T}_\tau H : \tau \in \mathbb{R}\}$ in $BC(\mathbb{R} : Y)$ is compact. Consider the mapping between spaces,

$$s : C(\mathbb{R} : K) \times L \to BC(\mathbb{R} : X),$$

defined by

$$(\phi, \tilde{H}) \mapsto (t \mapsto \tilde{H}(t)(\phi(t))).$$

The compactness of L makes the family of mappings $\tilde{H}(t)$ equicontinuous. Now the map s is continuous, for we apply the equicontinuity to

$$\tilde{H}(t)(\phi(t)) - \tilde{H}'(t)(\phi'(t)),$$

which we can split into

$$\tilde{H}(t)(\phi(t)) - \tilde{H}(t)(\phi'(t)) + \tilde{H}(t)(\phi'(t)) - \tilde{H}'(t)(\phi'(t)),$$

whereas, because of equicontinuity,

$$s(\phi, \tilde{H}) - s(\phi', \tilde{H}') = t \mapsto (\tilde{H}(t)(\phi(t)) - \tilde{H}'(t)(\phi'(t)))$$

implies that s is continuous. Also, we have for the translate

$$\mathcal{T}_\tau(s(\phi, H)) = s(\mathcal{T}_\tau \phi, \mathcal{T}_\tau \mathcal{H}),$$

and as $\{\mathcal{T}_\tau \phi : \tau \in \mathbb{R}\}$ is contained in a compact subset L' of $C(\mathbb{R}, X)$, we have that the translates $\mathcal{T}_\tau(s(\phi, H))$ are contained in the compact subset $s(L' \times L)$ of $BC(\mathbb{R}, X)$. We conclude that if ϕ is almost-periodic, $s(\phi, H)$ is almost-periodic.

□

We formulate another basic result to produce the framework for theorems 15.5 and 15.6.

Theorem 15.8
Consider Eq. (15.2) with the basic conditions; assume that X is an associated separable Hilbert space and that $-i\mathcal{A}$ is self-adjoint and generates a denumerable, complete orthonormal set of eigenfunctions. If $f(z, t, 0)$ is almost-periodic, $F(z, t, 0) = T(-t)f(T(t)z, t, 0)$ is almost-periodic and the average $F_0(z)$ exists uniformly for z in compact subsets of D. Morover, a solution starting in a compact subset of D will remain in the interior of D on the timescale $1/\varepsilon$.

Proof
For $z \in X$, we have $z = \sum_k z_k e_k$, and it is well-known that the series $T(t)z = \sum_k e^{-i\lambda_k t} z_k e_k$ (λ_k the eigenvalues) converges uniformly and is in general almost-periodic. From Theorem 15.7 it follows that $t \mapsto F(z, t, 0)$ is almost-periodic with average $F_0(z)$. The existence of the solution in a compact subset of D on the timescale $1/\varepsilon$ follows from the usual contraction argument.

□

Remarks

1. Apart from the introduction of suitable spaces and norms, the proofs of the averaging Theorems 15.5 and 15.6 run, with the results formulated in Theorems 15.7 and 15.8, along the line of proofs using "local averages" in the cases of ordinary differential equations; see for instance Sanders and Verhulst (1985). We shall not reproduce them here.
2. That the average $F^0(z)$ exists uniformly is very important in the cases where the spectrum $\{\lambda_k\}$ accumulates near a point that leads to "small denominators". Because of this uniform existence, such an accumulation does not destroy the approximation.

15.9.3 Application to Hyperbolic Equations

A straightforward application is now to consider semilinear initial value problems of hyperbolic type,

$$u_{tt} + Au = \varepsilon f(u, u_t, t, \varepsilon), \quad u(0) = u_0, u_t(0) = v_0, \tag{15.7}$$

where A is a positive self-adjoint linear operator on a separable Hilbert space and f satisfies the basic conditions. This is outlined in Buitelaar (1993) and for an interesting application in Buitelaar (1994). We briefly mention two examples.

Example 15.13
Consider the cubic Klein-Gordon equation

$$u_{tt} - u_{xx} + u = \varepsilon u^3, \ t > 0, 0 \le x \le \pi,$$

with boundary conditions $u(0, t) = u(\pi, t) = 0$.

In this case, the operator $A = -\frac{\partial}{\partial x^2} + 1$ and a suitable domain for the eigenfunctions is $\{u \in W^{1,2}(0, \pi) : u(0) = u(\pi) = 0\}$. Here $W^{1,2}(0, \pi)$ is the Sobolev space consisting of functions $u \in L_2(0, \pi)$ that have first-order generalised derivatives in $L_2(0, \pi)$. The eigenvalues are $\lambda_n = \sqrt{n^2 + 1}, n = 1, 2, \cdots$ and the spectrum is nonresonant. For general initial conditions, we obtain by averaging an $o(1)$-approximation on the timescale $1/\varepsilon$ (Theorem 15.5).

Example 15.14
Consider the nonlinear wave equation

$$u_{tt} - u_{xx} = \varepsilon u^3, \ t > 0, 0 \le x \le \pi,$$

with boundary conditions $u(0, t) = u(\pi, t) = 0$. For the operator $A = -\frac{\partial}{\partial x^2}$, we have eigenfunctions in $\{u \in W^{1,2}(0, \pi) : u(0) = u(\pi) = 0\}$. The eigenvalues are $\lambda_n = n, n = 1, 2, \cdots$ and the spectrum is resonant. For general initial conditions, we obtain by averaging an $O(\varepsilon)$-approximation on the timescale $1/\varepsilon$ (Theorem 15.6).

In all problems, we have to check the basic conditions for the nonlinearities and the possible resonances in the spectrum. The initial conditions have to be elements of the corresponding Sobolev space.

In many problems, the initial conditions are expressed as a finite sum of eigenfunctions. As shown in Chapter 13, this generally leads to improved approximations and also to a more tractable analysis, as it restricts the number of possible resonances.

Epilogue

"The time has come," the Walrus said,
"To talk of many things:
Of shoes - and ships - and sealing wax -
Of cabbages - and kings -
And why the sea is boiling hot -
And whether pigs have wings."
From "The Walrus and the Carpenter"
in "Through the Looking-glass" by Lewis Carroll

This book focuses on examples and methods, on solving problems and action. Such an approach has the advantage of showing what analysis can do and how it works. Hopefully it inspires one to tackle new problems. The danger, on the other hand, is the widespread misunderstanding that perturbation theory is "a bag of tricks" with no other merit than that the tricks work. A book full of examples may obscure the underlying theory and background. One should realise that nearly all of the perturbation results discussed in this book are part of established mathematics with rigorous formulations and appropriate proofs of validity.

The discussion about whether one should emphasise examples and techniques or mathematical foundations has been going on for a long time in applied mathematics. It is sometimes called the controversy between "doers" and "provers". I believe that both attitudes in applied mathematics are one-sided and impose strong limitations on scientific achievement.

A severe restriction in research to proving the validity of methods will in the end lead to a lack of inspiration to explore new fields and mathematical problems. When pursued exclusively, it may even lead to "artificial generalisations" without a clear purpose.

The danger in which "doers" find themselves is even more serious. Following your intuition is fine, but one should be aware that reality can be counterintuitive. Many unexpected results, in expansions, local variables, and timescales, were surprises, and surprises are exactly what scientists are looking for. Interesting in such cases is the interplay of exploring examples and studying mathematical foundations. Examples are the theory of Hamiltonian systems and slow manifold theory discussed in earlier chapters. No progress could have been achieved there without many analytic and numerical calculations hand in hand with basic mathematical theory that was developed in the same period of time.

Finally, a word about "synthesis". Is a problem solved when we have a lot of formulas describing solutions, numerical output, and a few relevant theorems? Is the accumulation of quantitative information enough to consider a problem solved? These are questions considered by Henri Poincaré, who answered them by asking in addition for qualitative insight. He asked for both quantitative and geometric descriptions of the solutions of problems and in this way he single-handedly developed the theory of geometric dynamics, including perturbation theory and topology.

In this modern tradition, there are a number of good examples; we mention the books by Guckenheimer and Holmes (1997) and Thompson and Stewart (2002). In our book on perturbation methods, there has been little space for this kind of synthesis. Both proofs of validity of perturbation techniques and qualitative, geometric insight have to be supplemented and are essential for a complete understanding of phenomena and problems.

Answers to Odd-Numbered Exercises

Chapter 2

2.1 $\|e^{-\frac{x}{\varepsilon}}\|$ on $[1,2]$ is respectively $O_s(e^{-\frac{1}{\varepsilon}}), O_s(\sqrt{\varepsilon}e^{-\frac{1}{\varepsilon}}), O_s(\frac{1}{\varepsilon}e^{-\frac{1}{\varepsilon}}), O_s(\frac{1}{\sqrt{\varepsilon}}e^{-\frac{1}{\varepsilon}})$.
For $x \geq \sqrt{\varepsilon}, |e^{-\frac{x}{\sqrt{\varepsilon}}}| \leq e^{-\frac{1}{\sqrt{\varepsilon}}}$.

2.3 a. $f(x) = O_s(g(x))$ for $x \to 0$. b. $f(x) = o(g(x))$ for $x \to \infty$.

2.5 $\delta(\varepsilon) = \varepsilon - \frac{1}{6}\varepsilon^3$; no, $\sin\varepsilon - \delta(\varepsilon) = o(\varepsilon^4)$.

2.7 a. From 2.5 $f(\varepsilon) = \delta(\varepsilon) + O(\varepsilon^5)$; from 2.6 $\varepsilon^{2n}e^{-\frac{1}{\varepsilon}}$ for $e^{-\frac{1}{\varepsilon}}\cos\varepsilon$.
b. The combined series converges to $f(\varepsilon)$.

2.9 a. For $0 < d \leq x, d$ independent of ε.
b. For $|x| \leq M, M$ independent of ε.

2.11 a. Yes. b. The procedure does not apply as predicted by the implicit function theorem; see the introduction to Chapter 10.

2.13 a. The stationary solutions satisfy $\cos E = 0, (1 - e^2)^{\frac{3}{2}}e = \frac{\varepsilon}{\beta}$.
b. Either e or $1 - e^2$ is small; we have $e_0 = \frac{\varepsilon}{\beta} + O(\varepsilon^2)$ and $e_0 = 1 - \frac{1}{2}(\frac{\varepsilon}{\beta})^{\frac{1}{3}} + O(\varepsilon^{\frac{4}{3}})$.

2.15 $0 < \int_0^1 e^{-\frac{x^2}{\varepsilon}} dx < \sqrt{\varepsilon}\int_0^{\infty} e^{-y^2} dy = \frac{1}{2}\sqrt{\pi\varepsilon}$.

Chapter 3

3.1 a. Expand e^{-t} in a Taylor series; c. $a \neq 1$ ($a = 1$ trivial):
$\Gamma(a,x) = \Gamma(a) - e^{-x}x^{a-1} + O(e^{-x}x^{a-2})$ as $x \to \infty$.

3.5 Transform the integrals by $s = t + \frac{1}{2}\varepsilon t^2, u = -t + \frac{1}{2}\varepsilon t^2$, respectively.

Chapter 4

4.1 $\phi(x) = -f(x) + f(0)e^{-\frac{x}{\sqrt{\varepsilon}}} + f(1)e^{-\frac{1-x}{\sqrt{\varepsilon}}} + \sqrt{\varepsilon}f''(0)e^{-\frac{x}{\sqrt{\varepsilon}}} + \sqrt{\varepsilon}f''(1)e^{-\frac{1-x}{\sqrt{\varepsilon}}} + O(\varepsilon)$.

4.3 a. Near $x = 0$ with $\xi = x/\varepsilon$ degeneration $-\frac{d}{d\xi} - 1$;
near $x = 1$ with $\eta = (1 - x)/\varepsilon$ degeneration $\eta\frac{d}{d\eta} + 1 - \eta$;
with $\zeta = (1 - x)/\varepsilon^2$ degeneration $(\zeta - 1)\frac{d}{d\zeta} + 1$.

b. $y(x) = \frac{\varepsilon^2}{x-1+\varepsilon^2} e^{\frac{1-x}{\varepsilon}}$.

Chapter 5

5.1 $\phi_\varepsilon(x) = -\frac{x^2}{\cos x} + \alpha e^{-\frac{x}{\sqrt{\varepsilon}}} + (\beta + \frac{1}{\cos 1})e^{-\sqrt{\cos 1}\frac{1-x}{\sqrt{\varepsilon}}} + \cdots$.
The approximation is formal.

5.3 a. $\phi_\varepsilon(x) = \frac{\beta}{3}(1+2x) + (\alpha - \frac{\beta}{3})e^{-\frac{x}{\varepsilon}} + \cdots$.

b. $\phi_\varepsilon(x) = \frac{\beta}{3}(1+2x) + (\alpha - \frac{\beta}{3})\frac{1}{(1+2x)^2}e^{-\frac{x+x^2}{\varepsilon}} + \cdots$;
asymptotically equivalent but different in the boundary layers.

5.5 a. Rescaling does not work. b. The general solution is
$Ax^{-1+\sqrt{1+\frac{1}{\varepsilon}}} + Bx^{-1-\sqrt{1+\frac{1}{\varepsilon}}}$. Because of the singularity, we cannot fit the boundary condition.

Chapter 6

6.1 a. $\phi_0 = 0, \phi_n = 0 (n \geq 1)$ assuming homogeneous boundary conditions.

b. $\tilde{\phi}_\varepsilon(x) = \frac{\alpha}{1-\alpha x/\sqrt{2\varepsilon}} + \frac{\beta}{1-\beta(1-x)/\sqrt{2\varepsilon}}$.

c. The asymptotic validity follows from the relation with the exact solution (Example 6.3).

6.3 Assuming a boundary layer for $x(t)$ near $t = 0$:
$x(t) = \frac{1}{2}(1+t) - \frac{\varepsilon}{2\varepsilon+t} + \cdots$, $y(t) = \frac{1}{4}(1+t)^2 + \cdots$.

6.5 a. Not necessarily, as we have to consider solutions outside the saddle loop (Fig. 6.2); b. idem.

Chapter 7

7.1 In the notation of Section 7.2
$\phi_0(x, y) = -y - \sqrt{1 - x^2}$, $\phi_1(x, y) = \frac{y}{(1-x^2)^{\frac{3}{2}}} + \frac{1}{1-x^2}$,

$\psi_0(\xi, \alpha) = 2\sin\alpha e^{-\xi\sin\alpha}$, $\psi_1(\xi, \alpha) = -(\xi^2\cos^2\alpha - 2\xi\sin\alpha)e^{-\xi\sin\alpha}$.

7.5 Regular expansion gives $\phi_0(x, y) = -x$, suggesting boundary layers near $y = 0, y = b$, resulting in $\phi_\varepsilon(x, y) = -x + xe^{-y/\sqrt{\varepsilon}} + xe^{-(b-y)/\sqrt{\varepsilon}} + \cdots$.

Chapter 8

8.1 a. $x(t) = \int_0^t e^{-\frac{s}{\varepsilon}} \sin(t - s) ds = \frac{\varepsilon}{1+\varepsilon^2}(\sin t - \cos t) + \frac{\varepsilon}{1+\varepsilon^2}e^{-\frac{t}{\varepsilon}}$.

b. Consider $t = O_s(1)$, transform $t = \varepsilon\tau$, and apply partial integration:
$x(t) = \int_0^t \frac{\varepsilon}{\varepsilon+s} \sin(t - s) ds = \varepsilon \int_0^{t/\varepsilon} \frac{1}{1+\tau} \sin(t - \varepsilon\tau) d\tau =$
$\varepsilon^2 \int_0^{t/\varepsilon} \ln(1 + \tau) \cos(t - \varepsilon\tau) d\tau = -\varepsilon t \ln\varepsilon + \varepsilon t \ln(\varepsilon + t) - \varepsilon t + \cdots$.

8.3 a. Yes. b. $\varepsilon = 0 \Rightarrow b_0(t) = \frac{1}{2}(x + \sin t), a_0(t) = \frac{11}{5}e^{2t} - \frac{1}{5}\cos t - \frac{2}{5}\sin t$ (slow manifold approximation). Higher order is derived from substitution of the O'Malley-Vasil'eva expansion with lowest-order boundary layer equation
$\frac{d\beta_0}{d\tau} = 2 - 2b_0(0) - 2\beta_0, \beta_0(0) = 3 - b_0(0)$.

8.5 For the energy at large time $E = \frac{1}{2} - m_r - \frac{1}{2}\varepsilon^2(1-m_r)^2 + \varepsilon^4 \cdots$ producing the critical mass $m_r = \frac{1}{2} - \frac{1}{8}\varepsilon^2 + \varepsilon^4 \cdots$.

8.7 Slow manifold approximation $y = \lambda + \varepsilon \cdots$, $x = a_0(t) + \varepsilon \cdots$ with $\ddot{a}_0 + a_0 + \lambda a_0^2 = 0$.

Chapter 9

9.1 (Based on Kevorkian and Cole, 1996, Chapter 3) a. $\alpha = 1, \beta = a$. b.
$\psi_l(x) = -\psi_r(x), \psi_r(x) = \frac{1}{2} \int g(x) e^{-x/2\varepsilon} dx$. c. $u(x,t) = \frac{1}{2} e^{-at/2\varepsilon} \int_{-t}^{t} V(x + s) e^{-s/2\varepsilon} I_0(g(s)) ds$ with I_0 the modified Bessel function of the first kind and order 0. d. $U_0 = 0, U_1 = \frac{g(x-\frac{t}{a})}{a}, W_1 = \frac{g(x)}{a} e^{-a\tau}$.

9.3 a. $u = 0$, $u = 1$; $u = 0$ unstable. b. $\frac{\partial u_0}{\partial t} = u_0(1 - u_0), u_0(x,0) = g(x)$, $\frac{\partial u_1}{\partial t} = u_{0xx} + (1 - 2u_0)u_1, u_1(x,0) = 0$; separation of variables gives $\int_q^{u_0} \frac{ds}{s(1-s)} = t + \int_q^{g(x)} \frac{ds}{s(1-s)}, 0 < q < 1$ or $u_0(x,t) = \frac{g(x)}{g(x)+(1-g(x))e^{-t}}$.
Note that there appear to be no boundary layers but that the validity of the approximation is restricted. For a discussion involving the use of the timescale $\tau = \varepsilon t$, see Holmes (1998).

Chapter 10

10.1 a. For a solution x_0, we have the requirement $3x_0^2 - (3+\varepsilon) \neq 0$ as $\varepsilon \to 0$; $x_0 = 1$ is a solution if $\varepsilon = 0$ so there is no Taylor series near $x_0 = 1$.
b. $1 \pm \frac{1}{\sqrt{3}} \varepsilon^{\frac{1}{2}} + \cdots, -2 - \frac{2}{9}\varepsilon + \cdots$.

10.3 a. $\dot{x}_0 = 1 - x_0, x_0(0) = x(0), \dot{x}_1 = -x_1 + x_0^2, x_1(0) = 0; x_0(t) = 1 + x(0)e^{-t}, x_1(t) = 1 + (x^2(0) + 2x(0)t - 1)e^{-t} - x^2(0)e^{-2t}$. b. Yes.

10.5 The periodic solution of the Van der Pol equation: $x(\theta) = 2\cos\theta + \varepsilon(\frac{3}{4}\sin\theta - \frac{1}{4}\sin 3\theta) + \varepsilon^2 \cdots$ with $\theta = (1 - \frac{1}{16}\varepsilon^2)t$.

Chapter 11

11.1 $\dot{r}_a = -\varepsilon r_a - \varepsilon \frac{3}{8} a r_a^3, \dot{\psi}_a = 0$; the approximation is defined by $r_a^2(t) = c(1 + \frac{3}{4} a r_a^2(t)) e^{-2\varepsilon t}, \psi_a(t) = \psi_0$, with c determined by the initial condition $r(0)$.

11.3 In the notation of Example 11.6 $\dot{y}_{1a} = -\frac{1}{2}\varepsilon y_{2a} + \frac{3}{8}\varepsilon a y_{2a}(y_{1a}^2 + y_{2a}^2), \dot{y}_{2a} = -\frac{1}{2}\varepsilon y_{1a} - \frac{3}{8}\varepsilon a y_{1a}(y_{1a}^2 + y_{2a}^2)$, which modifies the Floquet diagram.

11.5 $\lim_{T\to\infty} \frac{1}{T} \int_0^T \sin t^2 dt = 0$ as $\int_0^\infty \sin t^2 dt = \frac{1}{2}\sqrt{\frac{\pi}{2}}$. We have $x(t) = 1 + O(\varepsilon)$ on the timescale $1/\varepsilon$.

11.7 $\dot{r}_a = \frac{1}{2}\varepsilon r_a(1 - 3r_a^2), \dot{\psi}_a = 0$, with behaviour as in the Van der Pol equation.

11.9 $\dot{r}_a = \frac{1}{2}\varepsilon r_a(1 - \frac{9}{4}r_a^2), \dot{\psi}_a = 0$. Following the remark in Section 11.3, we have a periodic solution with approximation $\frac{2}{3}\cos(t + \psi_0)$.

11.11 Stationary solutions of the averaged system $\psi_a = 0, \pi, \beta_0 r_a - \frac{3}{4}\gamma r_a^3 \pm h = 0$ with one or three real solutions and Jacobian condition $\pm\frac{3}{2}\gamma r_a^3 + h \neq 0$. Depending on the parameters, the solutions have a centre (two imaginary eigenvalues) or a saddle (one positive and one negative eigenvalue) character.

Chapter 12

12.1 $\frac{r_a^2(t)}{1+\frac{3}{8}r_a^2(t)} = \frac{r_0^2}{1+\frac{3}{8}r_0^2} e^{-\varepsilon t}$.

12.3 Resonance manifolds $x = 0$ and $x = 1$; approximation outside the resonance manifolds $x_a(t) = x(0) + \varepsilon t$ so if $x(0) < 1$ we will run into resonance. Putting $\phi_1 - \phi_2 = \psi$, we can apply separation of variables to find the integral $\frac{1}{2}x^2 - \frac{1}{3}x^3 = \varepsilon\psi - \varepsilon\cos\psi + c$ with c determined by the initial conditions.

12.5 Putting $\psi_1 = \phi_1 - \phi_2$ near $x = 1$ and $\psi_2 = \phi_1 + \phi_2$ near $x = -1$, the first-order approximation in the resonance manifolds is described by the equation $\ddot{\psi} - \varepsilon\cos\psi = 0$.

12.7 The first-order approximation for the amplitude is given by $\dot{r}_a = -\frac{1}{2}\varepsilon r_a$, which implies attraction towards zero. When averaging over the angle, we conclude that this approximation is valid for all time. Averaging the second-order terms cannot change this.

12.9 The solutions from the system for (v_1, v_2) are stable, so the higher-order tongues must be of size $O(\varepsilon^2)$.

12.11 b. Amplitude-phase coordinates and second-order averaging produces with $\chi = 2(\psi_{1a} - \psi_{2a})$

$$\dot{r}_{1a} = \varepsilon^2(\frac{1}{12}a_1 a_2 - \frac{1}{2}a_2^2)r_{1a}r_{2a}^2\sin\chi,$$

$$\dot{\psi}_{1a} = -\varepsilon^2\frac{5}{12}a_1^2 r_{1a}^2 - \varepsilon^2(\frac{1}{2}a_1 a_2 + \frac{1}{3}a_2^2)r_{2a}^2 + \varepsilon^2(\frac{1}{12}a_1 a_2 - \frac{1}{2}a_2^2)r_{2a}^2\cos\chi,$$

$$\dot{r}_{2a} = -\varepsilon^2(\frac{1}{12}a_1 a_2 - \frac{1}{2}a_2^2)r_{1a}^2 r_{2a}\sin\chi,$$

$$\dot{\psi}_{2a} = -\varepsilon^2(\frac{1}{2}a_1 a_2 + \frac{1}{3}a_2^2)r_{1a}^2 - \varepsilon^2\frac{5}{12}a_2^2 r_{2a}^2 + \varepsilon^2(\frac{1}{12}a_1 a_2 - \frac{1}{2}a_2^2)r_{1a}^2\cos\chi.$$

c. $r_{1a}^2 + r_{2a}^2 = 2E_0$, which approximates the energy manifold; if $\frac{1}{6}a_1 a_2 \neq a_2^2$, we have as a second integral $r_{1a}^2(r_{2a}^2(-\alpha - \cos\chi) + \gamma) = c$ with $\alpha = (\frac{5}{6}a_1^2 - 2a_1 a_2 - \frac{1}{2}a_2^2)/(\frac{1}{3}a_1 a_2 - 2a_2^2)$, $\gamma = 5(a_1^2 - a_2^2)E_0/(a_1 a_2 - 5a_2^2)$, E_0, c determined by the initial conditions. The second integral describes the foliation into tori of the energy manifold.

d. $\chi = 0, 2\pi, 0 < \frac{a_2}{3a_2 - a_1} < 1$ and $\chi = \pi, 3\pi, 0 < \frac{7a_2}{9a_2 - 5a_1} < 1$.

Chapter 13

13.1 $\dot{E}_n = \varepsilon E_n(1 - \frac{3}{16}(n^2 + 1)E_n - \frac{1}{4}(m^2 + 1)E_m)$,
$\dot{E}_m = \varepsilon E_m(1 - \frac{3}{16}(m^2 + 1)E_m - \frac{1}{4}(n^2 + 1)E_n)$, $n \neq m$; periodic solutions are the normal modes $(E_n, 0), (0, E_m)$ and $E_n = \frac{16}{7(n^2+1)}, E_m = \frac{16}{7(m^2+1)}$. (This solution is unstable.)

13.3 $u = u_n \sin nx + u_m \sin mx, n \neq m$ produces

$$\ddot{u}_n + n^2 u_n = \varepsilon(\dot{u}_n - \frac{3}{4}\dot{u}_n^3 + \frac{3}{4}\delta_{m,3n}\dot{u}_n^2\dot{u}_m - \frac{3}{2}\dot{u}_n\dot{u}_m^2 + \frac{1}{4}\delta_{3m,n}\dot{u}_m^3)$$

and a similar equation for u_m with the roles of n and m interchanged; $\delta_{p,q} = 1$ if $p = q$, and $\delta_{p,q} = 0$ if $p \neq q$; putting $m \neq 3n, n \neq 3m, u_n = r_n \cos(nt + \phi_n), u_m = r_m \cos(mt + \phi_m)$, etc., we find after averaging

$$\dot{r}_{na} = \frac{1}{2}\varepsilon r_{na}\left(1 - \frac{9}{16}n^2 r_{na}^2 - \frac{3}{4}m^2 r_{ma}^2\right), \; \dot{\phi}_{na} = 0,$$

$$\dot{r}_{ma} = \frac{1}{2}\varepsilon r_{ma}\left(1 - \frac{9}{16}m^2 r_{ma}^2 - \frac{3}{4}n^2 r_{na}^2\right), \; \dot{\phi}_{ma} = 0.$$

There are two normal modes and one general position periodic solution.
13.5 To the truncated system are added $\varepsilon\beta\dot{u}_1$ and $\varepsilon\beta\dot{u}_2$; for the trivial solution, one gets the well-known "lifting" of the stability tongue off the parameter axis. In general, there are many bifurcations possible; see again the study of Rand (1996).

Chapter 14
14.1 a.

$$\frac{\partial^2 u_0}{\partial x^2} = \frac{\partial^2 u_0}{\partial t^2}, \; u_0(x,0) = f(x), u_{0t}(x,0) = 0,$$

$$\frac{\partial^2 u_1}{\partial x^2} = \frac{\partial^2 u_1}{\partial t^2} + \frac{\partial u_0}{\partial t}, \; u_1(x,0) = 0, u_{1t}(x,0) = 0.$$

b.

$$\frac{\partial^2 u_0}{\partial \xi \partial \eta} = 0, \frac{\partial^2 u_1}{\partial \xi \partial \eta} = -\frac{1}{4}\frac{\partial u_0}{\partial \xi} + \frac{1}{4}\frac{\partial u_0}{\partial \eta}.$$

c. Use that $u_0 = \frac{1}{2}(f(\xi) + f(\eta))$.
14.3 To the transformed equation is added $-\varepsilon\omega u_\theta$; only the secularity condition for $a(\tau)$ changes and becomes $\frac{da}{d\tau} + a = 0$.

Literature

1. ABLOWITZ, M.J. AND BENNEY, D.J. (1970), *The evolution of multi-phase modes for nonlinear dispersive waves*, Stud. Appl. Math. 49, pp. 225–238.
2. ALYMKULOV, K. (1986), *Periodic solutions of a system obtained by a singular perturbation of a conservative system*, Dokl. Akad. Nauk 186 (5); transl. in Sov. Math. Dokl. 33, pp. 215–219.
3. ANDERSEN, G.M. AND GEER, J.F. (1982), *Power series expansions for the frequency and period of the limit cycle of the van der Pol equation*, SIAM J. Appl. Math. 42, pp. 678–693.
4. ANOSOV, D.V. (1960), *On limit cycles in systems of differential equations with a small parameter in the highest derivatives*, Mat. Sb. 50 (92), pp. 299–334; translated in AMS Trans.. Ser. 2, vol. 33, pp. 233–276.
5. ARIS, R. (1975), *The Mathematical Theory of Diffusion and Reaction in Permeable Catalysts*, 2 vols., Oxford University Press, Oxford.
6. ARNOLD, V.I. (1965), *Conditions for the applicability, and estimate of the error, of an averaging method for systems which pass through states of resonance during the course of their evolution*, Sov. Math. 6, pp. 331–334.
7. ARNOLD, V.I. (1978), *Mathematical Methods of Classical Mechanics*, Springer-Verlag New York.
8. ARNOLD, V.I. (1982), *Geometric Methods in the Theory of Ordinary Differential Equations*, Springer-Verlag, New York.
9. BAKRI, T, NABERGOJ, R., TONDL, A., AND VERHULST, F. (2004), *Parametric excitation in nonlinear dynamics*, Int. J. Nonlinear Mech. 39, pp. 311–329.
10. BAMBUSI, D. (1999), *On long time stability in Hamiltonian perturbations of nonresonant linear pde's*, Nonlinearity 12, pp. 823–850.
11. BENNEY, D.J. AND NEWELL, A.C. (1967), *The propagation of nonlinear wave envelopes*, J. Math. Phys. 46, pp. 133–139.
12. BOBENKO, A.I. AND KUKSIN, S. (1995), *The nonlinear Klein-Gordon equation on an interval as a perturbed Sine-Gordon equation*, Comments Meth. Helvetici 70, pp. 63–112.
13. BOGAEVSKI, V.N. AND POVZNER, A. (1991), *Algebraic Methods in Nonlinear Perturbation Theory*, Applied Mathematical Sciences 88, Springer-Verlag, New York.

14. BOGOLIUBOV, N.N. AND MITROPOLSKY, YU.A. (1961), *Asymptotic Methods in the Theory of Nonlinear Oscillations*, Gordon and Breach, New York.

15. BOGOLIUBOV, N.N. AND MITROPOLSKY, YU.A. (1963), *The method of integral manifolds in nonlinear mechanics*, in *Contributions to Differential Equations*, vol.2, pp. 123–196, Interscience-John Wiley, New York .

16. BOREL, E. (1901), *Lecons sur les Séries Divergentes*, Gauthier-Villars, Paris.

17. BOREL, E. (1902), *Lecons sur les Séries A Termes Positifs*, Gauthier-Villars, Paris.

18. BOURGAIN, J. (1996), *Construction of approximative and almost periodic solutions of perturbed linear Schrödinger and wave equations*, Geom. Funct. Anal. 6, pp. 201–230.

19. BOURLAND, F.J. AND HABERMAN, R. (1989), *The slowly varying phase shift for perturbed, single and multi-phased strongly nonlinear dispersive waves*, Physica 35D, pp. 127–147.

20. BOURLAND, F.J. AND HABERMAN, R. (1990), *Separatrix crossing: time-invariant potentials with dissipation*, SIAM J. Appl. Math. 50, pp. 1716–1744.

21. BUCKMASTER, J.D. AND LUDFORD, G.S.S. (1983), *Lectures on mathematical combustion*, CBSM-NSF Conf. Appl. Math. 43, SIAM, Philadelphia.

22. BUITELAAR, R.P. (1993), *The method of averaging in Banach spaces*, Thesis, University of Utrecht.

23. BUITELAAR, R.P. (1994), *On the averaging method for rod equations with quadratic nonlinearity*, Math. Methods Appl. Sc. 17, pp. 209–228.

24. BUTUZOV, V.F. AND VASIL'EVA, A.B. (1983), *Singularly perturbed differential equations of parabolic type*, in *Asymptotic Analysis II*, Lecture Notes in Mathematics 985 (Verhulst, F., ed.) Springer, Berlin, pp. 38–75.

25. CARRIER, G. AND PEARSON, C. (1968), *Ordinary Differential Equations*, Blaisdell, Waltham, MA.

26. CARY, J.R., ESCANDE, D.F. AND TENNYSON, J.L. (1986), *Adiabatic invariant change due to separatrix crossing*, Phys. Rev. A 34, pp. 4256–4275.

27. CHANG, K.W. AND HOWES, F.A. (1984), *Nonlinear Singular Perturbation Phenomena: Theory and Application*, Applied Mathematical Sciences 56, Springer-Verlag, New York.

28. CHIKWENDU, S.C. AND KEVORKIAN, J. (1972), *A perturbation method for hyperbolic equations with small nonlinearities*, SIAM J. Appl. Math. 22, pp. 235–258.

29. CHIKWENDU, S.C. AND EASWARAN, C.V. (1992), *Multiple scale solution of initial-boundary value problems for weakly nonlinear wave equations on the semi-infinite line*, SIAM J. Appl. Math. 52, pp. 946–958.

30. CHOW, S.-N. AND HALE, J.K. (1982), *Methods of Bifurcation Theory*, Grundlehren der Mathematischen Wissenschaften 251. Springer-Verlag, New York-Berlin

31. CHOW, S.-N. AND SANDERS, J.A. (1986), *On the number of critical points of the period*, J. Diff. Eq. 64, pp. 51–66.

32. CLASS, A.G., MATKOWSKY, B.J., AND KLIMENKO, A.Y. (2003), *A unified model of flames as gasdynamic discontinuities*, J. Fluid Mech. 491, pp. 11–49.

33. COCHRAN, J.A. (1962), *Problems in Singular Perturbation Theory*, Stanford University, Stanford, CA.

34. CODDINGTON, E.A. AND LEVINSON, N. (1952), *A boundary value problem for a nonlinear differential equation with a small parameter*, Proc. Am. Math. Soc. 3, pp. 73–81.

35. DE BRUIJN, N.G. (1958), *Asymptotic Methods in Analysis*, North-Holland, Amsterdam, reprinted Dover, New York (1981).

36. DE JAGER, E.M. AND JIANG FURU (1996), *The Theory of Singular Perturbations*, Elsevier, North-Holland Series in Applied Mathematics and Mechanics 42, Amsterdam.

37. DE GROEN, P.P.N. (1977), *Spectral properties of second order singularly perturbed boundary value problems with turning points*, J. Math. Anal. Appl. 57, pp. 119–149.

38. DE GROEN, P.P.N. (1980), *The nature of resonance in a singular perturbation problem of turning point type*, SIAM J. Math. Anal. 11, pp. 1–22.

39. DIMINNIE, D.C. AND HABERMAN, R. (2002), *Slow passage through homoclinic orbits for the unfolding of a saddle-center bifurcation and the change in the adiabatic invariant*, Physica D 162, pp. 134–52.

40. DOELMAN, A. AND VERHULST, F. (1994), *Bifurcations of strongly non-linear self-excited oscillations*, Math. Methods Appl. Sci. 17, pp. 189–207.

41. DORR, F.W., PARTER, S.V., AND SHAMPINE, L.F. (1973), *Application of the maximum principle to singular perturbation problems*, SIAM Rev. 15, pp. 43–88.

42. DU BOIS-REYMOND, P. (1887), *Théorie générale des fonctions*, Hermann, Paris.

43. ECKHAUS, W. (1979), *Asymptotic Analysis of Singular Perturbations*, North-Holland, Amsterdam.

44. ECKHAUS, W., (1994), *Fundamental concepts of matching*, SIAM Rev. 36, pp. 431–439.

45. ECKHAUS, W. AND DE JAGER, E.M. (1966), *Asymptotic solutions of singular perturbation problems for linear differential equations of elliptic type*, Arch. Rat. Mech. Anal. 23, pp. 26–86.

46. ECKHAUS, W., VAN HARTEN, A., AND PERADZYNSKI, Z. (1985), *A singularly perturbed free boundary problem describing a laser sustained plasma*, SIAM J. Appl. Math. 45, pp. 1–31.

47. ECKHAUS, W. AND GARBEY, M. (1990), *Asymptotic analysis on large timescales for singular perturbations of hyperbolic type*, SIAM J. Math. Anal. 21, pp. 867–883.

48. ERDÉLYI, A. (1956), *Aymptotic Expansions*, Dover, New York.

49. EULER, LEONHARD (1754), *Novi commentarii ac. sci. Petropolitanae 5*, pp. 205–237; in *Opera Omnia*, Ser. I, 14, pp. 585–617.

50. EVAN-EWANOWSKI, R.M. (1976), *Resonance Oscillations in Mechanical Systems*, Elsevier, Amsterdam.

51. FATOU, P. (1928), *Sur le mouvement d'un système soumis á des forces á courte période*, Bull. Soc. Math. 56, pp. 98–139.

52. FEČKAN, M. (2000), *A Galerkin-averaging method for weakly nonlinear equations*, Nonlinear Anal. 41, pp. 345–369.

53. FEČKAN, M. (2001), *Galerkin-averaging method in infinite-dimensional spaces for weakly nonlinear problems*, in Progress in Nonlinear Differential Equations and Their Applications 43 (Grossinho, H.R., Ramos, M., Rebelo, C., and Sanches, L., eds.), Birkhäuser Verlag, Basel.

54. FENICHEL, N. (1971), *Persistence and smoothness of invariant manifolds for flows*, Indiana Univ. Math. J. 21, pp. 193–225.

55. FENICHEL, N. (1974), *Asymptotic stability with rate conditions*, Indiana Univ. Math. J. 23, pp. 1109–1137.

56. FENICHEL, N. (1977), *Asymptotic stability with rate conditions, II*, Indiana Univ. Math. J. 26, pp. 81–93.

57. FENICHEL, N. (1979), *Geometric singular perturbations theory for ordinary differential equations*, J. Diff. Eq. 31, pp. 53–98.
58. FIFE, P.C. (1988), *Dynamics of Internal Layers and Diffusive Interfaces*, CBSM-NSF Conf. Appl. Math. 53, SIAM, Philadelphia.
59. FINK, J.P., HALL, W.S., AND HAUSRATH,A.R. (1974), *A convergent two-time method for periodic differential equations*, J. Diff. Eq. 15, pp. 459–498.
60. FLAHERTY, J.E. AND O'MALLEY, R.E. (1980), *Analysis and numerical methods for nonlinear singular singularly perturbed initial value problems*, SIAM J. Appl. Math. 38, pp. 225–248.
61. FLATTO, L. AND LEVINSON, N (1955), *Periodic solutions of singularly perturbed equations*, J. Math. Mech. 4, pp. 943–950.
62. FRAENKEL, L.E. (1969), *On the method of matched asymptotic expansions, parts I, II and III*, Proc. Cambridge Philos. Soc. 65, pp. 209–284.
63. GEEL, R. (1981), *Linear initial value problems with a singular perturbation of hyperbolic type*, Proc. R. Soc. Edinburgh Section (A) 87, pp. 167–187 and 89, pp. 333–345.
64. GRASMAN, J. (1971), *On the Birth of Boundary Layers*, Mathematical Centre Tract 36, Amsterdam.
65. GRASMAN, J. (1974), *An elliptic singular perturbation problem with almost characteristic boundaries*, J. Math. Anal. Appl. 46, pp. 438–446.
66. GRASMAN, J. (1987), *Asymptotic Methods of Relaxation Oscillations and Applications*, Applied Mathematical Sciences 63, Springer-Verlag, New York.
67. GRASMAN, J. AND MATKOWSKY, B.J. (1977), *A variational approach to singularly perturbed boundary value problems for ordinary and partial differential equations with turning points*, SIAM J. Appl. Math. 32, pp. 588–597.
68. GRASMAN, J. AND VAN HERWAARDEN, O.A. (1999), *Asymptotic Methods for the Fokker-Planck Equation and the Exit Problem in Applications*, Springer-Verlag, New York.
69. GREBENIKOV, E.A. AND RYABOV, YU.A. (1983) (Russian ed. 1979), *Constructive Methods in the Analysis of Nonlinear Systems*, Mir, Moscow.
70. GUCKENHEIMER, J. AND HOLMES, P. (1997), *Nonlinear Oscillations, Dynamical Systems and Bifurcations of Vector Fields*, Applied Mathematical Sciences 42, Springer-Verlag, New York.
71. HABERMAN, R. (1978), *Slowly varying jump and transition phenomena associated with algebraic bifurcation problems*, SIAM J. Appl. Math. 37, pp. 69–106.
72. HABERMAN, R. (1991), *Standard form and a method of averaging for strongly nonlinear oscillatory dispersive traveling waves*, SIAM J. Appl. Math. 51, pp. 1638–1652.
73. HABERMAN, R. AND BOURLAND, F.J. (1988), *Variation of wave action: modulations of the phase shift for strongly nonlinear dispersive waves with weak dissipation*, Physica D 32, pp. 72–82.
74. HALE, J.K. (1963), *Oscillations in Nonlinear Systems*, McGraw-Hill, New York, reprinted Dover, New York (1992).
75. HALE, J.K. (1967), *Periodic solutions of a class of hyperbolic equations containing a small parameter*, Arch. Rat. Mech. Anal. 23, pp. 380–398.
76. HALE, J.K. (1969), *Ordinary Differential Equations*, Wiley-Interscience, New York.
77. HARDY, G.H. (1910), *Orders of Infinity*, Cambridge University Press, Cambridge.

78. HEIJNEKAMP, J.J., KROL, M.S., AND VERHULST, F. (1995), *Averaging in nonlinear transport problems*, Math. Methods Appl. Sciences 18, pp. 437–448.
79. HENRARD, J. (1993), *The adiabatic invariant in classical mechanics*, Dynamics Reported 2, pp. 117–235, Springer-Verlag, Heidelberg.
80. HINCH, E.J. (1991), *Perturbation Methods*, Cambridge University Press, Cambridge.
81. HIRSCH, M., PUGH, C., AND SHUB, M. (1977), *Invariant Manifolds*, Lecture Notes in Mathematics 583, Springer-Verlag, Berlin.
82. HOLMES, M.H. (1998), *Introduction to Perturbation Methods*, Texts in Applied Mathematics 20, Springer-Verlag, New York.
83. HOLMES, P.J., MARSDEN, J.E. AND SCHEURLE, J. (1988), *Exponentially small splittings of separatrices with application to KAM theory and degenerate bifurcations*, Contemp. Math. 81, pp. 213–244.
84. HOPPENSTEADT, F. (1967), *Stability in systems with parameters*, J. Math. Anal. Appl. 18, pp. 129–134.
85. HOPPENSTEADT, F. (1969), *Asymptotic series solutions for nonlinear ordinary differential equations with a small parameter*, J. Math. Anal. Appl. 25, pp. 521–536.
86. HÖRMANDER, L. (1983), *The Analysis of Linear Partial Differential Operators*, vol. 2, Springer-Verlag, New York.
87. HOWES, F.A. (1978), *Boundary-interior layer interactions in nonlinear singular perturbation theory*, Mem. Am. Math. Soc. 203.
88. HUET, D. (1977), *Perturbations singulières de problèmes elliptiqes*, in Lecture Notes in Mathematics 594 (Brauner, C.M., Gay, B., and Mathieu, J., eds.), Springer-Verlag, Berlin.
89. HUT, P. AND VERHULST, F. (1981), *Explosive mass loss in binary stars: the two time-scale method*, Astron. Astrophys. 101, pp. 134–137.
90. HUVENEERS, R.J.A.G. AND VERHULST, F. (1997), *A metaphor for adiabatic evolution to symmetry*, SIAM J. Appl. Math. 57, pp. 1421–1442.
91. IL'IN, A.M. (1999), *The boundary layer*, in *Partial Differential Equations V, Asymptotic Methods for Partial Differential Equations* (Fedoryuk, M.V., ed.), Encyclopaedia of Mathematical Sciences 34, Springer-Verlag, New York.
92. JACKSON, E.A.(1978), *Nonlinearity and irreversibility in lattice dynamics*, Rocky Mountain J. Math. 8, pp. 127-196.
93. JACKSON, E.A. (1991), *Perspectives of Nonlinear Dynamics*, 2 vols., Cambridge University Press, Cambridge.
94. JEFFREYS, H. (1962), *Asymptotic Approximations*, At the Clarendon Press, Oxford.
95. JONES, C.K.R.T. (1994), *Geometric singular perturbation theory*, in *Dynamical Systems*, Montecatini Terme 1994 (Johnson, R., ed.), Lecture Notes in Mathematics 1609, pp. 44–118, Springer-Verlag, Berlin.
96. JONES, C.K.R.T. AND KOPELL, N. (1994), *Tracking invariant manifolds with differential forms in singularly perturbed systems*, J. Diff. Eq. 108, pp. 64–88.
97. JONES, D.S. (1997), *Introduction to Asymptotics, a Treatment Using Nonstandard Analysis*, World Scientific, Singapore.
98. KAPER, T.J. (1999), *An introduction to geometric methods and dynamical systems theory for singular perturbation problems*, in Proceedings Symposia Applied Mathematics 56: *Analyzing Multiscale Phenomena Using Singular Perturbation Methods*, (Cronin, J. and O'Malley, Jr., R.E., eds.). pp. 85–131, American Mathematical Society, Providence, RI.

99. KAPER, T.J. AND JONES, C.K.R.T. (2001), *A primer on the exchange lemma for fast-slow systems*, IMA Volumes in Mathematics and its Applications 122: *Multiple-Time-Scale Dynamical Systems*, (Jones, C.K.R.T., and Khibnik, A.I., eds.). Springer-Verlag, New York.

100. KAPLUN, S. AND LAGERSTROM, P.A. (1957), *Asymptotic expansions of Navier-Stokes solutions for small Reynolds numbers*, J. Math. Mech. 6, pp. 585–593.

101. KELLER, J.B. AND KOGELMAN, S. (1970), *Asymptotic solutions of initial value problems for nonlinear partial differential equations*, SIAM J. Appl. Math. 18, pp. 748–758.

102. KEVORKIAN, J.K. (1961), *The uniformly valid asymptotic representation of the solution of certain nonlinear differential equations*, California Institute of Technology, Pasadena.

103. KEVORKIAN, J.K. (1987), *Perturbation techniques for oscillatory systems with slowly varying coefficients*, SIAM Rev. 29, pp. 391–461.

104. KEVORKIAN, J.K. AND COLE, J.D. (1996), *Multiple Scale and Singular Perturbation Methods*, Springer-Verlag, New York.

105. KOKOTOVIĆ, P., KHALIL, H.K., AND O'REILLY, J. (1999), *Singular Perturbation Methods in Control*, Classics in Applied Mathematics, SIAM, Philadelphia (orig. ed. Academic Pres, New York, 1986).

106. KRANTZ, S.G. AND PARKS, H.R. (2002), *The Implicit Function Theorem, History, Theory and Applications*, Birkhäuser, Boston.

107. KROL, M.S. (1989), *On a Galerkin-averaging method for weakly non-linear wave equations*, Math. Methods Appl. Sci. 11, pp. 649–664.

108. KROL, M.S. (1991), *On the averaging method in nearly time-periodic advection-diffusion problems*, SIAM J. Appl. Math. 51, pp. 1622–1637.

109. KRUPA, M. AND SZMOLYAN, P. (2001), *Extending geometric singular perturbation theory to nonhyperbolic points - fold and canard points in two dimensions*, SIAM J. Math. Anal. 33, pp. 286–314.

110. KRYLOFF, N. AND BOGOLIUBOV, N. (1935), *Méthodes approchées de la mécanique non linéaire dans leur application dans l'étude de la perturbation des mouvements périodiques et de divers phénomènes de résonance s'y rapportant*, Acad. Sci. Ukraine 14.

111. KUKSIN, S. (1991), *Nearly Integrable Infinite-Dimensional Hamiltonian Systems*, Lecture Notes in Mathematics 1556, Springer-Verlag, Berlin.

112. KUZMAK, G.E. (1959), *Asymptotic solutions of nonlinear second order differential equations with variable coefficients*, J. Appl. Math. Mech. (PMM) 10, pp. 730–744.

113. LAGERSTROM, P.A. AND CASTEN, R.G. (1972), *Basic concepts underlying singular perturbation techniques*, SIAM Rev. 14, pp. 63–120.

114. LAGRANGE, J.L. (1788), *Mécanique Analytique* (2 vols.), Paris (reprinted by Blanchard, Paris, 1965).

115. LANGE, C.G. (1983), *On spurious solutions of singular perturbation problems*, Stud. in Appl. Math. 68, pp. 227–257.

116. LARDNER, R.W. (1977), *Asymptotic solutions of nonlinear wave equations using the methods of averaging and two-timing*, Q. Appl. Math. 35, pp. 225–238.

117. LAUWERIER, H.A. (1974), *Asymptotic Analysis*, Mathematical Centre Tracts 54, Amsterdam.

118. LEBOVITZ, N.R. AND SCHAAR, R.J. (1975), *Exchange of stabilities in autonomous systems*, Stud. in Appl. Math. 54, pp. 229–260.

119. LEBOVITZ, N.R. AND SCHAAR, R.J. (1977), *Exchange of stabilities in autonomous systems-II. Vertical bifurcation.*, Stud. in Appl. Math. 56, pp. 1–50.

120. LELIKOVA, E.F. (1978), *A method of matched asymptotic expansions for the equation* $\varepsilon \Delta u - au_z = f$ *in a parallelepiped*, Diff. Urav. 14, pp. 1638–1648; transl. in Diff. Eq. 14, pp. 1165–1172.

121. LEVINSON, N. (1950), *The first boundary value problem for* $\varepsilon \Delta u + A(x,y)u_x + B(x,y)u_y + C(x,y)u = D(x,y)$ *for small* ε, Ann. Math. 51, pp. 428–445.

122. LIDSKII, B.V. AND SHULMAN, E.I. (1988), *Periodic solutions of the equation* $u_{tt} - u_{xx} = u^3$, Functional Anal. Appl. 22, pp. 332–333.

123. LIONS, J.L. (1973), *Perturbations singulières dans les Problèmes aux Limites et en Contrôle Optimal*, Lecture Notes in Mathematics 323, Springer-Verlag, Berlin.

124. LOCHAK, P. AND MEUNIER, C. (1988), *Multiphase Averaging for Classical Systems*, Applied Mathematical Sciences 72, Springer-Verlag, New York.

125. LOCHAK, P., MARCO, J.-P., AND SAUZIN, D. (2003), *On the Splitting of Invariant Manifolds in Multidimensional Near-Integrable Hamiltonian Systems*, Memoirs of the AMS 775, American Mathematical Society, Providence, RI.

126. LUKE, J.C. (1966), *A perturbation method for nonlinear dispersive wave problems*, Proc. R. Soc. London Ser. A, 292.

127. LUTZ, R. AND GOZE, M. (1981), *Nonstandard Analysis*, Lecture Notes in Mathematics 881, Springer-Verlag, Berlin.

128. MACGILLIVRAY, A.D. (1997), *A method for incorporating trancendentally small terms into the method of matched asymptotic expansions*, Stud. in Appl. Math. 99, pp. 285-310.

129. MAGNUS, W. AND WINKLER, S. (1966), *Hill's Equation*, John Wiley, New York (reprinted by Dover, New York, 1979).

130. MAHONY, J.J. (1962), *An expansion method for singular perturbation problems*, J. Austr. Math. Soc. 2, pp. 440–463.

131. MATKOWSKY, B.J. AND SIVASHINSKY, G.I. (1979), *An asymptotic derivation of two models in flame theory associated with the constant density approximation*, SIAM J. Appl. Math. 37, pp. 686–699.

132. MAUSS, J. (1969), *Approximation asymptotique uniforme de la solution d'un problème de perturbation singulière de type elliptique*, J. Mécanique 8, pp. 373–391.

133. MAUSS, J. (1970), *Comportement asymptotique des solutions de problèmes de perturbation singulière pour une equation de type elliptique*, J. Mécanique 9, pp. 523–596.

134. MCLAUGHLIN, D.W. AND SCOTT, A.C. (1978), *Perturbation analysis of fluxon dynamics*, Phys. Rev. A 18, pp. 1652–1680.

135. MISHCHENKO, F.F. AND ROSOV, N.KH. (1980), *Differential Equations with Small Parameters and Relaxation Oscillations*, Plenum Press, New York.

136. MITROPOLSKY, YU.A. (1981), private communication at seminar, Mathematics Institute, Kiev.

137. MITROPOLSKY, Y.A., KHOMA, G., AND GROMYAK, M. (1997), *Asymptotic Methods for Investigating Quasiwave Equations of Hyperbolic Type*, Kluwer, Dordrecht.

138. MORSE, P.M. AND FESHBACH, H. (1953), *Methods of Theoretical Physics*, McGraw-Hill, New York.

139. MURDOCK, J. (2003), *Normal Forms and Unfoldings for Local Dynamical Systems*, Springer-Verlag, New York.

140. NAYFEH, A.H. (1973), *Perturbation Methods*, Wiley-Interscience, New York.

141. NEISHTADT, A.I. (1986), *Change of an adiabatic invariant at a separatrix*, Fiz. Plasmy 12, p. 992.

142. NEISHTADT, A.I. (1991), *Averaging and passage through resonances*, Proceedings International Congress Mathematicians 1990, Math. Soc. Japan, Kyoto, pp. 1271–1283, Springer-Verlag, Tokyo; see also the paper and references in Chaos 1, pp. 42–48 (1991).

143. OLVER, F.W.J. (1974), *Introduction to Asymptotics and Special Functions*, Academic Press, New York.

144. O'MALLEY, JR., R.E. (1968), *Topics in singular perturbations*, Adv. Math. 2, pp. 365–470.

145. O'MALLEY, JR., R.E. (1971), *Boundary layer methods for nonlinear initial value problems*, SIAM Rev. 13, pp. 425–434.

146. O'MALLEY, JR., R.E. (1974), *Introduction to Singular Perturbations*, Academic press, New York.

147. O'MALLEY, JR., R.E. (1976), *Phase-plane solutions to some singular perturbation problems*, J. Math. Anal. Appl. 54, pp. 449–466.

148. O'MALLEY, JR., R.E. (1991), *Singular Perturbation Methods for Ordinary Differential Equations*, Applied Mathematical Sciences 89, Springer-Verlag, New York.

149. PALS, H. (1996), *The Galerkin-averaging method for the Klein-Gordon equation in two space dimensions*, Nonlinear Anal. 27, pp. 841–856.

150. PARLETT, B.N. (1998), *The Symmetric Eigenvalue Problem*, Classics in Applied Mathematics 20, SIAM, Philadelphia.

151. PERKO, L.M. (1969), *Higher order averaging and related methods for perturbed periodic and quasi-periodic systems*, SIAM J. Appl. Math. 17, pp. 698–724.

152. POINCARÉ, H. (1886), *Sur les intégrales irrégulières des équations linéaires*, Acta Math. 8, pp. 295–344.

153. POINCARÉ, H. (1892, 1893, 1899), *Les Méthodes Nouvelles de la Mécanique Céleste*, 3 vols., Gauthier-Villars, Paris.

154. POPOVIĆ, N. AND SZMOLYAN, P. (2004), *A geometric analysis of the Lagerstrom model*, J. Diff. Eq. 199, pp. 290–325.

155. PRANDTL, L. (1905), *Uber Flüssigheitsbewegung bei sehr kleine Reibung*, Proceedings 3rd International Congress of Mathematicians, Heidelberg, 1904, (Krazer, A., ed.), pp. 484–491, Leipzig. (Also publ. by Kraus Reprint Ltd, Nendeln/Liechtenstein (1967).)

156. PRANDTL, L. AND TIETJENS, O.G. (1934), *Applied Hydro- and Aeromechanics*, McGraw-Hill, New York.

157. PRESLER, W.H AND BROUCKE, R. (1981a), *Computerized formal solutions of dynamical systems with two degrees of freedom and an application to the Contopoulos potential. I. The exact resonance case*, Comput. Math. Appl. 7, pp. 451–471.

158. PRESLER, W.H AND BROUCKE, R. (1981b), *Computerized formal solutions of dynamical systems with two degrees of freedom and an application to the Contopoulos potential. II. The near-resonance case*, Comput. Math. Appl. 7, pp. 473–485.

159. PROTTER, M.H. AND WEINBERGER, H.F. (1967), *Maximum Principles in Differential Equations*, Prentice-Hall, Englewood Cliffs, NJ.

160. RAFEL, G.G. (1983), *Applications of a combined Galerkin-averaging method*, Lecture Notes in Mathematics 985 (Verhulst, F., ed.), Springer-Verlag pp. 349-369, Berlin.

161. RAND, R. H. (1994), *Topics in Nonlinear Dynamics with Computer Algebra*, Gordon and Breach, New York.

162. RAND, R.H. (1996), *Dynamics of a nonlinear parametrically-excited pde: 2-term truncation*, Mech. Res. Commun. 23, pp. 283–289.

163. RAND, R.H., NEWMAN, W.I., DENARDO, B.C., AND NEWMAN, A.L. (1995), *Dynamics of a nonlinear parametrically-excited partial differential equation*, Design Engineering Technical Conferences, vol. 3, pt A, ASME, New York (also: Chaos 9, pp. 242–253 (1999)).

164. ROSEAU, M. (1966), *Vibrations nonlinéaires et théorie de la stabilité*, Springer-Verlag, Berlin.

165. SANCHEZ HUBERT, J. AND SANCHEZ PALENCIA, E. (1989), *Vibration and Coupling of Continuous Systems: Asymptotic Methods*, Springer-Verlag, New York.

166. SANDERS, J.A. AND VERHULST, F. (1985), *Averaging Methods in Nonlinear Dynamical Systems*, Applied Mathematical Sciences 59, Springer-Verlag, New York.

167. SCOTT, A. (1999), *Nonlinear Science: Emergence and Dynamics of Coherent Structures*, Oxford University Press, Oxford.

168. SHIH, S.-D. AND R.B. KELLOGG (1987), *Asymptotic analysis of a singular perturbation problem*, SIAM J. Math. Anal. 18, pp. 1467–1511.

169. SHIH, S.-D. (2001), *On a class of singularly perturbed parabolic equations*, Z. Angew. Math. Mech. 81, pp. 337–345.

170. SHISKOVA, M.A. (1973), *Examination of a system of differential equations with a small parameter in the highest derivatives*, Dokl. Akad. Nauk 209 (3), pp. 576–579, transl. in Sov. Math. Dokl. 14, pp. 483–487.

171. SHIVAMOGGI, B.K. (2003), *Perturbation Methods for Differential Equations*, Birkhäuser, Boston.

172. SHTARAS, A.L. (1989), *The averaging method for weakly nonlinear operator equations*, Math. USSR Sb. 62, pp. 223–242.

173. SMITH, D.R.. (1985), *Singular Perturbation Theory: An Introduction with Applications*, Cambridge University Press, Cambridge.

174. STAKGOLD, I. (2000), *Boundary Value Problems of Mathematical Physics*, 2 vols., MacMillan, New York (1968), reprinted in Classics in Applied Mathematics 29, SIAM, Philadelphia.

175. STEWART, G.W. AND SUN, J. (1990), *Matrix Perturbation Theory*, Academic Press, Boston.

176. STIELTJES, TH. (1886), Ann. de l'Ecole Norm. Sup. 3 (3), pp. 201–258.

177. STRAUSS, W.A. (1992), *Partial Differential Equations: An Introduction*, John Wiley, New York.

178. STROUCKEN, A.C.J. AND VERHULST, F. (1987), *The Galerkin-averaging method for nonlinear, undamped continuous systems*, Math. Methods Appl. Sci. 9, pp. 520–549.

179. SZMOLYAN, P. (1992), *Analysis of a singularly perturbed traveling wave problem*, SIAM J. Appl. Math. 52, pp. 485–493.

180. THOMPSON, J.M.T. AND STEWART, H.B. (2002), *Nonlinear Dynamics and Chaos* (2nd ed.), John Wiley and Sons, New York.

181. TIKHONOV, A.N. (1952), *Systems of differential equations containing a small parameter multiplying the derivative* (in Russian), Mat. Sb. 31 (73), pp. 575–586.

182. TING, L. (2000), *Boundary layer theory to matched asymptotics*, Z. Angew. Math. Mech. (special issue on the occasion of the 125th anniversary of the birth of Ludwig Prandtl) 80, pp. 845–855.

183. TONDL, A., RUIJGROK, T., VERHULST, F. AND NABERGOJ, R. (2000), *Autoparametric Resonance in Mechanical Systems*, Cambridge University Press, New York.

184. TRENOGIN, V.A. (1970), *Development and applications of the asymptotic method of Liusternik and Vishik*, Usp. Mat. Nauk 25, pp. 123–156; transl. in Russ. Math. Surv. 25, pp. 119–156.

185. TUWANKOTTA, J.M. AND VERHULST, F. (2000), *Symmetry and resonance in Hamiltonian systems*, SIAM J. Appl. Math. 61, pp. 1369–1385.

186. VAINBERG, B.R. (1989), *Asymptotic Methods in Equations of Mathematical Physics*, Gordon and Breach, New York.

187. VAINBERG, B.R. AND TRENOGIN, V.A. (1974), *Theory of Branching of Solutions of Non-linear Equations*, Noordhoff, Leyden (transl. of Moscow ed., 1969).

188. VAN DEN BERG, I (1987), *Nonstandard Asymptotic Analysis*, Lecture Notes in Mathematics 1249, Springer-Verlag, Berlin.

189. VAN DEN BROEK, B. (1988), *Studies in nonlinear resonance, applications of averaging*, Thesis, University of Utrecht.

190. VAN DEN BROEK, B. AND VERHULST, F. (1987), *Averaging techniques and the oscillator- flywheel problem*, Nieuw Arch. Wiskunde 4th Ser. 5, pp. 185–206.

191. VAN DER AA, E. AND KROL, M.S. (1990), *A weakly nonlinear wave equation with many resonances*, in Thesis of M.S. Krol, University of Utrecht.

192. VAN DER BURGH, A.H.P. (1975), *On the higher order asymptotic approximations for the solutions of the equations of motion of an elastic pendulum*, J. Sound Vibr. 42, pp. 463–475.

193. VAN DER BURGH, A.H.P. (1979), *On the asymptotic validity of perturbation methods for hyperbolic differential equations*, in *Asymptotic Analysis, from Theory to Application* (Verhulst, F., ed.) Lecture Notes in Mathematics 711, pp. 229–240, Springer-Verlag, Berlin.

194. VAN DER CORPUT, J.G. (1966), *Neutralized values I, II*, Kon. Nederl. Akad. Wetensch. Proc. Ser. A 69, pp. 387–411.

195. VAN DYKE, M. (1964), *Perturbation Methods in Fluid mechanics*, Academic press, New York; annotated edition, Parabolic Press (1975), Stanford, CA.

196. VAN HARTEN, A. (1975), *Singularly Perturbed Non-linear 2nd order Elliptic Boundary Value Problems*, Thesis, University of Utrecht.

197. VAN HARTEN, A. (1978), *Nonlinear singular perturbation problems: proofs of correctness of a formal approximation based on a contraction principle in a Banach space*, J. Math. Anal. Appl. 65, pp. 126–168.

198. VAN HARTEN, A. (1979), *Feed-back control of singularly perturbed heating problems*, in Lecture Notes in Mathematics 711 (Verhulst, F., ed.), Springer-Verlag, Berlin, pp. 94–124.

199. VAN HARTEN, A. (1982), *Applications of singular perturbation techniques to combustion theory*, in Lecture Notes in Mathematics 942 (Eckhaus, W., and de Jager, E.M., eds.), Springer-Verlag, Berlin, pp. 295–308.

200. VAN HORSSEN, W.T. (1988), *An asymptotic theory for a class of initial-boundary value problems for weakly nonlinear wave equations with an application to a model of the galloping oscillations of overhead transmission lines*, SIAM J. Appl. Math. 48, pp. 1227–1243.

201. VAN HORSSEN, W.T. AND VAN DER BURGH, A.H.P. (1988), *On initial boundary value problems for weakly nonlinear telegraph equations. Asymptotic theory and application*, SIAM J. Appl. Math. 48, pp. 719–736.

202. VAN HORSSEN, W.T. (1992), *Asymptotics for a class of semilinear hyperbolic equations with an application to a problem with a quadratic nonlinearity*, Nonlinear Anal. 19, pp. 501–530.

203. VAN INGEN, J.L. (1998), *Looking back at forty years of teaching and research in Ludwig Prandtl's heritage of boundary layer flows*, Z. Angew. Math. Mech. 78, pp. 3–20.

204. VASIL'EVA, A.B. (1963), *Asymptotic behaviour of solutions to certain problems involving nonlinear differential equations containing a small parameter multiplying the highest derivatives*, Russ. Math. Surv. 18, pp. 13–84.

205. VASIL'EVA, A.B., BUTUZOV, V.F. AND KALACHEV, L.V. (1995), *The Boundary Function Method for Singular Perturbation Problems*, SIAM Studies in Applied Mathematics 14, SIAM, Philadeplhia.

206. VERHULST, F. (1975), *Asymptotic expansions in the perturbed two-body problem with application to systems with variable mass*, Celes. Mech. 11, pp. 95–129.

207. VERHULST, F. (1976), *Matched asymptotic expansions in the two-body problem with quick loss of mass*, J. Inst. Math. Appl. 18, pp. 87–98.

208. VERHULST, F. (1988), *A note on higher order averaging*, Int. J. Non-Linear Mech. 23, pp. 341–346.

209. VERHULST, F. AND HUVENEERS, R.J.A.G. (1998), *Evolution towards symmetry*, Regular and Chaotic Dynamics 3, pp. 45–55.

210. VERHULST, F. (1998), *Symmetry and integrability in Hamiltonian normal forms*, Proceedings of a Workshop on Symmetry and Perturbation Theory (Bambusi, D. and Gaeta, G., eds.), Quaderni CNR 54, Firenze.

211. VERHULST, F. (1999), *On averaging methods for partial differential equations*, in *Symmetry and Perturbation Theory*, SPT 98 (Degasperis, A. and Gaeta, G., eds.), pp. 79–95, World Scientific, Singapore.

212. VERHULST, F. (2000), *Nonlinear Differential Equations and Dynamical Systems*, Universitext, Springer-Verlag, New York.

213. VON KÁRMÁN, TH. AND BIOT, M.A. (1940), *Mathematical Methods in Engineering*, McGraw-Hill, New York.

214. VISHIK, M.I. AND LIUSTERNIK, L.A. (1957), *On elliptic equations containing small parameters in the terms with higher derivatives*, Dokl. Akad. Nauk, SSR, 113, pp. 734–737.

215. WARD, M.J. (1992), *Eliminating indeterminacy in singularly perturbed boundary value problems with translation invariant potentials*, Stud. Appl. Math. 91, pp. 51-93.

216. WARD, M.J. (1999), *Exponential asymptotics and convection-diffusion-reaction models*, in Proceedings Symposia Applied Mathematics 56, pp. 151–184 (Cronin, J., and O'Malley, Jr., R.E., eds.), American Mathematical Society, Providence, RI.

217. WASOW, W. (1965), *Asymptotic Expansions for Ordinary Differential Equations*, Interscience, New York.

218. WASOW, W. (1984), *Linear Turning Point Theory*, Springer-Verlag, New York.
219. WHITHAM, G.B. (1970), *Two-timing, variational principles and waves*, J. Fluid. Mech. 44, pp. 373–395.
220. WHITHAM, G.B. (1974), *Linear and Nonlinear Waves*, John Wiley, New York.
221. WILKINSON, J.H. (1965), *The Algebraic Eigenvalue Problem*, Oxford University Press, Oxford.
222. YAKUBOVICH, V.A. AND STARZHINSKII, V.M. (1975), *Linear Differential Equations with Periodic Coefficients* (vols. 1 and 2), John Wiley, New York (Israel Program for Scientific Translations).

Index

Texts in Applied Mathematics